Evolutionary Ethics

The International Library of Essays on Evolutionary Thought
Series Editor: Neil Levy

Titles in the Series:

Philosophy of Evolutionary Biology: Volume I
Stefan Linquist

Evolutionary Psychology: Volume II
Stefan Linquist and Neil Levy

Evolutionary Ethics: Volume III
Neil Levy

The Evolution of Culture: Volume IV
Stefan Linquist

Intelligent Design and Religion as a Natural Phenomenon: Volume V
John S. Wilkins

Evolutionary Ethics
Volume III

Edited by

Neil Levy

Florey Neuroscience Institutes, Australia, and Oxford Centre for Neuroethics

Routledge
Taylor & Francis Group

LONDON AND NEW YORK

First published 2010 by Ashgate Publishing

2 Park Square, Milton Park, Abingdon, Oxfordshire OX14 4RN
711 Third Avenue, New York, NY 10017

Routledge is an imprint of the Taylor & Francis Group, an informa business

First issued in paperback 2018

Wherever possible, these reprints are made from a copy of the original printing, but these can themselves be of very variable quality. Whilst the publisher has made every effort to ensure the quality of the reprint, some variability may inevitably remain.

British Library Cataloguing in Publication Data
Evolutionary ethics.
  Volume 3. – (International library of essays on
  evolutionary thought)
  1. Ethics, Evolutionary.
  I. Series II. Levy, Neil, 1967-
  171.7-dc22

Library of Congress Control Number: 2009938975

ISBN 978-0-7546-2758-6 (hbk)
ISBN 978-1-138-37848-3 (pbk)

Contents

Acknowledgements   vii
Series Preface   ix
Introduction   xi

PART I   PRECURSORS TO MORALITY

1   Jessica C. Flack and Frans B.M. de Waal, (2000) '"Any Animal Whatever":
    Darwinian Building Blocks of Morality in Monkeys and Apes', *Journal of
    Consciousness Studies*, **7**, pp. 1–29.   3

PART II   MECHANISMS

2   Philip Kitcher (1998), 'Psychological Altruism, Evolutionary Origins, and Moral
    Rules', *Philosophical Studies*, **89**, pp. 283–316.   35
3   Elliott Sober and David Sloan Wilson (2000), 'Summary of: "Unto Others: The
    Evolution and Psychology of Unselfish Behavior"', *The Journal of Consciousness
    Studies*, **7**, pp. 185–206.   69
4   Brian Skyrms (2000), 'Game Theory, Rationality and Evolution of the Social
    Contract', *The Journal of Consciousness Studies*, **7**, pp. 269–84.   91
5   Robert H. Frank (1987), 'If *Homo Economicus* Could Choose His Own Utility
    Function, Would He Want One with a Conscience?', *The American Economic
    Review*, **77**, pp. 593–604.   107

PART III   ALTRUISM

6   Neven Sesardic (1995), 'Recent Work on Human Altruism and Evolution',
    Ethics, **106**, pp. 128–57.   121
7   Stephen Stich (2006), 'Evolution, Altruism and Cognitive Architecture: A
    Critique of Sober and Wilson's Argument for Psychological Altruism', *Biology
    and Philosophy*, **22**, pp. 267–81.   151

PART IV   META-ETHICS

8   Tamler Sommers and Alex Rosenberg (2003), 'Darwin's Nihilistic Idea:
    Evolution and the Meaninglessness of Life', *Biology and Philosophy*, **18**,
    pp. 653–68.   169
9   Michael Ruse (1986), 'Evolutionary Ethics: A Phoenix Arisen', *Zygon*, **21**,
    pp. 95–112.   185

10   Richard Joyce (2000), 'Darwinian Ethics and Error', *Biology and Philosophy*, **15**,
     pp. 713–32.																			203
11   Sharon Street (2006), 'A Darwinian Dilemma for Realist Theories of Value',
     Philosophical Studies, **127**, pp. 109–66.												223

PART V   NORMATIVE ETHICS

12   Robert J. Richards (1986), 'A Defense of Evolutionary Ethics', *Biology and
     Philosophy*, **1**, pp. 265–93.															283
13   Keith Sutherland and Jordan Hughes (2000), 'Is Darwin Right?', *Journal of
     Consciousness Studies*, **7**, pp. 63–86. (Including Response by Larry Arnhart.)		313

Name Index																			*337*

Acknowledgements

The editor and publishers wish to thank the following for permission to use copyright material.

American Economic Association for the essays: Robert H. Frank (1987), 'If *Homo Economicus* Could Choose His Own Utility Function, Would He Want One with a Conscience?', *The American Economic Review*, **77**, pp. 593–604. Copyright © 1987 American Economic Association.

Imprint Academic for the essays: Jessica C. Flack and Frans B.M. de Waal, (2000) '"Any Animal Whatever": Darwinian Building Blocks of Morality in Monkeys and Apes', *Journal of Consciousness Studies*, **7**, pp. 1–29. Copyright © 2000 Imprint Academic; Elliott Sober and David Sloan Wilson (2000), 'Summary of: "Unto Others: The Evolution and Psychology of Unselfish Behavior"', *The Journal of Consciousness Studies*, **7**, pp. 185–206. Copyright © 2000 Imprint Academic; Brian Skyrms (2000), 'Game Theory, Rationality and Evolution of the Social Contract', *The Journal of Consciousness Studies*, **7**, pp. 269–84. Copyright © 2000 Imprint Academic; Keith Sutherland and Jordan Hughes (2000), 'Is Darwin Right?', *Journal of Consciousness Studies*, **7**, pp. 63–86. (Including Response by Larry Arnhart.) Copyright © 2000 Imprint Academic.

John Wiley and Sons for the essay: Michael Ruse (1986), 'Evolutionary Ethics: A Phoenix Arisen', *Zygon*, **21**, pp. 95–112. Copyright © 1986 John Wiley and Sons.

Springer for the essays: Philip Kitcher (1998), 'Psychological Altruism, Evolutionary Origins, and Moral Rules', *Philosophical Studies*, **89**, pp. 283–316. Copyright © 1998 Springer; Stephen Stich (2006), 'Evolution, Altruism and Cognitive Architecture: A Critique of Sober and Wilson's Argument for Psychological Altruism', *Biology and Philosophy*, **22**, pp. 267–81. Copyright © 2006 Springer; Tamler Sommers and Alex Rosenberg (2003), 'Darwin's Nihilistic Idea: Evolution and the Meaninglessness of Life', *Biology and Philosophy*, **18**, pp. 653–68. Copyright © 2003 Springer; Richard Joyce (2000), 'Darwinian Ethics and Error', *Biology and Philosophy*, **15**, pp. 713–32. Copyright © 2000 Springer; Sharon Street (2006), 'A Darwinian Dilemma for Realist Theories of Values', *Philosophical Studies*, **127**, pp. 109–66. Copyright © 2006 Springer; Robert J. Richards (1986), 'A Defense of Evolutionary Ethics', *Biology and Philosophy*, **1**, pp. 265–93. Copyright © 1986 Springer.

The University of Chicago Press for the essay: Neven Sesardic (1995), 'Recent Work on Human Altruism and Evolution', *Ethics*, **106**, pp. 128–57. Copyright © 1995 The University of Chicago Press.

Every effort has been made to trace all the copyright holders, but if any have been inadvertently overlooked the publishers will be pleased to make the necessary arrangement at the first opportunity.

Series Preface

The theory of evolution is one of science's great achievements. Though to those outside science, it may seem that the theory is controversial, within science there is no controversy at all about its basic form. Moreover, the theory of evolution plays a pivotal role in guiding new research. 'Nothing in biology makes sense except in the light of evolution', Theodosius Dobzhansky famously wrote; the theory of evolution unifies disparate subfields of biology and generates testable predictions for each. The success of the theory and its explanatory fecundity for biology cannot be doubted. But might the theory also be capable of illuminating phenomena outside the direct purview of biology?

The volumes in this series are dedicated to exploring this question. They bring together some of the best writings of the past two decades which explore the relevance of evolution and evolutionarily-inspired thought to arenas of human life beyond the merely biological. Volumes focus on whether it is productive and illuminating to attempt to understand our most distinctive achievements and our most intimate features as evolved phenomena. Is the content of moral systems explained by evolution? To what extent are the processes of selection and reproduction that explain changes in gene frequencies also at work in explaining the reproduction of ideas? Can evolution shed light on why we think as we do, perceive as we do, even feel as we do? Might even our idea of God – and perhaps with it the perennial temptation to reject evolution in the name of religion – be explained by evolutionary thought?

Answering these questions requires not only a detailed grasp of the phenomena we aim to explain – the contours of religious thought, the features of morality, and so on – but also an understanding of the theory we aim to apply to the field. Though the theory of evolution is not itself controversial within science, there are lively controversies about its details. One volume of this theory is devoted to writings which illuminate these controversies and deepen our understanding of the mechanisms of evolution. It is only if we have an appreciation of how evolution works that we can begin to assess attempts to extend its reach to culture, to the mind, to morality and to religion.

The volumes are edited by experts in the philosophy of biology and include sensitive and thoughtful discussions of the material they contain. Naturally, in selecting the papers for inclusion, and given the large amount of high quality thought on the philosophy of biology, and on each of the topics covered by these volumes, it was necessary to make some hard choices. Each editor has chosen to focus on particular controversies within the field covered by their volume; on each topic, a range of views is canvassed (including the views of those who deny that evolution can contribute much to the understanding of non-biological features of human beings).

Evolution is our story; in coming to understand it, we come to understand ourselves. Readers of these volumes should be left with a deepened appreciation for the power and ambition of evolutionary thought, and with a greater understanding of what it means to be an evolved being.

NEIL LEVY
Florey Neuroscience Institutes, Australia and University of Oxford, UK

Introduction

The possibility of attempting to ground ethics, explain it, or explain it away, by reference to evolutionary theory has attracted thinkers at least since Darwin's day. There have been several flowerings of evolutionary ethics in the history of thought. After the success of sociobiology in the 1970s, the research programme became stagnant for a decade; recently, however, evolutionary proposals have become a focus for debates over meta-ethics (the philosophical discipline concerned with the foundations and nature of morality). This volume brings together some of the most interesting work in evolutionary ethics from the past twenty-five years. After a brief preface setting out the scope of evolutionary ethics, this introduction will survey recent developments in the field, situating them in the context of the history of evolutionary ethics. It will then briefly describe the papers contained in this volume.

The Aims and Scope of Evolutionary Ethics

'Evolutionary ethics' is an umbrella term that can be applied to several, closely linked but importantly different, projects. It can refer to an *explanatory* project, which has the aim of explaining the sources of moral judgments in human beings (and perhaps in other animals), to a *prescriptive* project, which has the aim of advancing normative claims on the basis of evolutionary considerations, or it might be a *meta-ethical* project, which has the aim of advancing debates over the nature of moral judgment and cognition by reference to evolutionary considerations. Evolutionary ethics is relatively uncontroversial as an explanatory project. Since we are evolved beings, all our phenotypic traits can be illuminated by evolutionary considerations, directly (when the trait is itself an adaptation) or indirectly (when the trait is a more or less mediated by-product of adaptations). Whether morality is an adaptation or a by-product, or a mix of both, and just how it, or its building-blocks, evolved, all that is hugely controversial. But it is not controversial that our evolutionary history helps explain why we are moral beings.

However, the meta-ethical project is controversial, and the normative project is hugely controversial. It is apparently coherent to believe that, though evolution explains why are capable of moral judgment, it tells us nothing about the *nature* of moral judgment, let alone what *content* moral judgments should have. But from the very beginning of evolutionarily inspired reflection on ethics, some thinkers have defended all three projects. For some, evolution provides communities and individuals with a guide to life. In this introduction, we shall survey the debates over these three projects, introducing the major themes and controversies reflected in the essays collected here.

The Rise and Fall and Rise of Evolutionary Ethics

The naturalistic tendency in moral philosopher has a very long history. Aristotle, in one way, and Hume, in another, both sought to give their ethical theories foundations in theories of human nature (indeed, Hume continues to serve as a model and inspiration for more recent naturalizers). But most contemporary thinkers seeking to naturalize morality today look to Darwin and to later evolutionary thinkers. Naturalizers, in this context, are thinkers who seek to explain – or, sometimes, to explain away – morality entirely in terms of events, processes and states that figure in the natural sciences.

Here, however, we are concerned more specifically with the *evolutionary* naturalizers: philosophers (and sometimes biologists) who attempt to explain, or explain away, ethics by reference to Darwinian and Neo-Darwinian natural selection.[1]

So understood, evolutionary ethics begins with Darwin himself. In both his major works, *The Origin of Species* and *The Descent of Man* (especially in the latter), Darwin makes scattered remarks of direct relevance to the claim that morality has an evolutionary origin. Most importantly, in *Descent* Darwin notices the problem of altruism, which has plagued evolutionary ethicists ever since, and offers the outline of a solution to it. The entire history of evolutionary ethics could be written in terms of this problem, and the changing fortunes of Darwin's solution. However, perhaps because Darwin had an aversion to unnecessary controversy, he did not play a conspicuous role in the debate over evolutionary ethics. Instead, the contemporary debate featured most prominently two of his most able champions. Between them, Herbert Spencer and Thomas Henry Huxley set the tone for much of the debate which was to dominate evolutionary approaches to ethics for the next century. Though their works are seldom read today, evolutionary ethicists continue to work in their shadow, whether they realize it or not.

Early evolutionary ethicists were concerned with both the explanatory project and the prescriptive project (unlike most thinkers today, who are more concerned with the explanatory project and the meta-ethical project). But all three questions were already clearly articulated by Huxley and Spencer, and the answers they gave laid down the pattern for their successors.

Huxley and Spencer were both enthusiastic evolutionists, but they diverged radically in the extent to which each believed that ethics could be directly explained in evolutionary terms. Huxley argued that ethics was an *autonomous* institution, independent of, indeed opposed to, mere biological nature. It is probably true, Huxley contends, that our moral sentiments are the product of natural selection. But, by the same token, so are our *im*moral sentiments. From a purely evolutionary point of view, there is no reason to prefer one set of sentiments to the other. They are each merely a set of motives which drives behaviour that might be expected to have survival value. Thus, we cannot derive ethical principles from natural selection itself:

---

[1]   A note on terminology. *Neo-Darwinism* is the product of the synthesis of Darwinian evolution with Mendelian genetics (which provided Darwinism with the mechanism of heredity it hitherto lacked). In this introduction, I use terms like 'evolution', 'Darwinism' and 'natural selection' to refer to the neo-Darwinian consensus view. According to it, evolution occurs through the process of random variation, differential reproduction and inheritance. There are (postulated) evolutionary mechanisms besides these (such as Lamarckian inheritance of acquired characteristics) but they are ignored here.

The thief and the murderer follow nature just as much as the philanthropist. Cosmic evolution may teach us how the good and the evil tendencies of man may have come about; but, in itself, it is incompetent to furnish any better reason why what we call good is preferable to what we call evil than we had before. Some day, I doubt not, we shall arrive at an understanding of the evolution of the aesthetic faculty; but all the understanding in the world will neither increase nor diminish the force of the intuition that this is beautiful and that is ugly. (Huxley, 1989, p. 138)

Thus, Huxley answers the meta-ethical question largely in the negative: at least by itself, evolution cannot tell us much about the meaning or reference of moral terms. 'Good' does not mean 'adaptive', or anything of that kind, for too many different kinds of actions and traits of character fit this description. Meta-ethics is relatively independent of biology.

Huxley is equally unequivocal with regard to the prescriptive question. If we were to follow nature's counsels, then we would glorify death, suffering and selfishness. Natural selection systematically rewards violence and greed. If it can serve as a model for us, it is only negatively, by showing us what *not* to do: 'Let us understand, once for all, that the ethical progress of society depends, not on imitating the cosmic process, still less in running away from it, but in combating it' (1989, p.141). Far from modelling morality on the process of natural selection, Huxley argued that we ought to oppose evolutionary processes in the name of morality.[2]

Spencer's views could not be more different. It was Spencer, and not Darwin, who coined the phrase that in many minds still encapsulates evolution – 'the survival of the fittest'. For him, evolution is an essentially *progressive* force, which systematically selects the best representatives of each species. If it is allowed to work without interference, therefore, natural selection improves the quality of species. We can therefore identify the *process* and the *product* of evolution with the good.

Spencer gives positive, and interrelated, answers to both the prescriptive and the meta-ethical questions. For him, 'good' just *means* 'highly evolved'; the study of evolution shows that 'the conduct to which we apply the name good, is the relatively more evolved conduct; and that bad is the name we apply to conduct which is relatively less evolved' (1883, p. 25).

If 'good' means 'highly evolved', then it is our moral obligation, in pursuing the good, to assist the process of evolution. Thus, Spencer's meta-ethical views have direct implications for his answer to the prescriptive question. We ought to allow natural selection to do its work. Spencer therefore opposes a short-term, sentimental, devotion to the welfare of the poor to a hard-headed, but ultimately also more ethical, concern to see that those who win out in the struggle for existence are those who are, genuinely, fit. He opposes organized charity, for instance, since it will tend to divorce fitness from reproductive success, while supporting private charities which aim to provide the 'deserving' poor with opportunities they might otherwise lack. We should intervene in social processes only to undo the effects of past interventions, or to compensate for sheer bad luck, Spencer argued. But we ought to resist the

---

[2]   I take it, however, that Huxley does not mean that 'the cosmic process' gives us an infallible – negative – guide to morality. If that were the case, then we would have an evolutionary analysis of goodness at hand: 'good' refers to all and only those actions and processes which are disfavoured by evolution. Huxley argues that both our moral and our immoral dispositions are the product of evolution. Accordingly, I take his point to be that evolution cannot offer us any way of distinguishing between them.

temptation to ameliorate the suffering of the undeserving, of those genetically destined to fall by the wayside.

Spencer's views, and similar positions, were massively influential in the early decades of the twentieth-century. The Social Darwinism he advocated, and the closely related eugenics movement, had thousands of adherents, from all parts of the political spectrum. Not just conservatives, but socialists and feminists, as well as the majority of mainstream scientists, supported these programmes, in whole or in part. They were politically successful, especially in Northern Europe, and in the United States. Eugenic policies, aiming at discouraging the immigration or birth of supposedly inferior individuals, became law in twenty-nine American states (Kevles, 1995). But it was the far more sinister eugenic policies of Nazi Germany which finally discredited the movement. After the war, eugenics went into a rapid decline. Its erstwhile intellectual leaders quickly distanced themselves from it or were marginalized, the legislation which had put its policies into practice was repealed. It was not, it seems, because eugenics had been shown to be false, because the theories which it had elaborated had been refuted, that it suffered this fate. It was because of its association with crimes of an unprecedented enormity that it lost favour.

Because of the link between views like Spencer's and eugenics, the vogue for finding normative implications in evolution largely passed. The belief that some individuals were naturally fitter than others came to be associated with racism and with a morally retrograde and scientifically unsound elitism. On more purely intellectual grounds, Moore's 'open question argument' – specifically formulated with Spencer in mind – seemed to defeat all attempts to identify goodness with being highly evolved (Moore 1903). It would, Moore argued, always be an open question whether any identification of moral terms with natural properties was correct. The direct route, from meta-ethics to normative principles, was apparently blocked.

If Spencer won the debate between him and Huxley during their lifetimes, in the sense that his views were far more influential at the time, then perhaps we can say that Huxley emerged the eventual victor. After the war, most moral philosophers sided with him in agreeing that ethics was an autonomous domain, even if our capacity to engage in it was itself a product of evolution. Even when a more specifically evolutionary ethics began to revive in the 1970s and '80s, spurred on by the success of sociobiology, the philosophers who engaged in it steered, for the most part, away from normative and motivational questions. Instead, they concerned themselves largely with questions that Huxley deemed appropriate, in particular with the question whether the fundamental bases of morality – whatever they turn out to be – might have a direct evolutionary origin.[3]

Gradually, however, their ambition grew. Indeed, the naturalizing impulse cannot appropriately be confined to genetic (in the historical, not Mendelian, sense) questions. If morality is the direct product of evolution, then we can explain its content in adaptive terms. Answering the genetic question in the affirmative will therefore have far-reaching implications for our account of moral motivation, for our meta-ethics and for normative questions as well. The debate between Spencer and Huxley is not over: if morality can be directly tied to evolution

---

3   Of course, given that natural selection is true, all human abilities and characteristics are the product, in some sense, of evolution. By asking whether morality is a *direct* product of evolution, I ask whether its evolutionary history can throw light on its nature and function, or whether we are better off seeking to understand it in, say, exclusively cultural terms.

(as Spencer claimed), it is a very different kind of phenomenon than the ethics which might emerge from processes of cultural and intellectual elaboration on a biological base, the kind of morality that Huxley defended. In particular, as we shall see, a neo-Spencerian morality seems far more likely to be subjectivist, non-realist and restricted than a Huxleyean morality. Before re-examining the meta-ethical and the prescriptive projects in the contemporary debate, let us now turn to the explanatory project.

The Problem of Altruism and The Evolution of Morality

Altruism is a puzzle for evolutionary ethicists, for fairly obvious reasons. Evolution systematically favours *phenotypic* traits – the bodily morphology and behaviour of organisms – which are adaptive. A phenotypic trait is adaptive just in case it increases the fitness of the organism; fitness, in turn, we define in term of the reproductive success of the organism. Fit phenotypic traits are thus reproduction-relevant: they are characteristics that enable the organism to compete successfully for reproduction-relevant resources, with conspecifics and with organisms from other species. Just which resources will be reproduction-relevant will vary from species to species, but they will always include the means of survival (food and shelter) and of attracting high-quality mates (assuming sexual reproduction). Thus, whether they are aware of it or not, all organisms are apparently engaged in a struggle for existence with every other.

Evolution can occur only if some organisms have traits that make them fitter than others, and if these traits are heritable. Offspring must tend to resemble their parents. Given that this is the case, traits that are fit will tend to spread in a population, since a fitter trait enables its possessor to have fitter offspring. The stage is now set for the problem of altruism. A behaviour is altruistic just in case it benefits others (whether they are members of the same species, or of other species) at the expense of the organism whose behaviour it is.[4] It follows from this that someone who acts altruistically reduces her own fitness, while boosting that of the recipient. Altruists will therefore, on average, have fewer descendants than non-altruists. We ought to expect altruism to be a casualty of the struggle for existence.

But altruism does seem to exist, among human beings and elsewhere in the natural world: people and other animals sometimes seem willing to put themselves out for others, without hope of recompense. This is the problem of altruism: given that we are the products of evolution, how did we come to be altruistic, to whatever extent? E.O. Wilson called this 'the central theoretical problem of sociobiology' (Wilson, 1975, p. 3). This problem is of first importance to evolutionary ethics, since it seems part of the definition of morality that it will require us sometimes to give at least *some* weight to the interests of others independently of our own interests: that is, morality requires altruism of us. Darwin's own solution to the problem of altruism remained the most influential up until the 1960s. It goes by the name of *group selection*. Though it is true that individuals who behave altruistically have lower fitness than other individuals who do not, internally altruistic *groups*

---

[4]   This is a stipulative and biologicized definition of a concept that has its place in ordinary moral discourse, but its main ingredients are defensible. We would not call an action altruistic if it did not (aim to) benefit someone other than the actor. If the actor believed that she would benefit from the act more than from alternatives, then it is not altruistic.

can outcompete internally selfish groups. That is, a group composed of individuals who have a disposition to aid one another might be expected to do better than a group composed of only selfish individuals:

> although a high standard of morality gives but a slight or no advantage to each individual man and his children over the other men of the same tribe, yet that an increase in the number of well-endowed men and an advancement in the standard of morality will certainly give an immense advantage to one tribe over another. A tribe including many members who, from possessing in a high degree the spirit of patriotism, fidelity, obedience, courage, and sympathy, were always ready to aid one another, and to sacrifice themselves for the common good, would be victorious over most other tribes; and this would be natural selection. (Darwin, 1871, p.166)

Thus, despite the fact that altruism is individually less fit than selfishness, selfish groups would go extinct, and altruism would spread.

Group selectionist explanations were routinely invoked by biologists from Darwin's day till the 1960s. But in that decade it was subjected to seemingly devastating criticism by a new generation of evolutionary biologists, especially John Maynard Smith (1964) and George C. Williams (1966). These biologists were more mathematically sophisticated then their predecessors, and utilized precise models to test group selectionist hypotheses. They found that group selection was vulnerable to *subversion from within*. On the assumption that altruistic acts benefit recipients more than they cost donors (so that altruism is a net benefit to the group), groups composed largely of altruists will indeed grow faster than those composed largely or exclusively of selfish individuals (just as Darwin predicted). But since selfish individuals are fitter than altruists *within* each group, the proportion of altruists in each group declines. Though (largely) altruistic groups might drive selfish groups to extinction, they are destined to become selfish themselves. So long as selfish organisms can arise within altruistic groups, whether by mutation or by immigration (and over the kind of time-spans with which we are concerned, such events are very likely), altruism seems destined to be driven to extinction.

We should not conclude from this that group selection is impossible. Under the right conditions, it can certainly be a powerful force. If the proportion of altruists within groups necessarily declines, then group selection requires that groups do not persist long enough for this factor to eliminate altruism. If altruistic groups break up and re-form, or establish colonies, *and* the successor groups have a higher proportion of altruists than the mother group, altruism can persist and even increase across the global population, so long as the formation of new altruistic groups occurs at a rate rapid enough to outrun the effects of subversion from within. In the 1970s, most biologists believed that these conditions would be met with so rarely that group selection could not be a powerful force in evolution, and was therefore extremely unlikely to be the source of altruism. More recently, group selectionism has made something of a comeback. Its most energetic defender is probably David Sloan Wilson, alone and together with Elliot Sober (Wilson and Sober 1994, reprinted here; Sober and Wilson 1998). Opponents allege, however, that this comeback is illusory, since it consists in important part, in reinterpreting supposedly individual selection processes as disguised instances of group selection. In any case, group selection is unlikely to be powerful enough to account for the evolution of altruism by itself.

In the wake of the decline of group selection, several alternative explanations for the evolution of altruism were developed. The most promising are *kin altruism* and *reciprocal altruism*. Kin altruism is one of the triumphs of the gene selectionist perspective upon evolution associated especially with George C. Williams (1966) and Richard Dawkins (1976). From this perspective, adaptations are for the ultimate benefit of the genes, not the individuals they 'build'. If this is the case, however, then 'altruistic' behavior by an organism can, under the right conditions, be a kind of genetic selfishness – in biological terms, such behaviours can boost *inclusive fitness* (fitness calculated from the genetic viewpoint). Other things being equal, sexually reproducing organisms share half their genes with full siblings, and with their offspring, one-quarter of their genes with nephews and nieces and grandchildren, and so on. Thus, by aiding a close relative, organisms can benefit the genes she shares with it. Genes which motivate kin altruism can therefore spread in a population. Evolutionary biologists have been able to use kin selection to explain the apparent altruism of the social insects. As a result of their unique biology, worker bees and ants share three-quarters of their genes with one another, while queens share only half their genes with their offspring. Workers therefore do better, from a genetic perspective, by aiding the queen to create more siblings of theirs than they would by having offspring of their own (Hamilton, 1972). The apparently selfless behaviour of bees in giving their lives for the good of the hive is explained in precisely the same manner: in benefiting her conspecifics, the bee benefits those genes she shares with them.

Kin altruism is restricted to relatives, but reciprocal altruism can cross almost any barriers, including within its ambit even members of other species. Reciprocal altruism, roughly, refers to the benefits given to a recipient in exchange for, or in the expectation of, a return (Trivers, 1971). Game theoretic modelling of exchanges between organisms has shown that this kind of altruism can evolve, even among organisms which lack the cognitive capacity to represent (or indeed to have) psychological states. Typically, interactions between potential cooperators are modelled as a prisoner's dilemma. The original prisoner's dilemma scenario is as follows:

> Two prisoners are interrogated separately by police. They are accused of committing a crime together, but the police do not have sufficient evidence to convict them of this crime. Each is offered the same deal: if she will confess her guilt, but agree to testify against her codefendant, she will be released on a good behaviour bond. If she stays silent, however, and her co-defendant accepts the deal, it will be him who is released, while she goes to jail for ten years. If both the accused confess, each will go to jail for five years. And if neither confess, the police will be unable to secure a conviction for either. However, they will be charged and convicted with some lesser crime – perhaps resisting arrest – and will receive a six-month jail sentence each.

Assuming that each agent is exclusively self-interested, and wants to minimize her time in jail, the rational strategy for each is to defect (that is, to betray her comrade). If agent A defects and B cooperates (by refusing to talk) then A will go free, while B gets ten years jail. But if B defects then A minimizes her jail term (halving it from ten years to five) is she defects as well. Of course, the situation is entirely symmetrical: what goes for A goes for B as well. Hence if both agents are rational, they will both defect, and both go to jail for five years each. Yet if they had cooperated with one another, they could each have avoided four and a half years jail time.

The exact numbers do not matter; it is the structure of the game and the relative payoffs that makes it a prisoner's dilemma. The payoffs must be as follows: the *temptation to defect* (the amount a player receives for defecting on a cooperator) must be greater than the reward for mutual cooperation, which must in turn be greater than the punishment for mutual defection, which must be greater than the *sucker's* payoff, the punishment for cooperating with a defector. Whenever we have a prisoner's dilemma structure, defect is the dominant strategy, which is to say that, whatever the other player does, each player is better off defecting.

If interactions have the form of a prisoner's dilemma, how can altruism evolve? Since each organism would do best by defecting, we should expect cooperation to go extinct. However, this melancholy conclusion can be avoided if the game is *iterated*: that is, if interactions between players are repeated (and it is unpredictable how many iterations there will be). Under those conditions, cooperation becomes a viable strategy. By cooperating, one player can signal his willingness to cooperate on condition that the other does likewise. The strategy of cooperating with strangers and with those who have cooperated in the past, while defecting on those with a reputation for defecting, has been shown to outcompete both unconditional cooperation and unconditional defection. Moreover, the strategy is *evolutionarily stable*: it is not vulnerable to subversion from within.

Does this strategy, which evolves because it serves the interests of the organisms, deserve to be called altruism at all? Some thinkers argue that neither it, nor kin altruism, is genuine altruism (Sesardic, Chapter 6). Anything which evolved to serve an individual's (or, perhaps worse, a gene's) interests cannot be genuinely altruistic. But this seems to run together two concerns better kept apart: the function of an adaptation and the details of its psychological implementation. Nothing evolves by natural selection unless it is able to enhance the ability of its vehicle to replicate. But it doesn't follow that the vehicle will be motivated by selfish concerns. Sexual desire evolves to enhance reproductive success, but it does not follow that sexual desire is the desire for offspring. Similarly, kin and reciprocal altruism evolved to serve the interests of organisms or their genes, but organisms need not be motivated by concern for their interests. Instead, reciprocal and kin altruism might lead to the development of, and be driven by, genuine sympathy for others, despite the fact that at a genetic level it is selfish.

Indeed, there are reasons to think that it is likely that these routes will be implemented by other-regarding motivations, rather than concern for one's own interests. If interactors are merely rational calculators of utility, they cannot convince one another of their disposition to cooperate once doing so is no longer in their interests – that is, once the other party has cooperated. So it is to each interactor's advantage to be able to show that it acts upon motives other than the calculation of utility. One way out of this dilemma might consist in demonstrating that the organism calculates long-term utility, but there are at least two problems with this solution. It may be too costly to implement, from a biological point of view, since the machinery for calculating long-term payoffs is complex, and it may be insufficient, in any case, since organisms will be less likely to interact with others when they know that these others might defect as soon as the payoffs from doing so cross a certain threshold.[5] Far

---

[5]   Moreover, it can be to an organism's advantage to possess dispositions to action, in certain circumstances, even though it would not be to its advantage actually to act upon those dispositions. If everyone knows that I shall stop at nothing to avenge a petty slight, then I may not be slighted at all, and that is to my advantage, though if I were slighted, and reacted in the threatened manner, I may pay a high price. For many examples of this kind, see Frank (1988).

better, then, for the organism to adopt the solution of manifesting the disposition to cooperate, on condition the other does, no matter what the payoffs. The disposition to altruism, now considered as a psychological state or a motive upon which organisms act, will, under the right conditions, boost inclusive fitness (Frank, 1988). So it is not true that altruism cannot emerge from reciprocal exchanges.

A great deal of subtle and important work by evolutionary biologists (Maynard Smith 1982; Dawkins 1976), game theorists (Skyrms, 1996; Vanderschaaf, 2000) and philosophers (Kitcher, 1993, Chapter 2; Sober, 1994; Sober and Wilson, 1998) has gone into elaborating these explanations for how altruism might emerge from the mechanisms of natural selection, even though those mechanisms ultimately reward "selfish" behavior. But these different theorists often have different targets in mind, when they seek to explain altruism. To avoid confusion, we need to adopt an important distinction Sesardic (Chapter 6) makes, between *psychological* altruism (altruism$_p$) and *evolutionary* altruism (altruism$_e$). An organism acts altruistically$_p$ if it acts with the intention of benefiting others at some cost to itself, whereas it acts altruistically$_e$ if it actually boosts the fitness of others relative to itself. Separating these kinds of altruism is essential if we are to understand just what claims different theorists are making.

Hence kin altruism is primarily an explanation of altruism$_e$. It seeks to explain how helping behavior can be selected for, via the notion of inclusive fitness. It can be agnostic on altruism$_p$, or downright sceptical concerning its existence. Certainly some theorists have tried to interpret kin altruism as consistent with psychological egoism. If organisms typically act to aid close relatives – say, their offspring – to relieve feelings of distress of their own, then kin selection may be psychologically egoistic. But kin selection might also be cited as the first step in a two-stage argument for altruism$_p$. This argument comes in two forms, defending a restricted and a general disposition to altruism$_p$. On the first, altruistic dispositions toward close kin are held to be the product of kin selection. This view is most convincingly articulated by Elliott Sober, in work on his own and with David Sloan Wilson (Sober, 1994; Sober and Wilson, 1998, Chapter 3). Essentially, Sober and Wilson argue that kin selection mechanisms will be driven by concern for the welfare of kin, rather than by egoistic desires because such concern is a more direct solution to the design problem under consideration, and therefore more reliable (on the plausible assumption that there are cases in which purely hedonistic desires will not motivate the organism to act in the ways which maximize its inclusive fitness). Therefore, kin selection will probably result in the formation of altruistic$_p$ dispositions and desires. But these desires will be restricted, in the sense that they will have as their target only (close) kin.

Thus kin selection might explain how we come to have altruistic desires directed at our kin. But this is far from an explanation of the kind of altruism necessary for morality. The extent to which morality requires us to treat all persons equally, regardless of ties of blood or affection, is a matter of ongoing debate, of course, but there can be no doubt that a morality worthy of the name requires *some* altruistic concern for those beyond the circle of family. We must be willing, at very least, to sacrifice our trivial interests for their most important ones. Might kin selection explain this more general altruism$_p$? Alexander Rosenberg (2000) speculates (I choose that term advisedly, to indicate his own lack of commitment to the hypothesis) that it might. If human beings have lived in relatively small groups for most of our evolutionary history (as most anthropologists believe) then we might have been selected for relatively indiscriminate altruistism$_p$. Mechanisms which would allow us to distinguish relatives from non-relatives would have a cost, a cost which would not be worth paying if we lived in bands which consisted almost entirely of kin and encounters with strangers were relatively rare. However, given enough time natural selection would evolve more discriminating mechanisms, Rosenberg suggests. (He obviously believes that humans have been around long enough for such indiscriminate altruism to have been eliminated.) Certainly, it does not seem uncontroversial to say that we tend to have stronger altruistic dispositions toward kin than toward strangers, which suggests that kin selection has not been implemented by way of indiscriminate altruism$_p$ in human beings. Perhaps, however, we have evolved a disposition to aid those who exhibit some characteristic which (in the environment of evolutionary adaptation) would have been a reliable marker for kinship. Palmer and Palmer (2002) suggest that accent and dialect might play this role. If altruistic dispositions are triggered by *any* similarity markers, then altruism$_p$ might be promoted by any information that allows us to empathize with others. But, if kin altruism is triggered by specific markers – whether of kinship, or of some other property that was, in the environment of evolutionary adaptation, reliably associated with kinship – then its range will of necessity be restricted. In that case, we shall have to turn to other mechanisms to explain the origins of our apparent tendency to possess (at least some) altruistic$_p$ tendencies toward almost all human beings.

Indeed, our altruistic$_p$ dispositions even extend beyond the bounds of our own species. This may seem mysterious from an evolutionary perspective, but it is not. Since reciprocal altruism is based on exchange, we can engage in it with any kind of organism so long as we are able to benefit one another. We have seen how reciprocal altruism might require the development of a cooperative disposition. We can now express this by saying that, though it is not in any obvious sense altruistic$_e$, it may nevertheless be the condition for the development of altruism$_p$, and this altruism$_p$ may, in turn, motivate altruistic$_e$ acts.

We now have some idea of how the dispositions which underlie morality might have come about. We are finally able to turn to our main subject: just what kind of morality might we expect to be a product of natural selection? Do some moral theories look more or less plausible in the light of the evolutionary story we have briefly sketched? In particular, does moral realism look less plausible in its light?

Evolutionary Meta-ethics

In what ways might the kinds of stories sketched by evolutionary biologists, game theorists and philosophers have the kinds of deflationary or eliminativist implications that some philosophers fear – or welcome – in them? The answer to this question will naturally depend on what we believe morality to consist in (that is, what properties an accurate analysis of the concept would impute to it; what morality is, or would be, if there was such a thing), such that evolution could threaten it.

We might best approach the question by asking what propositions a full and complete analysis of our concept of morality would contain, which might plausibly be threatened by the evolutionary hypothesis. There are at least five relevant possibilities:

(1) Morality might commit us to the existence of Platonic moral facts, which are ontologically independent of human beings or other rational beings.

(2) Morality might commit us to the existence of moral facts which, while not ontologically independent of the existence of any rational beings, are binding on all such beings, and which therefore cannot vary across space or time.

(3) Morality might commit us to the existence of objective properties, the truth conditions of which do not essentially contain references to the subjective states of the beings upon whom they are binding.

(4) Morality might commit us to the existence of moral facts which are such that their existence ensures that it is rational for us to behave morally, in general or (more strongly) on each particular occasion.

(5) Morality might require us to give some weight to the interests of all parties affected by our acts, regardless of their relationship to us.

Some of these claims are, prima facie at least, much more plausible than others. Thus, how threatening to moral realism an evolutionary explanation of morality will be depends upon which – if any – of these claims it is taken to undermine.

Evolutionary Expressivism

Evolutionary ethicists often take their hypotheses to undermine (3). In fact, they claim, the subjective states of the person making a moral judgments figure in its truth-conditions. The thought underlying this view is seldom explicitly developed, but the idea seems to be something like this: if morality is an adaptation, then we will very likely discover that its essential building blocks did not spring into existence with *homo sapiens*, but exist in other species as well, especially those closely related to us (Flack and de Waal, Chapter 1). Indeed, evolutionary hypotheses for the development of morality trace the development of the dispositions and behaviors thought to underlie it in social insects and bats, fish and monkeys (Wilkinson, 1990; Ridley, 1996). Clearly, however, these organisms are not moved to act by truth-assessable representations of the world. Instead, their proto-moral behavior must be driven by instinct or by feeling. It is this proto-morality which human beings inherit from simpler creatures, and which we go on to elaborate into complex intellectual systems. But no matter how intricate a superstructure we build upon these inherited foundations, it remains the case that morality is essentially subjective. We are motivated to act as we do by feelings, not by beliefs. To think otherwise is to insert a gap in nature, a sudden leap or saltation (to use Steven Jay Gould's useful term), where in fact there is none. It is to fail to see that we are continuous with the rest of the animal kingdom, in our morality as much as anywhere else (Waller, 1996, 1997; McShea and McShea, 1999).

Thus, evolutionary expressivists conclude, moral judgments do not really state facts about the world outside us at all. Instead, they express our feelings, our evolved sentiments. Evolution gives crucial support to expressivism. Something very like this view is defended by Ruse (1998) and Waller (1996, 1997).

Arguably, evolutionary expressivism has *normative* implications. It implies that claim (5) above is false. If our moral judgments express our evolved sentiments, then the content of those judgments is limited by the range of sentiments which we have evolved to feel. If we believe, further, as Ruse and Waller both do, that evolution has selected for altruistic$_p$ dispositions only with regard to close kin and perhaps others who bear markers reliably associated with close kin in the environment of evolutionary adaptation, then we shall conclude that the range of true moral judgments of which we are capable is much narrower than it is usually taken to be. For Ruse, for instance, it is a mistake to believe that morality requires us to give much weight to the needs of those who are distant from us. Since morality is 'rooted in our feelings' (Ruse, 1998, p. 241), but we are likely to feel much more strongly for kin than for the distant needy, morality cannot require impartiality or indiscriminate altruism of us.

However, it is a mistake to think, as Ruse and Waller sometimes seem to, that the only alternative to accepting the normative and meta-ethical positions for which they argue is to reject a substantially evolutionary explanation of morality. We can coherently deny either (a) that evolution can be expected to give rise to altruism$_p$ that is essentially limited to kin, or (b) that because emotions played a crucial role in the evolutionary history of morality, such emotions must figure in the truth-conditions of moral judgments today, or (c) that, on the assumption that evolution gives rise to relatively restricted altruism$_p$, it follows that we have correlatively restricted moral obligations. Claim (a) is denied by several philosophers (for example, Kitcher, Chapter 2), but evaluating it would take us too far afield, into the realm of game theory. Here we shall concentrate on claims (b) and (c).

Claims (b) and (c) together entail the Spencerian claim: that morality is identical with our evolved dispositions. Claim (b) holds that the emotions which figure in the evolutionary history of morality continue to figure in its contemporary truth-conditions, and claim (c) holds that our obligations extend just as far as do these sentiments. Since the evolutionary models of morality have it arising out of the interactions of relatively unsophisticated organisms, the implicit claim is that human rationality adds nothing significant to morality.

But why think this? The most powerful argument in its favour seems to be a redundancy argument. Proto-morality – the core of morality we share with vampire bats, cleaner fish and the other primates – is largely or entirely a subjective phenomenon, driven by instinct and desire. This fact makes the objectivist hypothesis redundant. Why postulate moral beliefs, when it is clear that the subjective core of morality is sufficient to explain moral behaviour?

Indeed, we might read the evolutionary expressivist as offering crucial support to a view which has attracted a great deal of attention in recent meta-ethics: the claim that moral considerations are explanatorily irrelevant. In Harman's (1977) original version of this claim, the moral properties of actions, events or characters were held to be irrelevant to the judgments we were disposed to make concerning them. Given our dispositions, we would make the same judgments regardless of their truth. This line of argument invites the response (roughly the one advanced by Sturgeon, 1985) that we have been given no reason to think that these dispositions do not track real properties, and without such an independent argument, no reason to reject their deliverances. From this perspective, we can view the evolutionary considerations as filling the gap Sturgeon sees in Harman's argument. By providing an explanation of our moral dispositions, which shows how we (might possibly) have come by them *for reasons that have nothing to do with morality*, it casts doubt on their reliability as trackers of truth. Indeed, the evolutionary history which gave rise to our moral sentiments systematically favours (what we call) selfishness. What more evidence do we need to be convinced that they are unlikely to track *real* properties of the world that are *really* moral?

Something rather like this reinforced redundancy argument is advanced by Joyce (2001), when he asks us to compare ourselves to John, who is certifiably paranoid. John believes that Sally is 'out to get him'. Now, it is possible that Sally is really out to get John, but knowing that John is paranoid leads us to think that his claim is unlikely to be true. John would judge that Sally is out to get him, no matter how she behaved. Similarly, Joyce claims, since we have been naturally selected to think that certain actions and events are right or wrong, our judgments are not sensitive to the truth of these claims. We would continue to make them, regardless of whether they were true. Hence, like John's claims about Sally, our moral judgments are unjustified, and therefore should be considered to be (probably) false.

How should we respond to this reinforced redundancy argument? The first thing we need to note is that it is very plausible to think that our evolved dispositions do track real properties (Rottschaefer and Martinsen, 1990). Indeed, Joyce himself concedes this point, noting that the dispositions which underlie morality would not be fitness-enhancing if they were not a response to real properties in the external world. So the argument goes through only if there are good reasons to doubt that these real properties are *really* moral.

Joyce has an independent argument against identifying the dispositional properties that trigger our moral sentiments with properly moral properties. However, since the argument is entirely conceptual, and not evolutionary, I relegate discussion of it to a footnote.[6] Suffice it to say that it is ultimately unconvincing. However, there is apparently another route to the same goal. It lies through that much-discussed scourge of evolutionary ethics the (so-called) naturalistic fallacy. We need to examine this alleged fallacy, both because it is widely taken to invalidate any substantively evolutionary ethics, and for its implications for the dispositional analysis of moral concepts.

---

[6]   The argument consists, essentially, in an attempt to undermine (4), the claim that we have reason to act morally. Construed literally, Joyce argues, moral claims make *categorical* demands on us; they require us to behave as they prescribe, regardless of our interests or desires. Such categorical requirements are, by their nature, reason-providing: If someone ought (morally) to φ, then she has a *reason* to φ, regardless of her interests or desires. But, Joyce argues, we can make no sense of a reason that is independent of our interests or desires. The broadest framework of reason-giving is the framework of practical rationality. It is also the only inescapable framework, because it, and it alone, is presupposed by any demand for reasons. But the demands of morality are not the demands of practical rationality, since the demands of the latter must be understood in terms of interests and desires. The demands of the only inescapable framework there is are hypothetical. So there cannot be the kinds of categorical demands to the existence of which morality commits us. Construed literally, moral discourse commits us to a kind of reason which is incoherent, so we are in error when we use it in this way.

Joyce is, no doubt, correct in holding that moral claims are appropriately analysed as categorically binding. It does not follow, however, that we must understand categoricality in terms of practical rationality. Indeed, if we did cash out moral 'oughts' in this fashion, we would miss their point. We are required to act morally, quite independently of our interests or desires. The moral ought can only be understood in irreducibly moral terms (Devitt, 2002). Thus, Joyce is right in claiming that moral judgments are categorically binding, and right to suggest that this entails that each person has a reason to act morally. But that reason must itself be understood morally. We ought to refrain from torturing innocent children because it is *wrong* to act in this manner, not because we necessarily have desires or interests which will be satisfied by so doing. Moral demands can only be grasped from inside the institution of morality. Cashing out its claims in non-moral terms is not merely contingently impossible, it is *constitutive* of morality that this is so. (For similar reasons, attempting to show that morality is objective on the grounds that accepting some moral system is in our evolutionary interests, as Campbell [1996] claims, misses the point of morality and therefore fails to justify it.)

The Naturalistic Fallacy

The naturalistic fallacy is often cited as a supposedly decisive objection to any evolutionary analysis of morality (Woolcock, 1999; Lemos, 1999). Someone commits the naturalistic fallacy when they attempt to define goodness in natural terms. Do evolutionary ethicists commit this fallacy? It is unlikely that they do, for the simple reason that it is far from clear that there is any such fallacy (Smith 1994, 2000).

Certainly, it is a mistake to think that any *evolutionary* analysis of the meaning of moral terms can capture the implicit commitments necessarily had by all competent language-users. If this were possible, then the meaning of such terms would have had to have undergone a significant alteration since the theory of evolution was formulated in the nineteenth century, and that is an extremely implausible suggestion (Joyce, Chapter 10). But a correct analysis of a concept doesn't have to be *a priori*. Indeed, the analyses of natural kind terms offered by the sciences are *a posteriori*, and far from obvious. It is no objection to the proposition that water is H_2O that competent speakers do not necessarily intend 'H_2O' by 'water'. Since the theory of evolution is a (well-confirmed) scientific hypothesis, we can expect that the kind of analyses it is capable of generating will also be *a posteriori*. Though we certainly don't *mean* 'likely to enhance inclusive fitness in the environment of evolutionary adaptation' (or whatever) by 'good', we might discover that actions and character traits that we are disposed to call good are *in fact* likely to enhance inclusive fitness. Nothing Moore or his supporters have said suggests that this cannot be the case.

That said, it needs to be recognized that the analyses of moral terms that evolutionary ethicists have offered have tended to be very implausible. Consider a representative analysis:

> the term 'evil' designates behaviours by one or members of a group (society) that, were it generalised, would reduce the long-term fitness (i.e., over many generations) of all members of the group. (Thompson, 2002, p. 246)

The problem with all such analyses is not that they commit any fallacy, it is that they are unconvincing as analyses, and they are unconvincing because of the revision of morality they would force upon us. Analyses do sometimes force us to revise our concepts. Consider the analysis of our concept 'fish'. When this concept was correctly analysed, we realized that not all organisms we were disposed to call fish had the features mentioned in the analysis. Whales and dolphins, for instance, breathe air and bear their young live. So we had to exclude whales and dolphins from the class of fish. This was a cost, in a sense, because the concept we were analysing was one which we had previously been disposed to apply to dolphins and tuna equally, but it was a cost we were prepared to pay, since we now realized that there were very significant differences between dolphin and tuna, such that, by applying different natural kind terms to them, we were doing a better job of cutting the world at its joints.

However, we are not prepared to pay the analogous cost in revising our moral terms. Consider how we would have to revise our morality if Thompson's analysis were to be accepted. If all and only acts which, if generalized, would reduce fitness in the long term are to be described as evil, then indiscriminate altruism is evil, on the plausible assumption (defended in Mackie, 1978) that indiscriminate altruism generalized provides the conditions for a rapid increase in organisms that play the 'defect' strategy in prisoner's dilemmas, and therefore lead to a fall in average fitness. Now, we can coherently debate whether indiscriminate altruism is good.

Perhaps it is better described as unwise. But whatever else it is, it seems radically implausible to suggest that it is evil.[7]

The point can be generalized: any analysis of moral terms that claims that they ought to be understood *directly* in evolutionary terms – which claims that 'good' is equivalent to 'highly evolved', as Spencer had, it or to having the properties that would be approved of by organisms which had evolved under ideal conditions, as Collier and Stingl (1993) suggest – fails, because it implies that certain propositions, to which we are more strongly attached than we are to any evolutionary analysis of goodness, are false. At least so long as the models for describing the evolution of morality are reasonably accurate (and these models or something very like them are accepted by all evolutionary ethicists), it is easy to think of actions and character traits that would meet the requirements of the suggested analysis, but which are not good, or which are good, yet would not be approved of. It is plausible to think that xenophobia is an adaptation which boosts (or, in the environment of evolutionary adaptation, boosted) inclusive fitness, but xenophobia is not good.

It is not because these analyses commit a fallacy that they are false. It is not an open question whether indiscriminate altruism is evil, or xenophobia good. These are closed questions, and the analyses which entail them fail for this banal reason. They are false, not fallacious. This does not imply that evolutionary considerations might not figure importantly in moral theory, perhaps even at levels other than the genetic. But it does imply that philosophers like Joyce are right in saying that we cannot identify moral properties with the properties we have evolved to feel certain kinds of dispositions toward. To put the same point in another way, it is a mistake to identify proto-morality – the animal base, as it were, of morality full-blown – with morality itself. On this point at least, we ought to side with Huxley against Spencer. We shall return to this point. For the moment, we need to pick up the main thread of our journey.

Evolutionary Error Theories

It is a mistake, we now know, to identify moral properties with the real properties toward which natural selection has made us sensitive. Morality is not a system of enlightened selfishness, as Spencer thought. Or, more carefully, our concept of morality is not of such a system. According to some evolutionary ethicists, we have here the basis for an error theory: nothing answers to our concept of morality. The question therefore arises how we came by such a concept. Why have we made such a glaring error?

There is, of course, an evolutionary explanation available of our alleged error. Since morality is fitness-enhancing, it is in our interests for us to be strongly motivated to act upon its dictates. Emotional pushes and pulls will frequently be sufficient to cause us to act appropriately, but morality will have an even more powerful grip on us if it is backed up by belief. Hence, not merely the dispositions to act morally, but also the disposition to believe that morality is objective, are adaptive. Both the subjective reality of morality, and the illusion

---

[7]   Harms' teleosemantic analysis of moral concepts fails for exactly the same reason. Harms argues that moral judgments are true just in case they are playing the role for which they were selected (Harms, 2000). But many true moral judgments are fitness-*reducing*, not fitness-enhancing, because morality has in some way broken free from the functions which account for its existence.

of objectivity, are the product of natural selection (Ruse, 1998). In fact, morality is nothing more than a 'collective illusion foisted upon us by our genes' (1998, p. 253).

Defending an error theory successfully depends on making good on two requirements. It demands, first, that a convincing analysis of the concept in question be given and, second, that it be shown that nothing corresponds to the concept. Mackie's error theory, for instance, analyses moral claims as committing those who make them to the existence of objective prescriptive facts: facts the recognition of which is necessarily motivating. Mackie (1977) argued that such 'queer' facts are metaphysical extravagances. The evolutionary error theory analyses morality as demanding more of us than we can give: it claims that our concept of morality is of a set of obligations which transcend the bounds of kin and reciprocity, but our moral dispositions are confined within these bounds. Our concept of morality commits us to (5), but (5) is false.

Those philosophers who are inclined to give dispositional analyses of moral concepts, however – with one or two explicitly evolutionary exceptions – do not take the dispositions in question to be simply evolved sentiments. On the contrary, philosophers who have taken this line explicitly argue that the dispositions in question need to be cultivated and trimmed (McDowell, 1985, 1995). For those of us who set great store by the method of reflective equilibrium, such a refined dispositional analysis seems much more convincing, indeed perhaps even inescapable, than the directly evolutionary analysis. Consider our brief discussion of directly evolutionary analyses of moral concepts. We rejected these analyses because they had wildly implausible implications. In other words, they clashed with our moral intuitions, suitably adjusted to accommodate our best moral and empirical theories. On the assumption that these intuitions are internalized, so that they are reliably expressed in our moral sentiments (an assumption which seems plausible given the way in which the responses of people and entire societies change across time), it was indeed to our dispositions to which we appealed in rejecting the directly evolutionary analyses.

The best dispositional accounts of morality therefore do not seem to be directly evolutionary. If these accounts are acceptable, then our moral judgments do not commit us to any error – not, at least, to any error identified by the evolutionary ethicists. If evolution truly undermines moral realism, we must look elsewhere for the reason.

Our discussion of the meta-ethical implication of evolutionary hypotheses concerning the origins and function of morality has so far focused on a set of issues which centre around (3), the claim that morality has truth-conditions which are independent of the subjective states of those who make them, and (5), that morality requires us to give some weight to the interests of all parties affected by our actions. We now turn to (2), the claim that moral facts cannot vary across space and time, and are therefore equally binding on all rational creatures.

Morality on Other Planets

We may not be alone in the universe. Intelligent life may have evolved on other planets. Perhaps, indeed, we are not the only moral animals in the universe. But what would alien morality look like? It might be the case, some evolutionary ethicists argue (Waller,1996; Ruse, 1998), that the kinds of actions which we regard as obligatory are rightly held to be immoral by some aliens. If their genetic constitution were different, to ours or if their evolution took a different path, then the illusion of objectivity under which they labour might attach to actions we regard as immoral. Surely this is sufficient to show that objective morality is an illusion? If there were an objective morality, then it would be binding upon all rational creatures (as Kant points out). But there is no such morality.

What are we to make of this argument? The contention that the contents of our morality is sensitive to the details of evolutionary history is plausible. What counts as harming and benefiting someone, most obviously, is in important part a function of their biology, which makes them vulnerable to certain dangers and in need of certain resources and opportunities. But this fact is surely not sufficient to establish the species-relativism of morality, for it is vulnerable to the same kinds of replies that are used to discredit many claims of descriptive cultural relativism. We establish any kind of interesting relativism only if we show that members of different groups have different *fundamental* obligations (Rachels, 1995; Moser and Carson, 2001). Clearly, the fact that Australians are required to drive on the lefthand side of the road, while Americans are required to drive on the right, does not establish any kind of interesting relativism. Similarly, the fact – if it is a fact – that we might have been required to eat one another's faeces (to use Ruse's own example) if evolution had taken a different path is not sufficient to establish the kind of species relativism that Ruse and Waller hope to demonstrate. It might be that at a high enough level of abstraction, the kind of morality which would emerge from the demands of cooperation would be the same in all possible worlds, consisting of injunctions to treat everyone impartially, to sacrifice one's own lesser interests for the greater interests of others, and so on. At least, nothing Ruse and Waller say shows that this is not the case.

However, Waller at least has a reply to this line of thought implicit in his work. Though it might be true that mere differences in the content of morality are not sufficient to establish the kind of species-relativity that would undermine the objectivity of morality, the *right* kind of differences would have this effect, and the right kind of differences might arise from different patterns of genetic inheritance. Biologists have convincingly explained the intensely social behavior of ants, bees and termites – in particular, their apparent readiness to sacrifice themselves for the nest – in terms of their unique system of reproduction. Worker bees, for instance, are more closely related to one another than to their queen, and more closely related to each other than we are to our children. So they boost their genetic fitness by 'farming' the queen, to make more siblings of theirs, and by sacrificing themselves for the nest. Now, Waller asks us to imagine that organisms with this kind of reproductive system evolved a degree of intelligence comparable to our, and as they did so came to possess a morality which is an outgrowth (at least in important part) of kin selection, just as ours probably is. Would they not have *fundamental* moral obligations to each other quite different from ours? They would find our emphasis on the individual and her rights 'not merely absurd, but morally odious' (Waller, 1996, p. 253). This thought experiment shows that morality is indeed species-relative, in the strong sense required to undermine moral objectivity.

Or does it? Actually, there are problems with understanding just who is the subject of moral obligations in Waller's thought experiment. It may be that individual intelligent ants do not have obligations toward one another because there are no *individual* intelligent ants, at least not in the sense in which we understand 'individual'. Eusocial organisms like ants and bees are in many ways better thought of as constituting a single super-organism, than as so many separate individuals. (One of the lessons of evolutionary biology is that many biological concepts have vague boundaries. 'Species' is one well-known example, but 'individual' may be equally vague. Is the Portugese Man-of-War one animal, or a colony of four different kinds of animals? There may be no unique defensible answer to this question.)

But, if eusocial organisms are not individuals in their own right, then the possibility that individual intelligent ants would reject our concept of individual rights does not demonstrate that moral concepts are species-relative, in the strong sense Waller requires. A nest of ants might be a rights-bearer, in the sense which concerns us, and all our moral concepts might apply to it. Indeed, it is not even clear that we can intelligibly attribute intelligence, as we understand it, to *individual* eusocial organisms. Like the Borg in *Star Trek*, they might constitute a single super-organism, with a single, distributed, mind. Waller therefore faces a dilemma: either the individual ants are not individuals at all, in the sense in which we are, and therefore the fact that they lack individual rights is no surprise, or they are individuals in our sense, and we have no reason to think them incapable of understanding and appreciating individual rights. On the first disjunct, their rights are appropriately (from our point of view) constrained by their genetic structure; on the second, they transcend that structure.

We might more appropriately regard the nest as the individual, the entity which has rights and obligations. In that case, the intelligent ants' lack of individual rights is no more interesting than is the fact that my skin cells lack rights against me.

Thus, Ruse and Waller fail to establish that morality is species-relative in any fundamental sense. Though it may certainly be wrong for us to treat the members of alien species in just the same ways we are required to treat one another, this is of no more meta-ethical significance than the fact that there are plants that are nutritious food sources for some animals and poisonous for others. There may be a single set of ethical obligations, consisting in injunctions to benefit and avoid harming others, though it may also be true that it will be difficult to know how to fulfil these obligations towards beings sufficiently unlike us in biological structure.

Spencer and Huxley Redux

By now it should be apparent that a number of apparently separate claims advanced by the deflationary evolutionary ethicists are actually interconnected. They claim (a) that a dispositional analysis of morality is correct, in the sense that, though it may not answer to our concept of morality (they are divided among themselves on this question), nevertheless it captures the reality of moral phenomena. They also hold (b) that the dispositions in question are the *direct* products of evolution, and therefore do not extend much beyond kin and those in a position to benefit us in return. Finally, they argue that (a) and (b) imply (c) that the truth-conditions of moral claims make essential reference to the subjective states of those who make them: since morality is founded on a set of dispositions we share with much simpler organisms, it is essentially a matter of feeling, and not belief.

We might call the conjunction of these three claims the neo-Spencerian position. Neo-Spencerians follow Spencer in identifying morality proper with the set of evolved dispositions, in opposition to Huxley and neo-Huxleyeans[8] who believe that morality is importantly different

---

[8]   Contemporary Huxleyeans – those who believe that culture or human cognitive abilities significantly transform our proto-moral dispositions – include, among philosophers, Dennett (1995, 2003), Rottschaefer (1998), McGinn (1979) and Singer (1981, 1999), and among biologists Williams (1995) and Dawkins (1976).

from proto-morality. We cannot, as we might have thought, put the Huxley–Spencer debate behind us. It continues to play itself out among evolutionary ethicists today.

As we have seen, the neo-Spencerians take their position to be better grounded scientifically. It makes human morality continuous with other animal behaviours, refusing – parsimoniously, as they see it – to postulate additional mechanisms to do the work in humans that instincts and feelings carry out in other animals. Irreducibly moral properties are mysterious and unscientific, they might claim. At very least, the burden of proof ought to be on those who reject the neo-Spencerian view, to show that additional mechanisms are necessary. Merely insisting that preferences cannot be moral reasons (Woolcock, 1993, 1999, 2000) just begs the question.

We have, I believe, already seen that we have decisive reason to reject the neo-Spencerian view. In the course of considering the deflationary evolutionary arguments, we have already taken it apart, plank by plank. Lest it seem that our arguments rest on an ignorant or superstitious affirmation of mysterious human powers over good scientific argument, I want briefly to scrutinize the scientific credentials of the neo-Spencerian position before I rehearse the arguments against it we have already sketched. How well does the neo-Spencerian view achieve its main aim, and explain the evolution of our moral sense?

The Spencerians face a major challenge in explaining how it is we come by the concept of morality at all. Although game-theoretical modelling demonstrates how our moral instincts might have evolved, the same models also lead us to expect that such instincts will not be the only products of repeated occasions for cooperation and conflict in the paleolithic. Group selectionist hypotheses are sometimes criticized on the grounds that though they might explain the evolution of genuine intragroup altruism$_p$, they would also tend to give rise to intergroup hostility and violence (Laland, Odling-Smee and Feldman, 2000). In fact, the same charge can be generalized. Whatever set of processes gave rise to our cooperative dispositions, they – or some other set of interactions – would also have given rise to selfish$_p$ dispositions, dispositions which lead us to favour our own lesser interests over the greater interests of others. Our entire complex human nature has evolutionary foundations, and it does not take much reflection for us to realize that that nature includes a great many a- and immoral dispositions, as well as moral ones.

The challenge for the Spencerian is therefore this: how did we come to group a certain set of dispositions together, giving only to them the name of the moral sentiments? It might be suggested that these emotions, and only these emotions, motivate cooperative behaviour, and that they therefore constitute a class that can easily be identified. But the class of dispositions which motivate (or might motivate) cooperative behaviour is not co-extensive with the class of moral dispositions. We can be motivated to cooperate for selfish$_p$ reasons, and we can have moral duties that are self-regarding. Sometimes, morality even requires us to act against our sympathetic impulses, as when we are cruel to be kind, or when our sympathetic impulse is out of place because the suffering which is its trigger is deserved punishment. Indeed, it is plausible to think that the linked notions of deserved punishment and undeserved suffering just *are* irreducibly moral notions that cannot be cashed out in nonmoral terms.

If morality is not to be entirely mysterious, however, we must be able to give some kind of explanation as to how we came to possess these irreducibly moral concepts. Here is one suggestion, that builds upon the work of evolutionary biologists. In a famous 1985 article, Trivers suggests that self-deception might have been a product of an 'arms race' between cheats and cheat-detectors in iterated prisoner's dilemmas. The thought runs as follows: given that the disposition to cooperate becomes widespread in a population, the defect strategy becomes profitable, and is maintained at some, relatively low, level by frequency-dependent selection. In this situation, most people will be disposed to cooperate with one another, but will also always be on the look-out for defectors. Cheats will therefore have to develop increasingly sophisticated means to hide their true intentions, which will prompt the development of ever better cheat-detection mechanisms. Trivers argues that cheats will be far better at hiding their intentions if they are hidden even from themselves, for then they will not have to fear giving them away inadvertently. Thus, cheats will come to believe themselves to be genuinely altruistic$_p$. Self-deception might thus be the product of natural selection, an offshoot of the profitable ability to deceive others (Trivers, 1985).

Trivers does not seem to have noticed, however, that the self-deception whose origin he traces itself requires that we come to have the notion of morality. To see this, consider what the conscious content of our beliefs must be, if we are to engage in such self-deception. We must believe, not merely that we are disposed to cooperate with one another, but that we are disposed to cooperate in the right *manner* for the right *reasons*: to cooperate with cooperators, in order to achieve morally permissible ends. In other words, Trivers-style self-deception requires that we possess the concept of morality, morality full-blown, not mere proto-morality. If his suggestion that self-deception is itself an evolved characteristic is correct, we must have acquired the concept of morality at the same time, or earlier, for this self-deception requires it. Thus, promoting our own selfish$_e$ concerns requires that we possess the idea of morality, on this hypothesis. If it is plausible, then the Spencerian view is self-defeating: the evolved dispositions cannot substitute for morality proper; instead, they require it.

This is precisely what we should expect, I suggest, given what we know about the evolution of morality. We know that it has foundations in proto-morality, in a set of evolved dispositions which, in the environment of evolutionary adaptation, were fitness-enhancing. But we know, too, that moral judgments have in some manner floated free of this subjective base. Plausibly, our moral emotions – what we can, in retrospect alone, recognize as such – are incoherent, in the state in which we inherit them, and need to be trimmed and altered, to be brought into something approaching wide reflective equilibrium. These dispositions made us sensitive to the needs and interests of others, but did so only partially and inconsistently, only to the extent to which this sensitivity was in our own interests. At the same time, however, it gave us the belief that our moral sensitivity should not be self-interested. This laid the groundwork for a thorough-going transformation of our inherited proto-morality, a transformation effected by our cognitive abilities (themselves products of natural selection).

Thus, we transformed our moral sentiments, rejecting some and changing others, so that the resulting morality, towards which we continue to work, bears only a passing resemblance to the set of inherited dispositions. On this view, Huxley stands vindicated. Our morality is the product of evolution, in the sense that we came to have its raw material and the powers with which we reworked this material, as a result of natural selection. It remains the case, however, that the resulting morality is able to turn against the very processes which gave birth to it. Huxley was right: morality lies not in natural selection, but, sometimes at least, in combating it.

Conclusion

Neo-Spencerians hold that evolutionary explanations of the origin of morality demonstrate that it consists, wholly or mainly, in a set of evolved dispositions, which extend little further than our kin. If this were true, we would have strong reasons to believe that a widespread concept of morality, according to which it consists (at minimum) in a set of obligations which require us to give some weight to the interests of everyone, is false. Morality would have a greatly restricted range and importance. Perhaps it would be eliminated altogether.

If neo-Spencerians were correct in their central claim, that morality is to be identified with proto-morality, then claim (5) above – that morality requires us to give some weight to (at minimum) all persons affected by our acts – would be false. But we have seen no reason simply to identify morality with proto-morality. In fact, the notion that rationality adds nothing essential to morality is highly implausible, in view of the fact that we hold all and only rational beings morally responsible. It is very plausible to think that an essential threshold is crossed when rationality is added to proto-morality (McGinn, 1979).

We have also seen that there are decisive *moral* reasons to reject the proposed analyses. They would force revisions on our concept of morality which are far greater than we are willing to countenance. Of course, the neo-Spencerians will reject our right to appeal to our concept of morality in this context. But unless they can make good on their identification of morality with proto-morality, we have no reason to take notice of them. Morality is proto-morality pruned and refined, proto-morality with at least the more glaring irrational elements eliminated. The intuitions it yields are far more likely to be truth-preserving than are those which the neo-Spencerian insists we ought to feel.

Though evolutionists, of all stripes, are no doubt right in insisting that morality has evolutionary origins, it cannot be identified with the set of dispositions which we have as a result of our evolutionary history. This initial set of dispositions need do no more than make us sensitive to the needs and interests of others, and give us the concept of a morality that is impartial. From that point on, the significant work is carried out by rationality, which prunes proto-morality to make it answer to the concept of morality. Our moral sentiments are gradually extended, and our self-interest trimmed in the process. There are no decisive arguments against moral realism to be found in evolution.

What relevance, finally, have the evolutionary hypotheses for our ethical and meta-ethical theories? The project of naturalizing ethics requires that we base our theories on the facts of human nature, as we know them, and of the social and physical world. A naturalistic ethics is concerned with what we are actually like. Evolutionary hypotheses can guide us, in suggesting hypotheses for further research: look for traits that would probably have been adaptive in the environment in which our ancestors evolved. But it cannot substitute for the examination of our actual characteristics. Indeed, theories of human nature, as it actually is, are themselves important constraints upon evolutionary theorizing. We go wrong, then, if we allow our theories to be driven by evolutionary hypotheses. They ought not to constitute a separate realm of theorizing about humanity and its nature, moral or otherwise. Instead, they constitute just one small part of the jigsaw, one more piece of evidence which, together with evidence from psychology and neurobiology, history and even literature, build up the picture of human nature.

Notes on the Essays

Precursors to Morality

If morality is an evolved phenomenon, we ought to expect to see its precursors and building blocks in other animals, especially the primates like chimpanzees with whom we share a relatively recent common ancestor. There is a great deal of attention being devoted to animal, and especially primate, morality today. Frans De Waal is perhaps the leading figure in this area, and Chapter 1 is a very good introduction to his work. Here, Jessica Flack and de Waal present evidence that non-human primates engage in reconciliation, mediation, and consolation, and that they apparently feel empathy and sympathy for one another. Nevertheless, Flack and de Waal do not attribute *morality* to non-human primates. For that, the organism must be capable of explicitly guiding their behaviour in the light of moral rules, they seem to think.

De Waal is also represented by a more recent short communication, from the prestigious scientific journal *Nature* (Brosnan and de Waal, 2003). Capuchin monkeys apparently have a sense of fairness which leads them to reject unequal pay for work.

Some people think that Flack and de Waal go too far in finding precursors of morality among non-human primates (for example, Bernstein, 2000). Others think that they do not go far enough, inasmuch as they confine their attention to *primates*. Marc Bekoff, for instance, thinks that the precursors of morality are to be found much more widely, in many group-living mammals, and perhaps beyond this. Gerald Wilkinson's fascinating essay (Wilkinson, 1990) shows that the characteristics needed for the evolution of reciprocal altruism – such as the ability to keep track of individuals and recall whether they have cooperated in the past – can be possessed even by organisms as unsophisticated cognitively as vampire bats.

Mechanisms

What needs to be added to proto-morality to yield morality? In his sophisticated sketch of the evolution of altruism, backed up with primatological and anthropological evidence, Philip Kitcher (Chapter 2) offers an answer. It is likely, he claims, that evolution leaves us with conflicting dispositions, to altruism and to selfishness. Since we do not experience the conflictual social life of other primates, we must have added something to these dispositions. Kitcher suggests it is a capacity for higher-order volitions: to side with some of our dispositions against others. Kitcher claims that, though natural selection is responsible for proto-morality, culture explains the rest.

No one has done more to defend group selection in contemporary biology than David Sloan Wilson. *Unto Others* (1998), written by Wilson and the prominent philosopher Elliott Sober, a summary of which appears here as Chapter 3, is the best application of group selection to morality in the contemporary literature. They point out that the question of what the unit of replication is – genes, individuals or groups – does not determine what the unit of selection is. If genes are the unit of replication, individuals may still benefit. So may groups. Moreover, kin selection is not, as selfish-gene theorists would have it, an alternative to group selection. It is an *instance* of it. Kin might form groups, even if they are distributed across space, inasmuch as they preferentially interact with one another. Since, however, human beings can choose with whom they interact, the way is open for group selection where the group is not limited to kin. Wilson and Sober claim to have found a 'smoking gun' of cultural group selection in the conflict between the Nuer and Dinka tribes in East Africa. The Nuer's superior group organization has caused a gradual erosion of the Dinka's territory and resources, with Dinka gradually defecting to the Nuer.

Wilson and Sober do not advocate replacing genetic or individual selection theories with group selection. Instead, they advocate a multilevel selection theory. Selection may happen on all three levels simultaneously.

Brian Skyrms' aim in Chapter 4 is to show that the kinds of strategies which can become widespread in a population as a result of evolution come decisively apart from those which rational choice theory would predict. Even strongly dominated strategies, strategies which lead to a smaller payoff no matter what other strategy is employed by other players, can, under the right conditions, persist indefinitely. Once again, the key to such counterintuitive outcomes is correlation – cooperators preferring to play against one another then against defectors.

Though it is easy to see why it is in an organism's interests to cooperate with cooperators, should we not expect natural selection to result in conditional cooperators: those who betray others whenever they can get away with it? Robert Frank (Chapter 5) presents a model to show that unconditional cooperation can evolve. The moral emotions, such as guilt, can serve as commitment devices; reliable signals of the intention to cooperate unconditionally. An organism capable of giving such a signal may be at an advantage, inasmuch as other organisms who give the signal will prefer to cooperate with fellow signallers rather than with those who either will not cooperate, or are incapable of giving the signal of the willingness to cooperate. Frank shows that unconditional cooperation can evolve, given certain starting conditions and relative payoffs for cooperation and defection. In many populations, the result will be a more-or-less stable polymorphism; that is, a population in which there are both cooperators and defectors. Cooperation will not go to fixation because when it is very common, the costs of detecting cheating are not worth paying; cheaters can take advantage of this fact and interact with cooperators, reaping the benefits of cooperation without paying the costs. But when cheating becomes common, cooperators will pay the cost of identifying cheats, and avoid them. Then the payoff for cooperating will rise, and cooperation will become more common.

Altruism

Neven Sesardic (Chapter 6) examines the so-called paradox of altruism: that natural selection will work to eliminate genuine altruism. The paradox is usually resolved by distinguishing between evolutionary and psychological altruism: organisms can be generally motivated to care for the welfare of others in ways that actually benefit themselves or their descendants. However, Sesardic argues that the paradox cannot be dealt with this easily: in fact, evolutionary and psychological altruism are likely connected. He explores further strategies for reducing or dissolving the paradox.

Stephen Stich (Chapter 7) carefully analyses Sober and Wilson's arguments for the claim that we can expect evolution to favour the development of altruistic desires for others' welfare, rather than instrumental desires which are motivated by the organisms concern for its own happiness. Stich finds only one of Sober and Wilson's arguments *prima facie* persuasive: the argument that if parental care were motivated by the belief that caring will increase the organism's happiness, it would be hostage to fortune: were the organism to find that caring was not conducive to its happiness, the belief would soon be abandoned. Beliefs are generally vulnerable in this way. But, Stich argues, not all belief-like states are vulnerable to empirical disconfirmation in this manner. There are also sub-doxastic states, which are cognitively impenetrable. However, it might be asked whether Stich actually establishes that desires which are instrumental are actually as or more plausible than desires that are intrinsic. We need to ask what representational content these sub-doxastic states have. It is far from clear that they match up to the categories of folk psychology which are at issue here.

Meta-ethics

Tamler Sommers and Alex Rosenberg (Chapter 8) argue that Darwinism (as they style it) leads to moral nihilism. There can be no intrinsic values or categorical obligations in our world, because such values and obligations 'make sense only against the background of purposes, goals, and ends which are not merely instrumental' (p. 169). It may be that Sommers and Rosenberg are putting a lot of weight on a particular interpretation of, 'intrinsic' and 'categorical', such that nothing counts as intrinsic or categorical unless it is part of the fabric of the unverse – ruling out response-dependent accounts by *fiat*. This suggested by their remark that 'Sacred to us … is not the same thing as sacred *tout court* or sacred *simpliciter*' (p. 183).

Sommers and Rosenberg claim that naturalism must reject Hume's law, showing how moral claims can be deduced from natural. It is not clear that they are right. Naturalists must *explain* how moral values come into existence, but it is not clear that their naturalistic story must *justify* these values. They can tell a response-dependent story, upon which it is our responses that justify those values. This response-dependent story need have no stronger relationship to naturalism than consistency with it. In other words, their claim that for naturalism to be vindicated 'Darwinism must underwrite morality and work to justify its claims' is false (p. 174).

Sommers and Rosenberg also suggest an epistemological deflationationism: 'if our best theory of why people believe P does not require that P is true, then there are no grounds to believe that P is true' (p. 183). Once again, however, some kind of response-dependent account might work here. We might claim that the extension of 'morally wrong', by virtue of its *meaning*, includes such things as unnecessary suffering.

Michael Ruse (Chapter 9) argues that earlier sociobiological views erred inasmuch as they committed the naturalistic fallacy, the (alleged) fallacy of inferring normative claims from purely factual statements. He aims to avoid this fallacy (though it is not entirely clear that he does so). But even if evolution has no *normative* implications, Ruse asserts, it does have *meta-ethical* implications: that is, implications for moral theory. Once we realize that morality is an evolved phenomenon, he argues, we come to see that its pretensions to objectivity are unfounded. Part of Ruse's evidence for this claim comes from pointing out that, had evolution taken a different turn, we might well feel obligated to perform repellent actions. This fact he takes to show that our actual obligations do not reflect timeless and human-independent facts. He is surely right. But the move from this claim to the claim that morality is an illusion is surely far too quick. There are moral realisms that do not claim that moral facts are timeless.

Between subjectivity and objectivity there is *intersubjectivity*: it is far from clear that this is not sufficient to vindicate some kind of moral realism.

Richard Joyce (Chapter 10) takes himself to be defending Ruse's error theory, according to which we are evolved to regard morality as categorically binding and objective, because creatures like us who regard moral imperatives as binding will be fitter than others. Joyce's paper is notable for explicitly considering a response-dependent (he calls it a dispositional) account of moral realism. He argues that such accounts fail because they cannot underwrite the notion of moral *requirement*. Moral claims are distinctive inasmuch as they are binding on everyone, regardless of their desires. Everyone, supposedly, has a reason to act morally. But in fact nothing can satisfy this condition: it is simply false that everyone has a reason to be moral. It is, Joyce readily concedes, possibly to offer a plausible naturalistic account of moral requirements and capacities, but no such account will be able to capture the requirement – what Joyce, in his recent book on the topic, calls morality's *practical clout* (Joyce, 2006). So an error theory, according to which moral claims are quite generally false, is vindicated. We are disposed, by virtue of our evolved psychological capacities and innate concepts, to believe in morality. But morality is an illusion.

Joyce does not recommend jettisoning morality, however. Instead, he advocates a moral fictionalism. Though it is irrational to believe in the objectivity of morality nevertheless it is in our best interests to behave as if we believed in its objectivity (for reasons very familiar from explorations of evolutionary ethics), and we may only be able to do that if we *accept* its objectivity, where 'acceptance' is a belief-like attitude, which disposes the agent to behaviour much like belief, but which is compatible with disbelief.

How plausible is Joyce's view? As I have argued elsewhere (Levy, 2006), its plausibility depends upon how we ought best to understand the practical clout that morality is supposed to have. Joyce understands it *motivationally*, claiming that it is part of the concept of morality that moral facts give us (putatively motivationally effective) *reasons to act*, regardless of the structure of our desires and beliefs. But there are other ways to understand the inescapability of moral claims. Consider Harman's (1977) example of Albert the cat-burner. As Joyce points out, no matter how carefully Albert reflects on his desire to burn cats, and no matter how much we idealize Albert (so long as we avoid begging any substantive questions), there is no reason to think that Albert will necessarily come to see that he has a (motivational) reason to refrain from cat-burning. But recall the claim about the irreducibility of moral claims urged above: the reason that moral transgressions remain wrong, regardless of what we believe, is that the facts upon which their wrongness supervenes do not vanish just because the rules change. What makes cat-burning wrong is not *our* responses, but the *cat's* (Railton, 1998); its suffering, and that suffering gives Albert a reason to refrain from cat-burning, whether he – or any else – is capable of grasping it. This is not a motivationally effective reason, to be sure, but practical authority comes in other flavours. Our evolved responses to cat suffering allow us to appreciate a fact about it – that it *is* suffering – that is inescapably reason-giving, not in the sense that any being, in any possible world with suffering cats, will necessarily be motivated to do anything about it, but just in the sense that a possible world that contains suffering cats contains *suffering*. That it is possible to act to decrease such suffering, or to avoid inflicting it, is a reason for so acting, though, if evolution had taken a different turn, we might well not have been capable of appreciating it.

Sharon Street's essay (Chapter 11) is one of the most sophisticated and detailed defences of the claim that evolutionary theory has anti-realist implications for meta-ethics. She considers a variety of realist positions, and sketches a Darwinian dilemma for them. Either they accept or reject the claim that evolutionary pressures make us better able to grasp moral facts. If they reject the claim, they leave wholly mysterious the relationship between our moral beliefs and moral facts. If, on the other hand, they admit that evolutionary pressures enable us to track moral facts, they face the problem that an evolutionary account according to which moral claims are *constituted* by evolved responses is so much more plausible than any tracking account.

But why is a properly evolutionary realism not an option? That is, why not think that the moral values constituted by our evolutionary responses are *real*? Street seems to think that a genuine realism about morality on which moral facts are constituted out of evolved disposition is ruled out by the realization that, had evolution taken a different turn, a different set of values would have resulted. Consider a possible future, in which we encounter aliens who have come to have different values as a result of their different evolutionary trajectory. They may therefore have moral claims that conflict with ours. But both moral claims cannot be simultaneously true. Hence the constitution view cannot support moral realism.

This is a powerful argument, but it is not entirely compelling. Street considers and rejects one possibility, which is to 'rigidify' our moral claims, in a similar way to which we rigidify, say, scientific claims. Water is H_2O in any possible world; we use 'water' to pick out the substance with that composition in any world, no matter what the inhabitants of the world call that substance and no matter what role, if any, it plays in their lives. Similarly, our moral terms – 'good', 'right', and so on – might pick out actions with certain natural properties, no matter what the attitude of alien beings to those actions. Street rejects this argument, on the grounds that the rigidifying move is equally available to aliens; they would have just as good grounds for applying their moral terms in their way as would we. True, but the rigidifying move is equally available to them with regard to chemical substances. Nevertheless, 'water' *does* refer uniquely to H_2O.

Normative Ethics

Deriving normative claims from factual claims is a fraught business, even if there is no naturalistic fallacy. Robert Richards (Chapter 12) makes one of the boldest attempts at such a derivation. He argues that, since we are evolved to promote community welfare, we *ought* to promote community welfare. Lemos (1999) offers effective criticisms of this argument, as well as of related attempts to bridge the is/ought gap.

Keith Sutherland and Jordan Hughes consider Arnhart's book-length defence of a Darwinian political theory (Chapter 13) that draws conservative implications from evolutionary biology. Arnhart claims that evolution supports an Aristotelian politics, according to which a good life is a life that allows for the fullest satisfaction of our most significant desires over a lifetime, and that only a minimal state can conduce to such a life. As Sutherland and Hughes point out, Darwinism seems to support conservatism because the political left, at least in the postwar period, was generally opposed to robust conceptions of human nature. I doubt, however, that taking our biology seriously entails political conservatism. It is worth pointing out that, contrary to their claim, Marx himself was not an opponent of human nature. In fact, he too accepted a view of human nature that was broadly Aristotelian. Moreover, the political conservatism presented here seems to turn on the claim that human nature is relatively fixed. But this is false, not because humans can transcend their biology – whatever that would amount to – but because humans are cultural animals *by nature*.

References

Bekoff, M. (2001), 'Social Play Behaviour: Cooperation, Fairness, Trust, and the Evolution of Morality', *Journal of Consciousness Studies*, **8**, pp. 81–90.

Bernstein, I.S. (2000), 'The Law of Parsimony Prevails: Missing Premises Allow any Conclusion', *Journal of Consciousness Studies*, **7** (1–2), pp. 31–4.

Brosnan, Sarah and de Waal, Frans, (2003), 'Monkeys Reject Unequal Pay', *Nature*, **425**, pp. 297–99.

Campbell, Richmond, (1996), 'Can Biology Make Ethics Objective?', *Biology and Philosophy*, **11**, pp. 21–31.

Collier, John and Stingl, Michael, (1993), 'Evolutionary Naturalism and the Objectivity of Morality', *Biology and Philosophy*, **8**, pp. 47–60.

Darwin, Charles (1871), *The Descent of Man, and Selection in Relation to Sex*, London: Murray.

Dawkins, Richard, (1976), *The Selfish Gene*, New York: Oxford University Press.

Dennett, Daniel C. (1995), *Darwin's Dangerous Idea: Evolution and the Meanings of Life*, New York: Simon & Schuster.

Dennett, Daniel C. (2003), *Freedom Evolves*, London: Allen Lane.

Devitt, Michael (2002) 'Moral Realism: A Naturalistic Perspective', *Croation Journal of Philosophy*, **2**, pp. 1–15.

Frank, Robert H. 1988. *Passions within Reason: the Strategic Role of the Emotions*, New York: Norton.

Hamilton, W.D. (1972), 'Altruism and Related Phenomena, Mainly in Social Insects', *Annual Reviews of Ecology and Systematics*, **3**, pp. 193–232.

Harman, Gilbert (1977). *The Nature of Morality: An Introduction to Ethics*, New York: Oxford University Press.

Harms, William F. (2000), 'Adaptation and Moral Realism', *Biology and Philosophy*, **15**, pp. 699–712.

Huxley, T.H. (1989), 'Evolution and Ethics', in James Paradis and George C. Williams (eds), *T.H. Huxley's Evolution and Ethics with New Essays on Its Victorian and Sociobiological Context*. Princeton: Princeton University Press, pp. 57–174.

Joyce, Richard (2000), 'Darwinian Ethics and Error', *Biology and Philosophy*, **15**, pp. 713–32.

Joyce, Richard (2001), *The Myth of Morality*, Cambridge: Cambridge University Press.

Joyce, Richard (2006), *The Evolution of Morality*, Cambridge, Mass. The MIT Press.

Kevles, Daniel J. (1995), *In the Name of Eugenics: Genetics and the Uses of Human Heredity*, Cambridge, MA: Harvard University Press.

Kitcher, Philip (1993), 'The Evolution of Human Altruism', *Journal of Philosophy*, **90**, pp. 497–516.

Laland, K.N., Odling-Smee, F.J. and Feldman, Marcus W. (2000), 'Group Selection: A Niche Construction Perspective', *Journal of Consciousness Studies*, **7**, pp. 221–25.

Lemos, John (1999), 'Bridging the Is/Ought Gap with Evolutionary Biology: Is This a Bridge Too Far?', *The Southern Journal of Philosophy*, **37**, pp. 559–77.

Levy, Neil (2006), 'What Evolves When Morality Evolves?' Studies In the History And Philosophy Of Science. Part C: *Studies In History And Philosophy Of Biological And Biomedical Sciences*, **37**, pp. 612–620.

Mackie, John L. (1977), *Ethics: Inventing Right and Wrong*, Harmondsworth: Penguin.

Mackie, John. L. (1978), 'The Law of the Jungle', *Philosophy*, **53**, pp. 455–64.

Maynard Smith, John (1964), 'Group Selection and Kin Selection', *Nature*, **201**, pp. 1145–47.

Maynard Smith, John (1982), *Evolution and the Theory of Games*, Cambridge: Cambridge University Press.

McDowell, John (1985), 'Values and Secondary Qualities', in Ted Honderich (ed.), *Morality and Objectivity: A Tribute to J.L. Mackie*, London Routledge & Kegan Paul, pp. 11–129.

McDowell, John (1995), 'Two Sorts of Naturalism' in Rosalind Hursthouse, Gavin Lawrence and Warren Quinn (eds), *Virtues and Reasons: Philippa Foot and Moral Theory, Essays in Honour of Philippa Foot*, Oxford: Oxford University Press, pp. 149–79.

McGinn, Colin (1979), 'Evolution and the Basis of Morality', *Inquiry*, **22**, pp. 81–99.

McShea, Robert J. and McShea, Daniel W. (1999), 'Biology and Value Theory', in Jane Maienschein and Michael Ruse (eds), *Biology and the Foundation of Ethics*, Cambridge: Cambridge University Press, pp. 307–37.

Moore, G.E. (1903), *Principia Ethica*, London: Cambridge University Press.

Moser, Paul K. and Carson, Thomas L. (2001), 'Introduction', in their *Moral Relativism*: A Reader, New York: Oxford University Press, pp. 1–21.

Palmer, Jack A. and Palmer, Linda K. (2002), *Evolutionary Psychology: The Ultimate Origins of Human Behavior*, Boston: Allyn & Bacon.

Rachels, James (1995), *The Elements of Moral Philosophy*, New York: McGraw-Hill.

Ridley, Matt (1996), *The Origin of Virtue*, London: Viking.

Rosenberg, Alexander (2000), 'The Biological Justification of Ethics: A Best-Case Scenario', in his *Darwinism in Philosophy, Social Science and Policy*, Cambridge: Cambridge University Press, pp. 118–36.

Rottschaefer, William A. (1998), *The Biology and Psychology of Moral Agency*, Cambridge: Cambridge University Press.

Rottschaefer, William A. and Martinsen, David (1990), 'Really Taking Darwin Seriously: An Alternative to Michael Ruse's Darwinian Metaethics', *Biology and Philosophy*, **5**, pp. 149–73.

Ruse, Michael (1998), *Taking Darwin Seriously: A Naturalistic Approach to Philosophy*, 2nd edn, Amherst: Prometheus Books.

Sesardic, Neven (1995), 'Human Altruism and Evolution', *Ethics*, **106**, pp. 128–157.

Singer, Peter (1981), *The Expanding Circle: Ethics and Sociobiology*, New York: Farrar, Straus & Giroux.

Singer Peter (1999), *A Darwinian Left: Politics, Evolution and Cooperation*, London: Weidenfeld & Nicolson.

Skyrms, Brian (1996), *Evolution of the Social Contract*, New York: Cambridge University Press.

Smith, Michael (1994), *The Moral Problem*, Oxford: Basil Blackwell.

Smith, Michael (2000), 'Moral Realism', in Hugh LaFollette (ed.), *The Blackwell Guide to Ethical Theory*, Oxford: Blackwell., pp. 15–37

Sober, Elliott (1994), 'Did Evolution Make us Psychological Egoists?', in his *From a Biological Point of View*, Cambridge: Cambridge University Press, pp. 8–27.

Sober, Elliott and Wilson, David S. (1998), *Unto Others: The Evolution and Psychology of Unselfish Behavior*, Cambridge, MA: MIT Press.

Spencer, Herbert (1883), *The Data of Ethics*, New York. D. Appleton & Co.

Sturgeon, Nicholas (1985), 'Moral Explanations', in David Copp and David Zimmerman (eds), *Morality, Reason and Truth*, Totowa, NJ: Rowman & Littlefield, pp. 49–78.

Thompson, Paul (2002), 'The Evolutionary Biology of Evil', *The Monist*, **85**, pp. 238–58.

Trivers, Robert (1971), 'The Evolution of Reciprocal Altruism', *Quarterly Review of Biology*, **46**, pp. 35–57.

Trivers, Robert (1985), *Social Evolution*, Menlo Park, CA: Benjamin/Cummings Publishing.

Vanderschaaf, Peter (2000), 'Game Theory, Evolution, and Justice', *Philosophy & Public Affairs*, **28**, pp. 325–58.

Waller, Bruce N. (1996), 'Moral Commitment without Objectivity or Illusion: Comments on Ruse and Woolcock', *Biology and Philosophy*, **11**, pp. 245–54.

Waller, Bruce N. (1997), 'What Rationality Adds to Animal Morality', *Biology and Philosophy*, **12**, pp. 341–56.

Wiggins, David (1976), 'Truth, Invention, and the Meaning of Life', *Proceedings of the British Academy*, **62**, pp. 331–78.

Wilkinson, Gerald. S. (1990), 'Food Sharing in Vampire Bats', *Scientific American*, February, pp. 76–82.

Williams, George C. (1966), *Adaptation and Natural Selection*, Princeton: Princeton University Press.

Williams, George C. (1995), 'Mother Nature Is a Wicked Old Witch!', in Matthew H. Nitecki and Doris V. Nitecki (eds), *Evolutionary Ethics*, Albany, NY: State University of New York Press, pp. 217–31.

Wilson, David and Sober, Elliott (1994), 'Re-Introducing Group Selection to the Human Behavioral Sciences', *Behavioral and Brain Sciences*, **17**, pp. 585–654.

Wilson, E.O. (1975), *Sociobiology: The New Synthesis*, Cambridge, MA: Harvard University Press.

Woolcock, Peter (1993), 'Ruse's Darwinian Meta-Ethics: A Critique', *Biology and Philosophy*, **8**, pp. 423–39.

Woolcock, Peter (1999), 'The Case Against Evolutionary Ethics Today', in Jane Maienschein and Michael Ruse (eds), *Biology and the Foundation of Ethics*, Cambridge: Cambridge University Press, pp. 276–306.

Woolcock, Peter (2000), 'Objectivity and Illusion in Evolutionary Ethics: Comments on Waller', *Biology and Philosophy*, **15**, pp. 39–60.

Part I
Precursors to Morality

[1]

'Any Animal Whatever'

Darwinian Building Blocks of Morality in Monkeys and Apes

Jessica C. Flack and Frans B.M. de Waal

To what degree has biology influenced and shaped the development of moral systems? One way to determine the extent to which human moral systems might be the product of natural selection is to explore behaviour in other species that is analogous and perhaps homologous to our own. Many non-human primates, for example, have similar methods to humans for resolving, managing, and preventing conflicts of interests within their groups. Such methods, which include reciprocity and food sharing, reconciliation, consolation, conflict intervention, and mediation, are the very building blocks of moral systems in that they are based on and facilitate cohesion among individuals and reflect a concerted effort by community members to find shared solutions to social conflict. Furthermore, these methods of resource distribution and conflict resolution often require or make use of capacities for empathy, sympathy, and sometimes even community concern. Non-human primates in societies in which such mechanisms are present may not be exactly moral beings, but they do show signs of a sense of social regularity that — just like the norms and rules underlying human moral conduct — promotes a mutually satisfactory modus vivendi.

Introduction

Any animal whatever, endowed with well-marked social instincts, the parental and filial affections being here included, would inevitably acquire a moral sense or conscience, as soon as its intellectual powers had become as well developed, or nearly as well developed, as in man.

Charles Darwin, *The Descent of Man* (1982 [1871], pp. 71–2)

Thomas Huxley, in his famous lecture, *Evolution and Ethics* (Huxley, 1894), advanced a view of human nature that has since dominated debate about the origins of morality. Huxley believed that human nature is essentially evil — a product of a nasty and unsympathetic natural world. Morality, he argued, is a human invention explicitly devised to control and combat selfish and competitive tendencies generated by the evolutionary process. By depicting morality in this way, Huxley was advocating that the search for morality's origins be de-coupled from evolution and conducted outside of biology.

Proponents of Huxley's dualistic view of nature and morality abound today. Among them is the evolutionary biologist, Richard Dawkins, who in 1976 (p. 3) wrote

> Be warned that if you wish, as I do, to build a society in which individuals cooperate generously and unselfishly towards a common good, you can expect little help from biological nature. Let us try to teach generosity and altruism, because we are born selfish.

Another well-known evolutionary biologist, George C. Williams (1988, p. 438), also reaffirmed, with minor variation, Huxley's position when he stated, 'I account for morality as an accidental capability produced, in its boundless stupidity, by a biological process that is normally opposed to the expression of such a capability'. And recently, the philosopher Daniel Dennett (1995, p. 481), although admitting that it is conceivable that perhaps the great apes, whales, and dolphins possess some of the requisite social cognition on which morality depends, wrote

> My pessimistic hunch is that the main reason we have not ruled out dolphins and whales as moralists of the deep is that they are so hard to study in the wild. Most of the evidence about chimpanzees — some of it self-censored by researchers for years — is that they are true denizens of Hobbes' state of nature, much more nasty and brutish than any would like to believe.

But if, as Dawkins suggests, the origins of morality — of the human sense of right and wrong used by society to promote pro-social behaviour — are not biological, then what is the source of strength that enabled humanity to escape from its own nature and implement moral systems? And from where did the desire to do so come? If, as Williams suggests, morality is an accidental product of natural selection, then why has such a 'costly' mistake not been corrected or eliminated by the very process that inadvertently created it? Our inability to answer these questions about the origins and consequences of moral systems is an indication that perhaps we need to broaden the scope of our search. After all, the degree to which the tendency to develop and enforce moral systems is universal across cultures (Midgley, 1991; Silberbauer, 1991), suggests that moral systems, contrary to Huxley's beliefs, do have biological origins and are an integral part of human nature.

Morality indeed may be an invention of sorts, but one that in all likelihood arose during the course of evolution and was only refined in its expression and content by various cultures. If, as we believe, morality arose from biological origins, then we should expect at a minimum that elements of it are present in other social species. And indeed, the evidence we will present in this paper suggests that chimpanzees and other social animals are not the 'true denizens of Hobbes' state of nature' they are surmised to be by Dennett. It may well be that chimpanzees are not moral creatures, but this does not mean that they do not have elements of moral systems in their societies. If we are to understand how our moral systems evolved, we must be open to the idea that the sets of rules that govern how non-human animals behave in their social groups provide clues to how morality arose during the course of evolution. These simple rules, which emerge out of these animals' social interactions, create an element of order that makes living together a possibility, and in a liberal sense, reflect elements of rudimentary moral systems. The order that these sets of rules create is vital to maintaining the stability of social systems and probably is the reason why human morality (whether or not an evolutionary accident) has not been eliminated by natural selection (Kummer, 1979). Garret Hardin (1983, p. 412) captured the essence of this argument in a statement about the importance of justice — 'The first goal of justice is to create a *modus vivendi* so that life can go on, not only in the next few minutes, but also indefinitely into the future.'

Had Huxley acknowledged that the origins of morality lay in biology but argued against searching within biology for the *specifics* of our moral systems, his case might have been more persuasive today. Such an argument would have at least fit the contemporary framework for addressing questions about why we are the way we are, which in the case of morality has been explored intensely (Nitecki and Nitecki, 1993, and contributions therein). Indeed, the only pertinent question seems to us: *To what degree* has biology influenced and shaped the development of moral systems? One way to determine the extent to which human morality might be the product of natural selection is to explore behaviour in other species that is analogous (similar traits that arose by convergent evolution due to the presence of similar selection pressures or evolutionary conditions), and perhaps homologous (traits that evolved in a common ancestor and that remain present in related species due to common phylogenetic descent) to our own.

Many non-human primates, for example, seem to have similar methods to humans for resolving, managing, and preventing conflicts of interests within their groups. Such methods, which include reciprocity and food sharing, reconciliation, consolation, conflict intervention, and mediation, are the very building blocks of moral systems in that their existence indicates, as Mary Midgley (1991, p. 12) wrote, 'a willingness and a capacity to look for shared solutions' to conflicts (see also Boehm, 2000). Furthermore, unlike strict dominance hierarchies, which may be an alternative to moral systems for organizing society, advanced methods of resource distribution and conflict resolution seem to require or make use of traits such as the capacity for empathy, sympathy, and sometimes even community concern. Conflict resolution that reflects concern for and possibly understanding of a predicament in which a fellow group member finds himself or herself provides for society the raw material out of which moral systems can be constructed.

Non-human primates in such societies may not be exactly moral beings, but they do show indications of a sense of social regularity that parallels the rules and regulations of human moral conduct (de Waal, 1996a; 1996b, chapter 3). In addition to conflict resolution, other key components or 'prerequisites' of morality recognizable in social animals are reciprocity, empathy, sympathy, and community concern. These components, which also include a sense of justice, and perhaps even the internalization of social norms, are fundamental to moral systems because they help generate connections among individuals within human and animal societies despite the conflicts of interests that inevitably arise. By generating or reinforcing connections among individuals, these mechanisms facilitate co-operative social interaction because they require individuals to make 'commitments' to behave in ways that later may prove contrary to independent individual interests (used throughout this paper in reference to those interests that are truly independent as well as in reference to those interests for which pursuit requires engaging in competition) that when pursued can jeopardize collective or shared interests (Frank, 1988; 1992).

Although many philosophers and biologists are sceptical that evolution can produce components of moral systems such as the capacity for sympathy and empathy or even the capacity for non-kin based co-operation that require the suspension of short

4   J.C. FLACK AND F.B.M. DE WAAL

term, independent interests, there also exists a tradition going back to Petr Kropotkin (1902) and, more recently, Robert Trivers (1971), in which the view has been that animals assist each other precisely because by doing so they achieve long term, collective benefits of greater value than the short term benefits derived from straightforward competition. Kropotkin specifically adhered to a view in which organisms struggle not necessarily against each other, but collectively against their environments. He strongly objected to Huxley's (1888) depiction of life as a 'continuous free fight'. Although some of Kropotkin's rationale was seriously flawed, the basic tenet of his ideas was on the mark. Almost seventy years later, in an article entitled 'The Evolution of Reciprocal Altruism', Trivers refined the concepts Kropotkin advanced and explained how co-operation and, more importantly, a system of reciprocity (called 'reciprocal altruism' by Trivers) could have evolved. Unlike simultaneous co-operation or mutualism, reciprocal altruism involves exchanged acts that, while beneficial to the recipient, are costly to the performer. This cost, which is generated because there is a time lag between giving and receiving, is eliminated as soon as a favour of equal value is returned to the performer (see Axelrod and Hamilton, 1981; Rothstein and Pierotti, 1988; Taylor and McGuire, 1988).

According to Richard Alexander (1987), reciprocity is essential to the development of moral systems. Systems of indirect reciprocity — a type of reciprocity that is dependent on status and reputation because performers of beneficent acts receive compensation for those acts from third parties rather than necessarily from the original receiver — require memory, consistency across time, and most importantly, a sense of social regularity or consensual sense of right and wrong (Alexander, 1987, p. 95). It is not yet clear whether systems of indirect reciprocity exist in non-human primate social groups, but certainly there is evidence from studies on food-sharing, grooming, and conflict intervention that suggest the existence of reciprocal systems and, at least among chimpanzees, a sense of social regularity (e.g. Cheney and Seyfarth, 1986; de Waal, 1991; 1996a; 1996b, chapter 3; 1997a; 1997b; Silk, 1992).

Food Sharing, Reciprocal Exchange, and Behavioural Expectations in Primates

Food sharing is known in chimpanzees (Nissen and Crawford, 1932; Kortlandt, 1962; Goodall, 1963; Nishida, 1970; Teleki, 1973; Boesch and Boesch, 1989; de Waal, 1989b; 1997a; Kuroda *et al.*, 1996), bonobos (Kano, 1980; Kuroda, 1984; Hohmann and Fruth, 1993; de Waal, 1992b), siamangs (Fox, 1984), orangutans (Edwards and Snowdon, 1980), and capuchin monkeys (Perry and Rose, 1994; Fragaszy, Feurerstein, and Mitra, 1997; de Waal, 1997b; Rose 1997). It is an alternative method to social dominance and direct competition by which adult members of a social group distribute resources among themselves. Most food sharing requires fine-tuned communication about intentions and desires in order to facilitate inter-individual food transfers. The food transfers typically observed are passive, involving selective relinquishment of plant and animal matter more frequently than active giving (de Waal, 1989b). Three non-exclusive hypotheses have been forwarded to explain the proximate reasons why one individual would voluntarily allow another to take food.

Richard Wrangham (1975) suggested that food possessors share with other group members in order to deter harassment and reduce the possibility that, as possessors,

they will become the recipients of aggression. This idea, known as the 'sharing-
-under-pressure' hypothesis, resembles Nicholas Blurton-Jones' (1987) 'tolerated-
-theft' model, according to which it is more common for possessors to let food be
taken from them than for them to actually give it away. Blurton-Jones reasoned that
possessors tolerate theft in order to avoid potentially risky fights.

The 'sharing-to-enhance-status hypothesis' has been used by Adriaan Kortlandt
(1972) and James Moore (1984) to explain male chimpanzee food sharing and the displays that frequently accompany the treatment of objects in the environment such as
captured prey. Both the act of sharing and the displays — for example, branch shak-
ing — draw attention to the food possessor in a way that may raise his or her status in
the group. Illustrative examples of this strategy can be found in Toshisada Nishida *et
al.*'s (1992) description of a chimpanzee alpha male in the wild who kept his position
through 'bribery' (i.e. selective food distribution to potential allies), and in de Waal's
(1982) account of a male contender for the alpha position in a zoo colony, who
appeared to gain in popularity by acquiring and distributing food to the group to
which the apes normally had no access.

A similar hypothesis was developed for human food distribution by Kristen
Hawkes (1990), an anthropologist, who suggested that men who provide food to
many individuals are 'showing off'. Showing off in this manner, according to
Hawkes, signals hunting prowess and generosity, two characteristics that may be
attractive to potential mates or potential political allies.

A third hypothesis — the reciprocity hypothesis — proposes that food sharing is
part of a system of mutual obligations that can involve material exchange, the
exchange of social favours such as grooming and agonistic support, or some combination of the two. For example, de Waal (1982) found that subordinate adult male
chimpanzees groom dominant males in return for an undisturbed mating session.
Suehisa Kuroda (1984) and de Waal (1987) found indications that adult male bonobos
exchange food with adolescent females in return for sex. The reciprocity hypothesis
thus differs significantly from the sharing-under-pressure hypothesis because it
addresses possessors and 'beggars' as potential long-term co-operators rather than
merely as present competitors who use sharing to appease one another. It differs from
the 'sharing-to-enhance-status' hypothesis because it emphasizes the co-operative
nature of the relationship between possessors and beggars and, consequently, emphasizes how sharing benefits both the possessor and the beggar rather than just the possessor. One advantage of the 'sharing-to-enhance-status' hypothesis is, however, that
it provides a testable proximate account of what social factors might motivate possessors to initially share with beggars, in that it suggests that possessors share because by
doing so they increase their social status in the group. In fact, the 'sharing-to-
-enhance-status' hypothesis, although a partial explanation of sharing, is useful if
considered in conjunction with the reciprocity hypothesis because it provides a proximate motivational explanation for why possessors allow some of their food to be
taken by others. Consequently, this hypothesis is not necessarily in conflict with the
reciprocity hypothesis, and may be an extension of it. Furthermore, the 'sharing-to-
-enhance-status' hypothesis, like the reciprocity hypothesis, involves the exchange of
favours between individuals using apparently equivalent, although unequal, curren-
cies: For example, a form of reciprocal exchange may emerge if A shares food with B,

6   J.C. FLACK AND F.B.M. DE WAAL

which makes A more popular with B resulting — as suggested by Hawkes — in agonistic support or matings.

De Waal (1989b; 1997a; 1997b) examined whether food itself is exchanged reciprocally over time or is shared in return for some social favour by investigating the
food sharing tendencies of brown capuchin monkeys and chimpanzees. Results of the
capuchin study indicated that female brown capuchins share food reciprocally. The
methodology used in this study differed substantially from the chimpanzee study
described next. The primary difference was that the capuchin's food sharing tendencies were examined in a dyadic context rather than in the presence of the entire group
as in the chimpanzee study. As shown in Figure 1, adult capuchins were separated
into pairs and placed into a test chamber divided into two sections by a mesh partition.
One capuchin was allowed continuous access to a bucket of attractive food. The individual with access to the food was free to monopolize all of it or could move close to
the mesh and share actively or passively by allowing his counterpart access to pieces
he had dropped. The situation was then reversed so that the second individual had
access to the attractive food (which was of a new type) and the first did not. The rate
of transfer between pairs of adult female capuchins was found to be reciprocal while
the rate of transfer between pairs of adult males was not. Males, however, were less
discriminating than females in terms of with whom they shared, and more generous in
the amount of food they shared. Although this study examined food-sharing in capuchins in an artificial environment created by the experimenters, the results were not
anomalous — food sharing among unrelated adults has been observed both among
capuchins in a colony at the Yerkes Regional Primate Centre as well as among wild
capuchins (Perry and Rose, 1994; Rose, 1997).

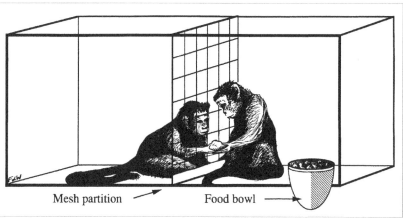

Mesh partition ⟶   Food bowl ⟶

Figure 1

Schematic drawing of the pair-test setup. One subject at a time receives food from a bowl attached
to the outside of the chamber. A mesh partition divides the test chamber, preventing direct access to
the food by the other subject. In a rare instance of active sharing, a male (right) hands a piece of food
to a female who reaches through the mesh to accept it. Both subjects visually monitor the transfer.
This drawing (by the second author) was made from an actual video still. From de Waal (1997b).

In order to study how chimpanzees share food in a social context, a situation was created in which a monopolizable food source was available to individuals in the social group. To accomplish this, a captive group of chimpanzees at the Field Station of the Yerkes Regional Primate Research Centre was provided with branches and leaves that were tightly bundled together so that the possibility existed for some group members to keep all of the food for themselves. Based on an analysis of nearly 7,000 recorded interactions over food, de Waal found that food exchanges between nine adult group members were quite balanced per dyad so that, on average, individuals A and B shared the same amount with each other. If individual A, however, shared a particular day with individual B, this did not necessarily result in B being more likely to share with A the following day. Grooming, on the other hand, did affect the likelihood to share when sharing and grooming occurred on the same day. For example, A was less likely to share with B if A had also groomed B the same day, but A was more likely to share with B if it had been B who had groomed A earlier that day. Other data indicating that food possessors actively resisted approaches by individuals who had not previously groomed them bolstered this result. Lastly, individuals who were reluctant to share their food had a greater chance of encountering aggression when they themselves approached a food possessor.

Although the chimpanzee food-sharing study confirmed one part of the prediction of Wrangham's and Blurton-Jones' hypotheses that there should be a negative correlation between rate of food distribution and frequency of received aggression, de Waal (1989b; 1996b, pp. 152–3) considers his results inconsistent with the tolerated theft model. He found that most aggression is directed not against the possessors of food, as the tolerated theft model predicts, but against beggars for food. In fact, even the lowest-ranking adult possessors are able to hold on to food unchallenged due to the 'respect of possession' first noted, with astonishment, by Goodall (1971, chapter 16), who wondered why the alpha male of her community failed to claim food possessed by others, and actually had to beg for it (Goodall, however, noted that the apparent respect for possession she observed among chimpanzees only applied to animal matter and not to bananas or other kinds of vegetable matter. This led her to suggest another somewhat different, although not mutually exclusive, explanation that focused more on the motivational state of the possessor and the corresponding response to this by the beggars). Respect of possession exists also in other primates, and was experimentally investigated by Sigg and Falett (1985), Kummer and Cords (1991), and discussed by Kummer (1991).

In fact, in de Waal's (1989) study the observed negative correlation between an individual's food distribution rate and the probability of aggression received concerned this individual as an *approacher* rather than as food possessor. This suggests either that food distributors respond to stingy individuals by sharing less with them than with others *or* that individuals, who for whatever reason, are more likely to be aggressively rebuffed when approaching food possessors, in turn become more reluctant themselves when they possess food to share with others.

Thus, the food sharing data are most in line with the reciprocity hypothesis. It is conceivable, though, that receipt of a favour (whether it be a service such as grooming or an object, such as food) positively influences an individual's social attitude so that this individual is willing to share indiscriminately with everyone else in its group (Hemelrijk, 1994). This so-called 'good mood' hypothesis, however, is not supported by the data (de Waal, 1997a), which show that if A receives grooming from B, A is only more likely to share with B but not with others in the group. The exchange between grooming and food is, therefore, partner-specific.

These studies on capuchins and chimpanzees address whether reciprocity is calculated or a by-product of frequent association and symmetrical relationships (de Waal and Luttrell, 1988; de Waal, 1997a). Calculated reciprocity is based on the capacity to keep mental note of favours given and received. It is a more sophisticated and cognitively complex (and consequently less easily accepted) form of reciprocity than symmetry-based reciprocity, which occurs when individuals preferentially direct favours to close associates. Since association is a symmetrical relationship characteristic (if A associates often with B, B does so often with A), the distribution of favours automatically becomes reciprocal (for a more in-depth discussion of what constitutes symmetry-based reciprocity, see de Waal, 1996b, p. 157). Although important, such symmetry-based reciprocity is not as cognitively demanding as calculated reciprocity, which was shown above to occur in chimpanzees and possibly female capuchins.

Calculated reciprocity — unlike symmetry-based reciprocity — raises interesting questions about the nature of expectations. The possibility that chimpanzees withhold favours from ungenerous individuals during future interactions, and are less resistant to the approaches of individuals who previously groomed them (de Waal, 1997a) suggests they have expectations about how they themselves and others should behave in certain contexts.

Other evidence to suggest that some primates have expectations about how others should behave comes from studies of patterns of conflict intervention. Chimpanzees and some species of macaques exhibit what appears to be calculated reciprocity in beneficial interventions, or the interference by a third party in an ongoing conflict in support of one of the two conflict opponents (de Waal and Luttrell, 1988; Silk, 1992). Thus, if A intervenes in favour of B, B is more likely to intervene in favour of A. In de Waal's and Luttrell's study, chimpanzees, but not macaques, also exhibited reciprocity in harmful interventions, suggesting the existence of a so-called 'revenge system'. In other words, in the chimpanzee group under study there existed a significant correlation between interventions given and received so that if A intervened against B, B was more likely to intervene against A in the future. This retaliatory pattern was not found in stumptail and rhesus macaque groups. Silk (1992), however, found evidence for a revenge system among males in bonnet macaque society in that the males in her study group appeared to monitor both the amount of aggression that they received from and directed at other males. Although Silk's data and numerous anecdotes suggest that macaques do have the capacity to engage in revenge, it is likely that revenge of this sort is not commonplace due to the greater risks in a macaque society (compared to a chimpanzee society) associated with directing aggression at dominants.

Revenge of another sort — indirect revenge — does, on the other hand, appear to be relatively common in at least one macaque species. Indirect revenge occurs when recipients of aggression redirect their aggression at the uninvolved juvenile or younger kin of their opponents. In this way, these often low ranking macaques are still able to 'punish' their attackers but are able to do so without much cost to themselves (Aureli, Cozzolino, Cordischi and Scucchi, 1992). For example, Aureli and colleagues found that Japanese macaque recipients of aggression were significantly more likely to attack the kin of their former opponents within one hour after the original conflict had occurred than if no conflict had occurred at all. One could argue, as the authors pointed out, that it is possible that this increase in aggression towards an opponent's kin after a conflict may be due to a general rather than selective aggressive tendency that is triggered by fighting and thus does not reflect a revenge system. Additional analyses revealed, however, that this hypothesis is not supported by the data — the relative probability that the original recipient of aggression would attack, following the conflict, the kin of a former opponent was significantly higher than the probability that the original recipient of aggression would attack following the conflict any group member subordinate to it. The existence of this form of revenge in macaque society suggests that a macaque's capacity to be vindictive is constrained by its rank in society rather than by its cognitive abilities.

These examples of retributive behaviour indicate that some form of calculated reciprocity is present in primate social systems. This kind of reciprocity and the kinds of responses seen by chimpanzees in the food-sharing study exemplify how and why prescriptive rules, rules that are generated when members of a group learn to recognize the contingencies between their own behaviour and the behaviour of others, are formed. The existence of such rules and, more significantly, of a set of expectations, essentially reflects a sense of social regularity, and may be a precursor to the human sense of justice (de Waal, 1991; Gruter, 1992; see also Hall, 1964; Nishida, 1994).

Trivers (1971) daringly labelled negative reactions to perceived violations of the social code, *moralistic aggression*. He emphasized that individuals who respond aggressively to perceived violations of the social code help reinforce systems of reciprocity by increasing the cost of not co-operating and, even more importantly, by increasing the cost of cheating, or failing to return a favour. When one individual cheats another, that individual exploits a relationship that is based on the benefits the partners previously obtained by co-operating. By doing so, the cheater benefits himself or herself at the partner's expense and destabilizes the system of reciprocity. Moralistic aggression, which often manifests itself as protest by subordinate individuals or punishment by dominant individuals, helps deter cheating. Consequently, it contributes to the creation of order, an element essential to the maintenance of the stability or integrity of social systems (de Waal, 1996a; Hardin, 1983). If unchecked, however, moralistic aggression can also lead to a spiral of spiteful retaliation that confers advantage on neither the original defector nor the moralistic aggressor, as is the case when those seeking retributive justice exacerbate conflicts to such a degree that feuds develop (Boehm, 1986; de Waal, 1996b, chapter 4).

Conclusion: Monkeys and apes appear capable of holding received services in mind, selectively repaying those individuals who performed the favours. They seem to hold negative acts in mind as well, leading to retribution and revenge. To what degree these reciprocity mechanisms are cognitively mediated is currently under investigation, but at least for chimpanzees there is evidence for a role of memory and expectation.

Conflict Resolution

Conflicts are inevitable in social groups. They may be generated by disagreement over social expectations or simply by competition over access to resources. Regardless of what triggers conflicts, group-living individuals need mechanisms for negotiating resolutions to them and for repairing the damage to their relationships that results once conflicts of interests have escalated to the point of aggression. One of the simplest ways that conflicts are regulated and resolved is through the establishment of clear-cut dominance relations (see Carpenter, 1942; Mendoza and Barchas, 1983; Bernstein, 1981; Bernstein and Ehardt, 1985; de Waal, 1996b; for a review, see Preuschoft and van Schaik, in press).

Primates in hierarchical social systems typically have many methods by which they communicate who is dominant and who is subordinate. Subordinate rhesus macaques, for example, bare their teeth in a ritualized expression and often present their hindquarters to an approaching dominant group member. Such displays signal to the dominant individual that the subordinate recognizes the type of relationship they share, which consequently eliminates any question of ambiguity or need for aggression and promotes harmony and stability at the group level (de Waal, 1986). Interestingly, it appears that the bared-teeth expression is a *formal* dominance signal in despotic species, such as the rhesus macaque, in that it is almost exclusively displayed by subordinate individuals (de Waal and Luttrell, 1985; Preuschoft, 1999). In more egalitarian and tolerant macaque species, such as Tonkean macaques, power asymmetries between individuals are less evident than in despotic species, like rhesus macaques. Coinciding with this difference in power is a difference in use of the bared-teeth expression, which in Tonkean macaques is neither ritualized nor formal but common to both subordinate and dominant individuals (Thierry, Demaria, Preuschoft and Desportes, 1989).

Strict dominance relationships are often an effective means by which conflicts can be negotiated. When conflicts persist despite dominance relationships, or in primate species where dominance relations are relaxed or almost absent, there must be alternative ways to work out problems and repair relationships (this does not, however, imply that the development of egalitarian social systems led to the development of conflict management devices or vice versa, only that generally the two go together). One of the most important of these post-conflict behaviours is *reconciliation*. Reconciliation, which is defined as a friendly reunion between former opponents not long after a confrontation, is illustrated in the following description of an agonistic interaction between two chimpanzees and the post-conflict behaviour that followed (de Waal, 1989c, p. 41):

> ... Nikkie, the leader of the group, has slapped Hennie during a passing charge. Hennie, a young adult female of nine years, sits apart for a while feeling with her hand the spot on the back of the neck where Nikkie hit her. Then she seems to forget about the incident; she lies down in the grass, staring into the distance. More than fifteen minutes later, Hennie slowly gets up and walks straight to a group that includes Nikkie and the oldest female, Mama. Hennie approaches Nikkie with a series of soft pant grunts. Then she stretches out her arm to offer Nikkie the back of her hand for a kiss. Nikkie's hand-kiss consists of taking Hennie's whole hand rather unceremoniously into his mouth. This contact is followed by a mouth-to-mouth kiss.

Reconciliation enables the immediate, negative consequences of aggression to be counteracted and reduces the tension-related behaviour of recipients of aggression (de Waal and van Roosmalen, 1979; Aureli and van Schaik, 1991; de Waal and Aureli, 1996; Aureli, 1997). Perhaps more importantly, though, reconciliation enables former opponents to restore their relationship (Kappeler and van Schaik,

1992) and indeed one can increase the rate of reconciliation by experimentally enhancing the value of the relationship, e.g. by making the food-intake of two individuals dependent on their co-operation (Cords and Thurnheer, 1993). This form of post-conflict behaviour has been demonstrated in many primate species, each of which has its own typical 'peacemaking' gestures, calls, facial expressions and rituals, including, for example, kissing and embracing (see de Waal and Yoshihara, 1983; Cords, 1988; de Waal and Ren, 1988; York and Rowell, 1988; Aureli, van Schaik and van Hooff, 1989; Judge, 1991; Ren *et al*, 1991; Kappeler, 1993). We label friendly post-conflict behaviour 'reconciliation' if we can demonstrate empirically that the former opponents are selectively attracted so that they tend to come together in this manner more than usual and more with each other than with individuals who had nothing to do with the fight. In order to determine the percentage of conflicts followed by reconciliation for individuals of a particular species, we compare the post-conflict period (PC) to a matched-control period (MC). We use the matched-control period because it enables us to determine whether the affiliation that takes place during the post-conflict period is triggered by the conflict, or if it is simply due to chance (for a detailed discussion of the PC/MC method, see Veenema, Das and Aureli, 1994). As seen in Figure 2, former stumptail macaque opponents affiliate considerably more in post-conflict periods than they do in the matched-control periods.

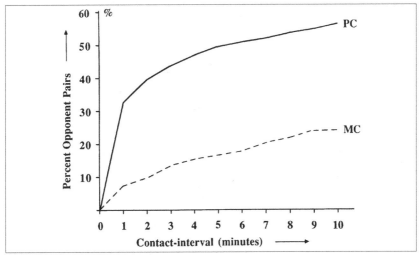

Figure 2

Cumulative percentage of pairs of opponents making their first nonagonistic body contact within a certain time interval. PC = post conflict observation, MC = matched control observation, N = number of pairs. From de Waal and Ren (1988).

Although reconciliation has been observed in most primate species and appears to be a universal method of repairing disturbed relationships, the degree to which it is used differs across primate species in a pattern that may reflect the level of integration and cohesion in a primate society. For example, analysis of post-conflict/matched-control (PC/MC) data from 670 pairs of former stumptail macaque opponents with PC/MC data from 573 pairs of former rhesus macaque opponents, revealed that stumptail monkeys reconciled on average significantly more often (i.e. 51.6 per cent) than rhesus monkeys (i.e. 21.1 per cent) (de Waal and Ren, 1988). In general, individuals in despotic species reconcile less frequently after conflicts than individuals in more tolerant and egalitarian species, most likely because the strict dominance hierarchies that are present in despotic species constrain the development of strong symmetrical relationships among group members (de Waal, 1989a).

Another way primates regulate and resolve conflicts of interests between group members is through conflict intervention. Although many studies have shown that interventions are related to coalition building and alliance formation, some interventions may have other functions as well (e.g. Reinhardt, Dodsworth and Scanlan, 1986; Bernstein and Ehardt, 1986; Boehm, 1994; Petit and Thierry, 1994; for reviews see Harcourt and de Waal, 1992). In some species, interventions by the highest ranking members of the social group end fights or at least reduce the severity of aggression that occurs during fights. The alpha male chimpanzee often plays such a role in fights involving females and/or juveniles in his group (de Waal, 1982). For example, if two juveniles are playing and a fight erupts, the alpha male just has to approach the area of the conflict to stop the fight. By doing so, he directly reduces the levels of aggression within the group, and also prevents the fight from escalating further by ending it before the juveniles' mothers intervene and possibly begin fighting themselves. Another example of this type of intervention in chimpanzees is given by the following excerpt from *Chimpanzee Politics* (1982, p. 124):

> On one occasion, a quarrel between Mama and Spin got out of hand and ended in biting and fighting. Numerous apes rushed up to the two warring females and joined in the fray. A huge knot of fighting, screaming apes rolled around in the sand, until Luit leapt in and literally beat them apart. He did not choose sides in the conflict, like the others; instead anyone who continued to fight received a blow from him.

This pattern of behaviour, often referred to as the 'control role' has been described in other species of primates as well, (e.g. Bernstein, 1964; Tokuda and Jensen, 1969; Reinhardt *et al.,* 1986), and is a type of arbitration. The most interesting types of interventions that fall under the control role heading are those that are impartial. Individuals who intervene without choosing sides seem to do so in order to restore peace rather than simply aid friends or family (de Waal and van Hooff, 1981; de Waal, 1982; Goodall, 1986). The ability to put one's own preferences aside in this manner is another indication that a rudimentary form of justice may exist in the social systems of non-human primates (Boehm, 1992; de Waal, 1996a; 1996b). On the other hand, one might argue that although such interventions ultimately may have the effect of restoring the peace and reducing overall levels of aggression in the group, it is possible that the intervener's intentions were simpler, and that he or she was motivated only by the desire to terminate an aversive stimulus — the noisy conflict — and not by any group-oriented motivation. Evidence from chimpanzees and macaques suggests, however, that this explanation for impartial interventions may be inadequate. Such interventions may indeed be motivated in part by a desire to terminate an aversive stimulus but, if that were the only motivation, we might expect the interventions to be severe and partial, and in favour of the individual with whom the intervener shares the best relationship. Furthermore, agonistic interventions, in particular, can trigger agonistic and nonagonistic involvement by other individuals and intense screaming by the target or targets of the intervention (Gouzoules, Gouzoules and

Marler, 1984). Thus, interventions often temporarily exacerbate the aversive stimuli that supposedly the intervener sought to suppress.

Another type of intervention that falls under the control role is the protective intervention. Interventions in this group include those that occur on behalf of recipients of aggression. When expressed in this form, the control role can be viewed as a way that lower ranking or weaker (either physically *or* socially) individuals are protected from higher ranking, stronger (either physically *or* socially) group members.

A particularly salient example of the power of control animals to protect recipients of aggression in some primate societies that demonstrates not only the degree to which a control animal can influence the outcome of a conflict, but also that other individuals in the group recognize this capacity in certain individuals, comes from an experiment in which the composition of a pigtail macaque group was manipulated (Flack and de Waal, unpublished data). During this experiment, the three highest-ranking male pigtail macaques were removed for the day once per week and confined to their indoor housing. We studied the patterns of conflict intervention and aggression that occurred in the group when the males were present and when the males were absent. During removal periods, although confined to their indoor housing, the males had vocal and very restricted visual access to the group in that they were separated from the group by only a two foot long tunnel (to the indoor housing), divided in the middle by a metal door that did not completely seal off the indoor area from the outdoor area.

Three very low-ranking females typically received moderate levels of aggression from other group members when the males were present, but the males often intervened in these conflicts and, when this occurred, the aggression that had been directed at the females usually subsided. During the period when the males were removed, the intensity and frequency of aggression directed at these low ranking females increased substantially, and severe biting (biting for more than five seconds in duration), in particular, was more common. Several times, the aggression directed at these females became so severe that the investigator was forced to intervene.

Over the course of the study period, however, the females discovered a way to deal with the increase in severe aggression that they received when the high-ranking males were absent. This began when one female started running into the tunnel that separated the indoor housing from the outdoor housing. Once in the tunnel, the female solicited support from the males locked inside. The males, who could hear the conflict, were always waiting at the door. The female being attacked would scream, bare her teeth, and stick her arm though a small space in the door as her attackers were rushing forward. Each time this occurred, one of the males confined to the indoor housing would emit a threat bark, and the female would scream, clearly distressed by both her attackers and the threat, but not move from her position. Her attackers, however, would jump back and cease their abusive behaviour. The female would then lie prone in the tunnel for the next fifteen minutes or so, and each time her attackers attempted to bite her, a male inside would emit a threat bark and the attack would cease. The other two females, too, learned that when they were severely attacked they could appeal to the confined high-ranking males in this way, and as the study period progressed, this behaviour increased in frequency. Notably, no individuals ever escaped from aggression by running into the tunnel when the males were present in

14   J.C. FLACK AND F.B.M. DE WAAL

the group, presumably because doing so prohibited escape and thus was very dangerous.

Interestingly, among chimpanzees, the individual who plays this control role need not be the alpha male of the group. The control animal may be any group member who the community permits — in the sense that none of the individuals in the social group protests the control animal's involvement in the conflict nor prohibits the particular control animal from 'playing' his or her role (de Waal, 1996b, chapter 3). Pascale Sicotte (1995) described a similar mechanism for resolving conflicts in bi-male groups of mountain gorillas in which females and infants sometimes interposed themselves between two fighting silverbacks. These interpositions, which occurred in 10–25 per cent of conflicts between silverbacks, involved more than just passive or chance interference in the agonistic dyad by a third party. Sicotte only included as interpositions those third party interventions in which a previously uninvolved individual interacted nonagonistically with at least one of the two males engaged in conflict, and in which the course of the initial agonistic interaction between the males was modified because one of the opponents directed its attention towards the third party. Notably, in one of the two groups Sicotte studied, he found that interpositions significantly increased the time between the end of the fight in which it occurred, and the start of the next fight.

One other important method of conflict resolution that has been identified in primate groups is mediation. Mediation occurs when a third party to a conflict becomes the bridge between two former opponents who cannot seem to bring themselves to reconcile without external help. It is characterized in the following example (de Waal and van Roosmalen, 1979, p. 62).

> Especially after serious conflicts between two adult males, the two opponents sometimes were brought together by an adult female. The female approached one of the males, kissed or touched him or presented towards him and then slowly walked towards the other male. If the male followed, he did so very close behind her (often inspecting her genitals) and without looking at the other male. On a few occasions the female looked behind at her follower, and sometimes returned to a male that stayed behind to pull at his arm to make him follow. When the female sat down close to the other male, both males started to groom her and they simply continued when she went off.

In the above example of mediation, a female, who apparently is trying to reunite two former opponents in her social group, seems to show community concern in that she apparently cares about resolving a conflict in which she had no part and, more importantly, about restoring a disturbed relationship that is not her own. Although such examples are rare in primates, and perhaps unlikely in any but apes, a very similar pattern of behaviour to that illustrated by the example above was observed in stumptail macaques (Flack, personal observation), suggesting that individuals in several species may have a sense of community concern that comes from having a stake in the quality of life within the group as a whole.

De Waal (1996b, p. 31) explained the evolution of community concern as follows:

> Inasmuch as every member benefits from a unified, cooperative group, one expects them to care about the society they live in, and to make an effort to improve and strengthen it similar to the way the spider repairs her web, and the beaver maintains the integrity of his dam. Continued infighting, particularly at the top of the hierarchy, may damage everyone's interests, hence the settlement of conflict is not just a matter of the parties involved,

ANY ANIMAL WHATEVER 15

it concerns the community as a whole. This is not to say that animals make sacrifices for their community, but rather that each and every individual has a stake in the quality of the social environment on which its survival depends. In trying to improve this quality for their own purposes, they help many of their group mates at the same time. A good example is arbitration and mediation in disputes; standard practice in human society — courts of law serve this function — but recognizable in other primates as well.

Two other patterns of behaviour in monkeys and apes illustrative of community concern are triadic reconciliation — the involvement of third parties in the reconciliation process — and the group-wide celebration that often follows the reconciliation of dramatic conflicts of chimpanzees (de Waal, 1992b; 1996b, chapter 4). These patterns of behaviour suggest that monkeys and apes devote time and energy to making sure their social group remains peaceful, perhaps because group members recognize the value that a harmonious coexistence can have to achieving shared interests. In this sense, community concern can be extremely beneficial to individuals within social groups — even if it requires subordinating independent interests (non-shared), at least on occasion, to community interests — so long as many common goals are shared among group members. This kind of community concern, however, does not require that monkeys and apes worry about how the community, as an abstract entity, is doing. It only requires that the individual works toward creating a community atmosphere that reflects his or her own best interests. Consequently, the evolution of community concern in individuals may not necessarily require group selection. Its evolution can most likely be explained using selection at the level of the individual (although it is possible that individual selection may perhaps provide only a partial explanation for the evolution of such behaviours — for more discussion of this matter, see Wilson and Sober, 1994, and this issue), particularly if we consider the other behaviours that probably co-evolved with community concern in order to mitigate the risks and reduce the short-term costs to individuals generated by investing in the community.

Punishment, for example, or the imposition of a cost or penalty by one individual (usually who has some authority or power over the other individual) on another for its behaviour (Bean, 1981), and perhaps indirect reciprocity and social norms, help offset the cost of placing community (shared) interests above individual (independent) interests. Punishment and indirect reciprocity facilitate the evolution of investment in community interests because they help deter cheating (as discussed earlier) and reinforce community-oriented behaviours by making it possible for reciprocity to become generalized — so that if one individual performs a favour for another, that favour may be returned by a third party (for discussion of punishment, see Boyd and Richerson, 1992; Clutton-Brock and Parker, 1995; for discussion of indirect reciprocity, see Alexander, 1987; Boyd and Richerson, 1989). As Axelrod (1986) discussed, an especially powerful mechanism by which cheating can be deterred emerges when punishment against defectors becomes linked with negative indirect reciprocity, or punishment of nonpunishers. This *metanorm,* as Axelrod calls it, makes the 'norm' against defection self-policing. Whether this occurs in non-human primates, however, remains to be empirically demonstrated.

There is, however, evidence from a recent study of play signalling patterns in juvenile chimpanzees that suggests primates may modify their behaviour in anticipation of punishment (Jeannotte, 1996). Lisa Jeannotte found that older play partners were significantly more likely to emit play signals, such as 'play face', during play bouts that occurred in proximity to adults, particularly those adults who were mothers of younger play partners and who were themselves young, than was the case when these adults were absent. These results suggest that an older juvenile play partner may increase its play signalling in the presence of a young mother to make clear that its interaction with the younger play partner is benign and does not warrant intervention or punishment.

Another mechanism that might enable the evolution of community concern is docility, or the receptivity to social influence that is common to social primates and that is useful for acquiring valuable information without the need for direct experience or evaluation (Simon, 1990). Thus, if certain values, like community concern, are fostered in a particular social environment, then an individual, simply due to its docile disposition, may adopt and become committed to those values even though at times those values or sentiments may encourage a course of action that is counter to an individual's independent (non-shared) interests. In this way, community concern evolves as a by-product of selection for docility (for a related argument about a similar mechanism, the 'conformist transmission', by which community concern might evolve, see Boyd and Richerson, 1985; 1992; Henrich and Boyd, 1998; see also Cronk, 1994 for a Marxian argument about how docility may actually make it possible in some moral systems for certain individuals to justify and perpetuate inequalities of power and access to resources).

Although at present, direct evidence for punishment, docility, and especially indirect reciprocity is scant or fragmented in the primate literature, such behavioural mechanisms by which collective action is facilitated do probably exist to varying degrees. The existence of these mechanisms, which are likely complementary rather than alternative methods that act in concert to produce stability in social systems (Henrich and Boyd, 1998), is indirectly supported by the presence of learned adjustment, succourant behaviour, empathy, and sympathy in non-human primates.

Conclusion: Despite inevitable conflicts of interests, a certain degree of stability must be maintained in primate societies so that individuals can realize their collective interests and make worthwhile their investments in sociality. Dominance relationships provide one simple way to regulate and order societies. Social systems with more level dominance relationships require additional mechanisms, however, such as reconciliation; consolation; impartial, protective, and pacifying interventions; and perhaps community concern. All of these mechanisms are present to varying degrees in monkeys and apes.

Empathy, Sympathy and Consolation

Although food-sharing, social reciprocity in general, and the different forms of conflict resolution seen in primates need not require a capacity for sympathy and empathy, it is likely that both are involved to some degree in all of these behaviours, and to a high degree in at least the more sophisticated forms of conflict intervention such as mediation. In order to help others, as the female in the mediation example was doing, individuals need to be concerned about and be able to understand others' needs and emotions.

Learned adjustment, which is common in primates, is a precursor to such behaviour in that it demonstrates the ability of monkeys and apes to change their behaviour as they become familiar with the limitations of those with whom they interact without requiring that these individuals understand why they should adjust their behaviour. Juvenile chimpanzees, for example, commonly restrict the degree of force they use in wrestling matches while playing with younger juveniles and infants (Hayaki, 1985). Monkeys and apes also adjust their behaviour in the presence of disabled group members (e.g. Fedigan and Fedigan, 1977; de Waal, 1996b, chapter 2; de Waal, Uno, Luttrell, Meisner and Jeannotte, 1996c). The adjustment may include increased social tolerance towards individuals who behave abnormally or intervention on behalf of disabled individuals who seem unaware of when they are involved in a dangerous predicament. Although the examples discussed above are most parsimoniously explained using learned adjustment, the possibility that cognitive empathy — the ability to comprehend the needs and emotions of other individuals — may provide a more accurate explanation for some of these behaviours needs to be explored in future studies.

An especially important area that needs investigation is how learned adjustment and cognitive empathy relate to the internalization of social norms. One study that begins to address this question is Jeannotte's previously mentioned study (1996) of play-signalling in juvenile chimpanzees in relation to social context and environment. Results of this study indicate that an older juvenile is not significantly more likely to play roughly when the age difference between it and its partner is small as opposed to large. Although this result seems to contradict findings from previous research that suggested that juvenile chimpanzees restrict the intensity of play when interacting with younger play partners, it does not necessarily do so. Jeannotte found that there was a strong correlation between the play intensity of one partner with that of the other. One likely explanation for this 'matching' is that it may be a consequence of restraint on the part of the older partner and escalation on the part of the younger partner (Hayaki, 1985).

Succourant behaviour, which includes care-giving and providing relief to distressed individuals who are not kin, is also an example of a category of behaviours that seem to require attachment to and concern for others and, in some cases, an understanding of other's needs and emotions (Scott, 1971). Succourant tendencies develop in primates early in life; even infants respond to tension generated by aggression by mounting one another or by mounting kin or even the individuals involved in the agonism. Although these infants probably are not helping to reduce tensions between the individuals involved in the agonism by responding in this way, they may be comforting themselves. This simple need to comfort oneself after or during a fight in which the infant itself was not involved suggests that the infant perceives distress in others and reacts vicariously to it by becoming distressed itself. This 'emotional contagion' (Hatfield, Cacioppo and Rapson, 1993) may be the mechanism underlying the development of succourance and suggests that primates do have the ability to empathize.

Non-human primates may be able to empathize with one another in that other group member's feelings and actions emotionally affect them, but are non-human primates also concerned about individuals who appear distressed? In other words, do they sympathize with or just react to individuals in their group who are distressed?

There is evidence suggesting that some primates do have concern for fellow group members; it comes from studies of consolation, or the appeasement of distressed individuals through affiliative gestures such as grooming and embracing by third parties following a fight (de Waal and van Roosmalen, 1979). The predominant and immediate effect of consolation — the alleviation of distress (de Waal and Aureli, 1996), is illustrated by the following example observed by Jane Goodall (1986, p. 361):

> An adult male challenged by another male often runs screaming to a third and establishes contact with him. Often both will then scream, embrace, mount or groom each other while looking toward the original aggressor. This . . . is how a victim tries to enlist the help of an ally. There are occasions, however, when it seems that the primary goal is to establish reassurance contact — as when fourteen-year-old Figan, after being attacked by a rival, went to hold hands with his mother.

In the above example, however, the recipient of aggression is *seeking* consolation. This type of consolation occurs in several primate species but may not require that the third party sympathizes with the recipient who approaches for reassurance (for other examples see Lindburg, 1973; de Waal and Yoshihara, 1983; Verbeek and de Waal, 1997). *Active* consolation, on the other hand, occurs when a third party approaches and affiliates with a recipient of aggression following a fight. Such action may require that the third party not only recognize the distress of the recipient of aggression but also be concerned enough about that individual to approach and appease it. Sometimes, for example, a juvenile chimpanzee will approach and embrace an adult male who has just lost a confrontation with his rival (de Waal, 1982).

There exist systematic data to support the conclusion that chimpanzees have the capacity to engage in active consolation (de Waal and Aureli, 1996). An analysis of 1,321 agonistic incidents among a captive group of seventeen chimpanzees housed in a large compound at the Field Station of the Yerkes Regional Primate Research Center revealed that significantly more affiliative contacts initiated by bystanders occurred immediately (within several minutes) after a conflict than after longer time intervals or in control periods not preceded by conflict. Furthermore, significantly more affiliative contacts initiated by bystanders occurred following serious incidents than mild incidents (this is particularly important because, if consolation occurs to alleviate distress, and if an individual's level of distress is proportional to the aggression intensity of the conflict in which it participated, then consolation should occur more frequently following serious aggressive incidents). And finally, bystanders initiated significantly more affiliative contact with the recipients of aggression than with the aggressors themselves.

In contrast to these findings for chimpanzees, researchers have been unable to quantitatively demonstrate active consolation in four macaque species (Aureli, 1992; Aureli and van Schaik, 1991; Judge, 1991; Aureli, Veenema, van Panthaleon van Eck and van Hoof, 1993; Aureli, Das, Verleur and van Hooff, 1994; Castles and Whiten, 1998; for a review, see de Waal and Aureli, 1996). This suggests that consolation may be limited to the great apes, possibly because it requires more sophisticated cognition than present in monkeys. Alternatively, it may be limited to apes because within their social systems, such behaviour is more advantageous and perhaps less costly than in monkey social systems in which approaching recipients of aggression, and areas of conflicts generally, can be dangerous due to the frequency with which aggression is redirected to bystanders in many monkey societies (de Waal and Aureli, 1996).

Conclusion: Moral sentiments such as sympathy, empathy, and community concern, engender a bond between individuals, the formation of which facilitates and is facilitated by co-operation. This bond is enabled by an individual's capacity to be sensitive to the emotions of others. Monkeys and apes are capable of learned adjustment, and have succourant tendencies, in that they comfort and console one another when distressed. But are they capable of genuine concern for others based on perspective-taking? There is some evidence to suggest that apes, like humans, are capable of cognitive empathy but its existence in monkeys remains questionable.

Implications of Primate Research for Understanding Human Morality

In the opening pages of a *Theory of Justice* (1971, p. 4), John Rawls elegantly states the central problem that plagues those human (and animal) societies in which implicit or explicit rules of conduct exist to make co-operation possible:

> . . . although a society is a co-operative venture for mutual advantage, it is typically marked by a conflict as well as by an identity of interests. There is an identity of interests since social cooperation makes possible a better life for all than any would have if each were to live solely by his own efforts. There is a conflict of interests since persons are not indifferent as to how the greater benefits of their collaboration are distributed, for in order to pursue their ends they each prefer a larger to a lesser share.

This problem, as identified above by Rawls, has in practice no true solutions. Furthermore, the research on the natural history and social behaviour of our non-human primate relatives illustrates how both our capacity and tendency to pursue our independent interests and our capacity and tendency to pursue shared interests are natural and important, at least from a biological point of view (see de Waal, 1992a). This suggests that morality was not *devised* to subjugate the independent interests of individuals. Rather, a moral system *emerged* out of the interaction of the two sets of interests, thus providing a way to express both. This conclusion should not be mistaken as justification for using natural selection as a model for what we ought to do or *not* do. What we ought to do and how we decide this is a separate question from why and how moral systems arose.

It is particularly important that in our pursuit of the origins and purpose of moral systems we resist the temptation to let our moral views frame, and thus obscure, how in the end we describe and explain the moral standards embodied in implicit social contracts. Even more importantly, we need to be careful not to hold up as moral systems only those that in our view wholly subjugate the independent interests of individuals in favour of those interests that are shared, simply because we value that these systems suppress conflict and deliver consensus. Doing so precludes from consideration those systems that from an operational standpoint are moral systems, but that may not fit perfectly our moral views of what is right, of worth, or of value.

Humans, nonetheless, may be the only truly moral creatures. Although one could argue that several elements of human morality are present in non-human primates — particularly in apes — there is no evidence at this time to suggest that non-human primates have moral systems that mirror the complexity of our own. In some species, individuals, by interacting every day, may create a kind of social contract that governs which types of behaviour are acceptable and tolerable and which are punishable —

yet these individuals have no way to conceptualize such decisions or abstract them from their context, let alone debate them amongst themselves. Consensus is only obvious in the absence of protest and prohibition.

Consensus achieved in this tacit manner, however, is not uncommon in our own species. This observation, in conjunction with the above research that suggests that an actual social contract of sorts arises out of interactions between group members in primate societies, makes plausible the idea that human morality is best understood as having arisen out of an implicit agreement among group members that enabled individuals to profit from the benefits of co-operative sociality.

Acknowledging that morality may have a social function and stressing that it may have emerged from such a social contract does not require that we accept this kind of 'actual' social contract as the medium through which we decide what is moral. Nor does it suggest that we revert to some form of Social Darwinism, an approach to deciding what we ought to do that was based not only on a misconceived, red-in-tooth-and-claw representation of natural selection, but worse, also on the idea that this interpretation of 'nature's way' should be used to guide (and justify) our own behaviour. Although we need to recognize that the social contract does in fact often represent the process by which we come to agree (as a group) what is acceptable, we also need to recognize that this is probably the case because the social contract is useful from an evolutionary perspective because it enables individuals in groups to reach consensus with minimal, if any, need for explicit co-ordination. Certainly, from a social perspective, one of the major limitations of any actual social contract is that such contracts do not necessarily produce the most 'moral' solutions to problems. The outcome such contracts produce is no more than a reflection of compromise, and of the behaviour that as a group we practice.

Another important observation about human behaviour made by David Hume (1739), Adam Smith (1759), and Edward Westermarck (1912) is that human morality is powerfully influenced by emotional responses and is not always governed by the abstract, intellectual rules upon which we have supposedly agreed. The primate research implicitly suggests that this emphasis on the role of emotions is both insightful and accurate — in primate groups individuals are motivated to respond to others based on the emotional reactions they have to one another's behaviour. That sympathy, based on empathy, seems to direct the emotional responses of some primates to others may reflect their ability to differentiate between self and other and, more significantly, to care for one another. Yet, the idea that emotion may be fundamental to morality contradicts what many philosophers — most significantly Immanuel Kant (1785) — have argued: That the human sense of right and wrong is more a consequence of rational processes than of emotional reactions. It would be quite erroneous, however, to equate moral emotions with a lack of rationality and judgement. The emotions discussed by Hume, Smith and Westermarck are actually very complex, involving retribution, reciprocity and perspective taking. The latter, as is now increasingly apparent from research into so-called Theory-of-Mind, involves complex mental abilities (for a review of Theory-of-Mind in non-human primates, see Heyes and Commentaries, 1998).

Darwin (1871; 1872), who was familiar with the thinking of Hume and Smith, advocated a perspective on human morality in line with these ideas in that he saw human nature as neither good nor bad but neutral. He recognized that moral systems enable individuals to reconcile what Hume saw as two sides to human nature — the dark, competitive side, which is dominated by greed and competition, and the 'sentimental,' co-operative side, which is marked by social instincts and compassion. To Darwin, this dualism in human nature arose from the evolution of two strategies (the individual and social) that together provided a method by which individuals can obtain limited resources. Thus, Darwin recognized that moral systems not only govern the expression and use of these strategies but also reflect their interaction.

Opposition to this more integrated view by some contemporary evolutionary biologists, such as Richard Dawkins and George Williams (see Introduction), leads us to propose that their views on morality be classified not as Darwinian but Huxleyan. For example, Dawkins recently reconciled human moral ideals with his interpretation of evolution by saying that we are entitled to throw out Darwinism ('in our political and social life we are entitled to throw out Darwinism, to say we don't want to live in a Darwinian world', *Human Ethology Bulletin,* March 1997). Because Darwin himself perceived absolutely no contradiction or dualism between the evolutionary process and human moral tendencies (e.g. de Waal, 1996b; Uchii, 1996), such views represent a considerable narrowing of what Darwin deemed possible.

The Kantian view of morality as an invention of reason supplemented with a sense of duty remained pervasive despite Darwin's insights. Perhaps primate research that suggests that morality is a consequence of our emotional needs and responses as well as of our ability to rationally evaluate alternatives is strong enough to warrant making room for a more integrated perspective of morality that acknowledges its biological basis and emotional component as well as the role of cognition. Perhaps Hume and Kant were both correct.

The foundations of morality may be built on our emotional reactions to one another but morality itself is no doubt also tempered and sometimes modified by two additional factors. First, morality may be modified by our ability to evaluate the situation generating these emotional reactions. Second, it may be tempered by our understanding of the consequences that our responses to the behaviour that elicited the emotional reaction have for ourselves and others. A problem, however, remains even after we acknowledge that what generates in each of us an understanding of what is good or virtuous is a combination of two factors: 1) the emotional reaction and intuition of each individual that jump-start the moral process, with 2) the cognitive-rational evaluations that enable the individual to determine what is right. The problem that remains despite this integration is how to translate the resultant conception in the individual of what is good to action at the community level. In human societies, as in animal societies, this is often achieved by some manifestation of the social contract. But as mentioned earlier, 'actual' social contracts, as rough compromises between competing agents, often with unequal powers and needs, may be unsatisfactory from a normative standpoint because they do not fully respect worth, value or rights.

Conclusion

Sympathy-related traits such as attachment, succourance, emotional contagion and learned adjustment in combination with a system of reciprocity and punishment, the ability to internalize social rules and the capacity to work out conflicts and repair relationships damaged by aggression, are found to some degree in many primate species, and are fundamental to the development of moral systems (Table 1).

All of these elements of moral systems are tools social animals — including humans — use to make living together a possibility. These capacities help keep in check the inevitable competition among group members due to conflicting interests. More importantly, however, sympathy-related traits and the capacity to work out con-

Table 1

It is hard to imagine human morality without the following tendencies and capacities also found in other species. These tendencies deserve to be called the four ingredients of morality:

Sympathy Related
Attachment, succourance, and emotional contagion.
Learned adjustment to and special treatment of the disabled and injured.
Ability to trade places mentally with others: cognitive empathy.*

Norm Related
Prescriptive social rules.
Internalization of rules and anticipation of punishment.*
A sense of social regularity and expectation about how one ought to be treated.*

Reciprocity
A concept of giving, trading, and revenge.
Moralistic aggression against violators of reciprocity rules.

Getting Along
Peacemaking and avoidance of conflict.
Community concern and maintenance of good relationships.*
Accommodation of conflicting interests through negotiation.

*It is particularly in these areas — empathy, internalization of rules, sense of justice, and community concern — that humans seem to have gone considerably further than most other animals.

flicts and repair relationships help promote cohesion, co-operation and social bonding, characteristics of a social group that may, from an evolutionary perspective, make group living a functionally effective strategy and, therefore, an attractive strategy in which individuals should invest resources. As the anthropologist Ruth Benedict wrote in 1934 (p. 251):

> One of the most misleading misconceptions due to this nineteenth-century dualism was the idea that what was subtracted from society was added to the individual and what was subtracted from the individual was added to society. . . . In reality, society and the individual are not antagonists. His culture provides the raw material of which the individual makes his life. If it is meagre, the individual suffers; if it is rich, the individual has the chance to rise to his opportunity.

ANY ANIMAL WHATEVER 23

At the end of this paper, in which we have discussed the possible evolutionary building blocks of human moral systems, it is essential to also point out the limitations of biological approaches to human morality. After all, we have in the past seen attempts to derive moral rules directly from nature, resulting in a dubious genre of literature going back to Ernest Seton's (1907) *The Natural History of the Ten Commandments*. Other biblical titles have followed, principally in the German language, spelling out how moral principles contribute to survival (e.g. Wickler, 1971). Much of this literature assumed that the world was waiting for biologists to point out what is Normal and Natural, hence worth being adopted as ideal. Attempts to derive ethical norms from nature, however, are highly questionable.

Our position is quite different. While human morality does need to take human nature into account by either fortifying certain natural tendencies — such as sympathy, reciprocity, loyalty to the group and family, and so on — or by countering other tendencies — such as within-group violence and cheating — it is in the end the society that decides, over a period of many generations, on the contents of its moral system. There is a parallel here with language ability: The capacity to develop and learn a very complex communication system such as language is naturally present in humans, but it is filled in by the environment resulting in numerous different languages. In the same way, we are born with a moral capacity, and a strong tendency to absorb the moral values of our social environment, but we are not born with a moral code in place. The filling in is done by the social environment often dictated by the demands of the physical environment (de Waal, 1996b).

Interestingly, moral development in human children hints at the same emphasis on conflict resolution and reciprocity (principles of 'fairness') as emphasized above for non-human primates. Instead of the traditional, Piagetian view of morality imposed upon the child by the all-knowing adults, increasingly it is thought that children develop moral rules in social interaction with each other, particularly during the resolution of conflict (e.g. Killen and Nucci, 1995; Killen and de Waal, in press).

At the same time that our moral systems rely on basic mental capacities and social tendencies that we share with other co-operative primates, such as chimpanzees, we also bring unique features to the table, such as a greater degree of rule internalization, a greater ability to adopt the perspective of others, and of course the unique capacity to debate issues amongst ourselves, and transmit them verbally, including their rationale. To communicate intentions and feelings is one thing, to clarify what is good, and why, and what is bad, and why, quite something else. Animals are no moral philosophers.

But, while there is no denying that we are creatures of intellect, it is also clear that we are born with powerful inclinations and emotions that bias our thinking and behaviour. It is in this area that many of the continuities with other animals lie. A chimpanzee stroking and patting a victim of attack or sharing her food with a hungry companion shows attitudes that are hard to distinguish from those of a person taking a crying child in the arms, or doing volunteer work in a soup kitchen. To dismiss such evidence as a product of subjective interpretation by 'romantically inspired naturalists' (e.g.: Williams, 1989, p.190) or to classify all animal behaviour as based on instinct and human behaviour as proof of moral decency is misleading (see Kummer, 1979). First of all, it is uneconomic in that it assumes different processes for similar behaviour in closely related species. Second, it ignores the growing body of evidence for mental complexity in the chimpanzee, including the possibility of empathy.

One wonders if, on the basis of external behaviour alone, an extraterrestrial observer charged with finding the only moral animal on earth would automatically end up pointing at *Homo sapiens*. We think it unlikely that human behaviour in all its variety, including the occasional horror, will necessarily strike the observer as the most moral. This raises of course the question how and whether morality sets us apart from the rest of the animal kingdom. The continuities are, in fact, quite striking, and need to weigh heavily in any debate about the evolution of morality. We do hesitate to call the members of any species other than our own 'moral beings', but we also believe that many of the tendencies and cognitive abilities underlying human morality antedate our species' appearance on this planet.

Acknowledgements

The authors thank Rudolf Makkreel and, especially, Leonard Katz for very helpful comments and suggestions. The first author thanks Meredith Small, Adam Arcadie, and Richard Baer, Jr. for patient and stimulating discussion about the evolution of morality.

References

Alexander, R.D. (1987), *The Biology of Moral Systems* (New York: Aldine de Gruyter).

Aureli, F. (1992), 'Post-conflict behaviour among wild long-tailed macaques (*Macaca- fascicularis*)', *Behavioural Ecology and Sociobiology*, **31**, pp. 329–37.

Aureli, F. (1997), 'Post-conflict anxiety in non-human primates: The mediating role of emotion in conflict resolution', *Aggressive behaviour*, **23**, pp. 315–28.

Aureli, F., Cozzolino, R., Cordischi, C., and Scucchi, S. (1992), 'Kin-oriented redirection among Japanese macaques — an expression of a revenge system?', *Animal Behaviour*, **44**, pp. 283–291.

Aureli, F., Das, M., Verleur, D., and van Hooff, J. (1994), 'Postconflict social interactions among Barbary macaques (*Macaca sylvanus*)', *International Journal of Primatology*, **15**, pp. 471–85.

Aureli, F., and van Hooff, J. (1993), 'Functional-aspects of redirected aggression in macaques', *Aggressive behaviour*, **19**, pp. 50–1.

Aureli, F., van Panthaleon van Eck, C. J., and Veenema, H. C. (1995), 'Long-tailed macaques avoid conflicts during short-term crowding', *Aggressive behaviour*, **21**, pp. 113–22.

Aureli, F., and van Schaik, C.P. (1991), 'Post-conflict behaviour in long-tailed macaques (*Macaca fascicularis*).2. Coping with the uncertainty', *Ethology*, **89**, pp. 101–14.

Aureli, F., van Schaik, C.P., and van Hooff, J. (1989), 'Functional-aspects of reconciliation among captive long-tailed macaques *(Macaca fascicularis)*', *American Journal of Primatology*, **19**, pp. 39–51.

Aureli, F., Veenema, H.C., van Panthaleon van Eck, C.J., and van Hooff, J. (1993), 'Reconciliation, consolation, and redirection in Japanese macaques *(Macaca fuscata)*', *Behaviour*, **124**, pp. 1–21.

Axelrod, R. (1986), 'An evolutionary approach to norms', *American Political Science Review*, **80**, pp. 1095–111.

Axelrod, R. and Hamilton, W.D. (1981), 'The evolution of co-operation', *Science*, **211**, pp. 1390–96.

Bean, P. (1981), *Punishment: A Philosophical and Criminological Inquiry* (Oxford: Rutherford).

Benedict, R. (1989 [1934]), *Patterns of Culture* (Boston: Houghton Mifflin Company).

Bernstein, I.S. (1964), 'Group social patterns as influenced by the removal and later reintroduction of the dominant male rhesus', *Psychological Reports*, **14**, pp. 3–10.

Bernstein, I.S. (1981), 'Dominance: The baby and the bathwater', *Behavioural and Brain Sciences*, **4**, pp. 419–58.

Bernstein, I.S., Ehardt, C. (1986), 'The influence of kinship and socialization on aggressive behaviour in rhesus monkeys (*Macaca fascicularis*)', *Animal Behaviour*, **34**, pp. 739–47,

Bernstein, I.S. and Ehardt, C.L. (1985), 'Intragroup agonistic behaviour in rhesus monkeys', *International Journal of Primatology*, **6**, pp. 209–26.

Blurton-Jones, N.G. (1987), 'Tolerated theft, suggestions about the ecology and evolution of sharing, hoarding, and scrounging', *Social Science Information*, **26**, pp. 31–54.

Boehm, C. (1986), 'Capital punishment in tribal Montenegro: Implications for law, biology, and Theory of Social Control', *Ethology and Sociobiology*, **7**, pp. 305–20.

Boehm, C. (1992), 'Segmentary warfare and the management of conflict: Comparison of east African chimpanzees and patrilineal-patrilocal humans', in *Coalitions and Alliances in Humans and Other Animals*, ed. A. H. Harcourt and F. B. M. de Waal, (Oxford: Oxford University Press), pp. 137–73.

Boehm, C. (1994), 'Pacifying interventions at Arnhem zoo and Gombe', in *Chimpanzee Cultures*, ed. R.W. Wrangham, W.C. McGrew, F.B.M. de Waal, and P.G. Heltne (Cambridge, MA: Harvard University Press), pp. 211–26.

Boehm, C. (2000), 'Conflict and the evolution of social control', *Journal of Consciousness Studies*, **7** (1–2), pp. 79–101.

Boesch, C. and Boesch, H. (1989), 'Hunting behaviour in wild chimpanzees in the Tai National Park', *American Journal of Physical Anthropology*, **78**, pp. 547–73.

Boyd, R. and Richerson, P.J. (1985), *Culture and the Evolutionary Process* (Chicago: University of Chicago).

Boyd, R. and Richerson, P.J. (1989), 'The evolution of indirect reciprocity', *Social Networks*, **11**, pp. 213–36.

Boyd, R. and Richerson, P.J. (1992), 'Punishment allows the evolution of cooperation (or anything else) in sizable groups', *Ethology and Sociobiology*, **13**, pp. 171–95.

Carpenter, C.R. (1942) 'Sexual behaviour of free-ranging rhesus monkeys, *Macaca mulatta*, 1: Specimens, procedures, and behavioural characteristics of oestrus', *Journal of Comparative Psychology*, **33**, pp. 113–42.

Castles, D.L., and Whiten, A. (1998), 'Post-conflict behaviour of wild olive baboons. I. Reconciliation, redirection and consolation', *Ethology*, **104**, pp. 126–47.

Cheney, D., Seyfarth, R. (1986), 'The Recognition of social alliances in vervet monkeys', *Animal Behaviour*, **34**, pp. 350–70.

Clutton-Brock, T.H. and. Parker, G.A. (1995), 'Punishment in animal societies', *Nature*, **373**, pp. 209–15.

Cords, M. (1988), 'Reconciliation of aggressive conflicts by immature long-tailed macaques', *Animal Behaviour*, **36**, pp. 1124–35.

Cords, M. and Thurnheer, S. (1993), 'Reconciling with valuable partners by long-tailed macaques', *Ethology*, **93**, pp. 315–25.

Cronk, L. (1994), 'Evolutionary theories of morality and the manipulative use of signals', *Zygon*, **29**, pp. 81–101.

Darwin, C. (1965 [1872]), *The Expression of Emotion in Man and Animals* (Chicago: University of Chicago Press).

Darwin, C. (1982 [1871]), *The Descent of Man, and Selection in Relation to Sex* (Princeton: Princeton University Press).

Dawkins, R. (1976), *The Selfish Gene* (Oxford: Oxford University Press).

Dawkins, R. (1997), 'Interview', *Human Ethology Bulletin*, **12**, (1).

Dennett, D.C. (1995), *Darwin's Dangerous Idea: Evolution and the Meaning of Life* (New York: Simon and Schuster).

Edwards, S. and Snowdon, C. (1980), 'Social behaviour of captive, group-living orang-utans', *International Journal of Primatology*, **1**, pp. 39–62.

Fedigan, L.M. and Fedigan, L. (1977), 'The social development of a handicapped infant in a free-living troop of Japanese monkeys', in *Primate Bio-social Development: Biological, Social and Ecological Determinants*, ed. S. Chevalier-Skolnikoff and R.E. Poirier, (New York: Garland), pp. 205–22.

Fox, G. (1984), 'Food transfer in gibbons', in *The Lesser Apes*, ed. H. Preuschoft, D.J. Chivers, W.Y. Brockelman, and N. Creel, (Edinburgh: Edinburg University Press), pp. 32–32.

Fragaszy, D.M., Feurerstein, J.M., and Mitra, D. (1997a), 'Transfers of food from adult infants in tuffed capuchins (*Cebus apella*)', *Journal of Comparative Psychology*, **111**, pp. 194–200.

Frank, R.H. (1988), *Passions Within Reason: The Strategic Role of the Emotions* (New York: Norton).

Frank, R. (1992), 'Emotion and the costs of altruism', in *The Sense of Justice: Biological Foundations of Law*, ed. R.D. Masters and M. Gruter, (Newbury, Calif.: Sage Publications), pp. 46–67.

Goodall, J. (1963), 'My Life Among Wild Chimpanzees', *National Geographic*, **124**, pp. 272–308.

Goodall, J. (1971), *In the Shadow of Man* (Boston: Houghton Mifflin).

Goodall, J. (1986), *The Chimpanzees of Gombe: Patterns of Behaviour* (Cambridge, Mass.: Belknap Press, Harvard University Press).

Gouzoules, S., Gouzoules, H., and Marler, P. (1984), 'Rhesus monkey (*Macaca mulatta*) screams: representational signaling in the recruitment of agonistic aid?' *Animal Behaviour,* **32**, pp. 182–93

Gruter, M. (1992), 'An ethological perspective on law and biology',in *The Sense of Justice: Biological Foundations of Law,* ed. R.D. Masters and. M. Gruter, (Newbury Park, Calif.: Sage Publications), pp. 95–105.

Hall, K.R.L. (1964), 'Aggression in monkey and ape societies', in *The Natural History of Aggression,* ed. J. Carthy and F. Ebling, (London: Academic Press), pp. 51–64.

Harcourt, A.H. and de Waal, F.B.M. (eds.) (1992), *Coalitions and Alliances in Humans and Other Animals* (Oxford: Oxford University Press).

Hardin, G. (1983), 'Is violence natural?', *Zygon,* **18**, pp. 405–13.

Hatfield, E., Cacioppo, J.T., and Rapson, R.L. (1993), 'Emotional contagion', *Current Directions in Psychological Science,* **2**, pp. 96–9.

Hawkes, K. (1990), 'Showing off: Tests of an hypothesis about men's foraging goals', *Ethology and Sociobiology,* **12**, pp. 29–54.

Hayaki, H. (1985), 'Social play of juvenile and adolescent chimps in the Mahale Mountains National Park, Tanzania', *Primates,* **26**, pp. 343–60.

Hemelrijk, C.K. (1994), 'Support for being groomed in longtailed macaques, *Macaca fasicularis*', *Animal Behaviour,* **48**, pp. 479–81.

Henrich, J., and Boyd, R. (1998), 'The evolution of conformist transmission and the emergence of between-group differences', *Evolution and Human Behaviour,* **19**, (4) pp. 215–41.

Heyes, C.M. and Commentators (1998), 'Theory of mind in non-human primates', *Behavioural and Brain Sciences,* **21**, pp. 101–48.

Hobbes, T. (1991 [1651]), *Leviathan* (Cambridge: Cambridge University Press).

Hohmann, G. and Fruth, B. (1993), 'Field observations on meat sharing among bonobos (*Pan paniscus*)', *Folia Primatologica,* **60**, pp. 225–9.

Hume, D. (1978 [1739]), *A Treatise of Human Nature* (Oxford: Oxford University Press).

Huxley, T.H. (1888), 'Struggle for existence and its bearing upon man', *Nineteenth Century,* Feb, 1888.

Huxley, T.H. (1989 [1894]), *Evolution and Ethics* (Princeton: Princeton University Press).

Jeannotte, L.A. (1996), *Play-signaling in juvenile chimpanzees in relationship to play intensity and social environment,* Unpublished Master's Thesis (Atlanta, Ga: Emory University).

Judge, P.G. (1991), 'Dyadic and triadic reconciliation in pigtail macaques (*Macaca nemestrina*)', *American Journal of Primatology,* **23**, pp. 225–37.

Kano, T. (1980), 'Social behaviour of wild pygmy chimpanzees (*Pan paniscus*) of Wamba: A preliminary report', *Journal of Human Evolution,* **9**, pp. 243–60.

Kant, I. (1947 [1785]), 'Groundwork for the metaphysics of morals', *The Moral Law,* (London: Hutchinson).

Kappeler, P.M. (1993), 'Reconciliation and post-conflict behaviour in ringtailed lemurs, *Lemur catta,* and redfronted lemurs, *Eulemur fulvus rufus*', *Animal Behaviour,* **45**, pp. 901–15.

Kappeler, P.M., and van Schaik, C.P. (1992), 'Methodological and evolutionary aspects of reconciliation among primates', *Ethology,* **92**, pp. 51–69.

Killen, M. and Nucci, L.P. (1995), 'Morality, autonomy and social conflict', in *Morality in Everyday Life: Developmental Perspectives,* ed. M. Killen and D. Hart, (Cambridge: Cambridge University Press), pp. 52–86.

Killen, M. and de Waal, F.B.M. (in press), 'The evolution and development of morality', in *Natural Conflict Resolution,* ed. F. Aureli and F.B.M. de Waal, (Berkeley: University of California Press).

Kortlandt, J. (1962) 'Chimpanzees in the wild', *Scientific American,* **206**, pp. 128–39.

Kortlandt, J. (1972) *New Perspectives on Ape and Human Evolution,* (Amsterdam: Stichting voor Psychobiologie).

Kropotkin, P. (1972 [1902]) *Mutual Aid: A Factor of Evolution,* (New York: New York University Press).

Kummer, H. (1980), 'Analogs of morality among non-human primates', in *Morality as a Biological Phenomenon,* ed. G.S. Stent, (Berkeley and Los Angeles: University of California Press), pp. 31–49.

Kummer, H. (1991), 'Evolutionary transformations of possessive behaviour', *Journal of Social Behaviour and Personality,* **6**, pp. 75–83.

Kummer, H. and M. Cords (1991), 'Cues of ownership in long-tailed macaques, *Macaca fascicularis*', *Animal Behaviour*, **42**, pp. 529–49.

Kuroda, S. (1984), 'Interaction over food among pygmy chimpanzees', in The *Pygmy Chimpanzee,* ed. R. Susman, (New York: Plenum), pp. 301–24.

Kuroda, S., Suzuki, S., and Nishihara, T. (1996), 'Preliminary report on predatory behaviour and meat sharing in tschego chimpanzees (*Pan troglodytes troglodytes*) in the Ndoki Forest, northern Congo', *Primates*, **37**, pp. 253–59.

Lindburg, D.G. (1973) 'Grooming as a social regulator of social interactions in rhesus monkeys', in *Behavioural Regulators of Behaviour in Primates*, ed. C. Carpenter, (Lewisburg: Bucknell University Press), pp. 85–105.

Mendoza, S.P. and Barchas, P.R. (1983), 'Behavioural processes leading to linear hierarchies following group formation in rhesus monkeys', *Journal of Human Evolution*, **12**, pp. 185–92.

Midgley, M. (1991), 'The origin of ethics', in *A Companion Guide to Ethics*, ed. P. Singer, (Oxford: Blackwell Reference), pp. 1–13.

Moore, J. (1984), 'The evolution of reciprocal sharing', *Ethology and Sociobiology*, **5**, pp. 5–14.

Nishida, T. (1970), 'Social behavior and relationship among chimpanzees of the Mahale Mountains', *Primates*, **11**, pp. 47–87.

Nishida, T. (1994), 'Review of recent findings on Mahale chimpanzees', in Chimpanzee Cultures, ed. R. Wrangham, W.C. McGrew, F.B.M. de Waal and P.G. Heltne (Cambridge, MA: Harvard University Press).

Nishida, T., and Turner, L.A. (1996), 'Food transfer between mother and infant chimpanzees of the Mahale Mountains National Park, Tanzania', *International Journal of Primatology*, **17**, pp. 947–68.

Nishida, T., Hasegawa, T., Hayaki, H., Takahata, Y., Uehara, S. (1992), 'Meat-sharing as a coalition strategy by an alpha male chimpanzee?', in *Topics in Primatology, Vo.l. 1, Human Origins*, ed. T. Nishida, W.C. McGrew, P. Maler, F.B.M. de Waal, (Tokyo: University of Tokyo Press) pp. 159–74.

Nissen, H.W. and Crawford, M.P. (1932), 'A preliminary study of food sharing behaviour in young chimpanzees', *Journal of Comparative Psychology*, **22**, pp. 383–419.

Nitecki, M.H., Nitecki, D.V. (1993), *Evolutionary Ethics* (Albany, NY: State University of New York Press).

Perry, S. (1997), 'Male-female social relationships in wild white-faced capuchins (*Cebus capucinus*)', *Behaviour*, **134**, pp. 477–510.

Perry, S. and Rose, L. (1994), 'Begging and food transfer of coati meat by white-faced capuchin monkeys *Cebus capucinus*', *Primates*, **35**, pp. 409–15.

Petit, O., and Thierry, B. (1994), 'Aggressive and peaceful interventions in conflicts in Tonkean macaques', *Animal Behaviour*, **48**, pp. 1427–36.

Preuschoft, S. (1999), 'Are primates behaviourists? Formal dominance, cognition, and free-floating rationales', *Journal of Comparative Psychology*, **113**, pp. 91–5.

Preuschoft, S. and van Schaik, C.P. (in press), 'Dominance and communication: Conflict management in various social settings', in *Natural Conflict Resolution*, ed. F. Aureli and F.B.M. de Waal, (Berkley, Calif.: University of California Press).

Rawls, J. (1971), *A Theory of Justice* (Oxford: Oxford University Press).

Reinhardt, V., Dodsworth, R. and Scanlan, J. (1986), 'Altruistic interference shown by the alpha female of a captive group of rhesus monkeys', *Folia primatologica*, **46**, pp. 44–50.

Ren, R., Yan, K., Su, Y. Gi, H. Liang, B., Bao, W. and de Waal, F.B.M. (1991), 'The reconciliation behavior of golden monkeys (*Rhinopithecus roxellanae*) in small groups', *Primates*, **32**, pp. 321–7.

Rose, L.M. (1997), 'Vertebrate predation and food-sharing in Cebus and Pan', *International Journal of Primatology*, **18**, pp. 727–65.

Rothstein, S. I. and Pierotti., R. (1988), 'Distinctions among reciprocal altruism, kin selection, and cooperation and a model for the initial evolution of beneficent behaviour', *Ethology and Sociobiology*, **9**, pp. 189–209.

Scott, J.P. (1971), *Internalization of Social Norms: A Sociological Theory of Moral Commitment* (Englewood Cliffs, NJ: Prentice-Hall).

Seton, E.T. (1907), *The Natural History of the Ten Commandments* (New York: Scribners).

Sicotte, P. (1995), 'Interpositions in conflicts between males in bimale groups of mountain gorillas', *Folia primatologica*, **65**, pp. 14–24.

Sigg, H. and J. Falett (1985), 'Experiments on respect of possession and property in hamadryas baboons (*Papio hamadryas*)', *Animal Behaviour*, **33**, pp. 978–84.

Silberbauer, G. (1991), 'Ethics in small scale societies', in *A Companion Guide to Ethics*, ed. P. Singer, (Oxford: Blackwell Reference), pp. 14–29.

Silk, J.B. (1992), 'The patterning of intervention among male bonnet macaques: Reciprocity, revenge and loyalty', *Current Anthropology*, **31**, pp. 318–24.

Simon, H.A. (1990), 'A mechanism for social selection and successful altruism', *Science*, **250**, pp. 1665–8.

Smith, A. (1937 [1759]), *A Theory of Moral Sentiments* (New York: Modern Library).

Taylor, C.E. and McGuire, M.T. (1988), 'Reciprocal altruism: Fifteen years later', *Ethology and Sociobiology*, **9**, pp. 67–72.

Teleki, G. (1973), 'Group response to the accidental death of a chimpanzee in Gombe National Park, Tanzania', *Folia primatologica*, **20**, pp. 81–94.

Thierry, B., Demaria, C., Preuschoft, S., and Desportes, C. (1989), 'Structural convergence between the silent bared-teeth display and the relaxed open-mouth display in the Tonkean macaque (*Macaca tonkeana*)', *Folia primatologica*, **52**, pp. 178–84.

Tokuda, K. and Jensen, G. (1969), 'The leader's role in controlling aggressive behaviour in a monkey group', *Primates*, **9**, pp. 319–22.

Trivers, R. (1971), 'The evolution of reciprocal altruism', *Quarterly Review of Biology*, **46**, pp. 35–57.

Uchii, S. (1996), 'Darwin and the evolution of morality', Paper presented at the Nineteeth Century Biology, International Fellows Conference, Centre for Philosophy of Science, University of Pittsburg, Published on internet: www.bun.kyoto-u.ac.jp/~suchii/D.onM.html

Veenema, H.C., Das, M., and Aureli, F. (1994), 'Methodological improvements for the study of reconciliation', *Behavioural Processes*, **31**, pp. 29–37.

Verbeek, P. and de Waal, F.B.M. (1997), 'Postconflict behaviour of captive capuchins in the presence and absence of attractive food', *International Journal of Primatology*, **18**, pp. 703–25.

de Waal, F.B.M. (1982), *Chimpanzee Politics: Power and Sex Among Apes* (London: Jonathon Cape).

de Waal, F.B.M. (1986), 'Integration of dominance and social bonding in primates', *Quarterly Review of Biology*, **61**, pp. 459–79.

de Waal, F.B.M. (1987), 'Tension regulation and nonreproductive functions of sex among captive bonobos (*Pan paniscus*)', *National Geographic Research*, **3**, pp. 318–35.

de Waal, F.B.M. (1989a), 'Dominance "style" and primate social organization', in *Comparative Socioecology: The Behavioural Ecology of Humans and Other Mammals*, ed. V. Standen and. R.A. Foley (Oxford: Blackwell), pp. 243–64.

de Waal, F.B.M. (1989b), 'Food sharing and reciprocal obligations among chimpanzees', *Journal of Human Evolution*, **18**, pp. 433–59.

de Waal, F.B.M. (1989c), *Peacemaking Among the Primates* (Cambridge, MA: Harvard University Press).

de Waal, F.B.M. (1991), 'The chimpanzee's sense of social regularity and its relation to the human sense of justice', *American Behavioural Scientist*, **34**, pp. 335–49.

de Waal, F.B.M. (1992a), 'Aggression as a well-integrated part of primate social relationships: Critical comments on the Seville Statement on Violence', in *Aggression and Peacefulness in Humans and Other Primates*, ed. J. Silverberg and J.P. Gray (New York: Oxford University Press), pp. 37–56.

de Waal, F.B.M. (1992b), 'Appeasement, celebration, and food sharing in the two *Pan* species', in *Topics in Primatology, Vol 1 Human Origins*, ed. T. Nishida, W.C. McGrew, P. Marler, F.B.M. de Waal, (Tokyo: Univeristy of Tokyo Press), pp. 37–50.

de Waal, F.B.M. (1993), 'Reconcilation among the primates: A review of empirical evidence and unresolved issues', in *Primate Social Conflict*, ed. W.A. Mason and S.P. Mendoza, (Albany: State University Press), pp. 111–44.

de Waal, F.B.M. (1996a), 'Conflict as Negotiation', in *Great Ape Societies*, ed. W.C. McGrew, L.F. Marchant, and T. Nishid, (Cambridge: Cambridge University Press), pp. 159–72.

de Waal, F.B.M. (1996b), *Good Natured: The Origins of Right and Wrong in Primates and Other Animals* (Cambridge, MA: Harvard University Press).

de Waal, F.B.M. (1997a), 'The chimpanzee's service economy: Food for grooming', *Evolution and Human Behaviour*, **18**, pp. 375–86.

de Waal, F.B.M. (1997b), 'Food transfers through mesh in brown capuchins', *Journal of Comparative Psychology,* **111**, pp. 370–78.

de Waal, F.B.M. and Aureli, F. (1996), 'Reconciliation, consolation, and a possible cognitive difference between macaques and chimpanzees',in *Reaching into Thought: The Minds of the Great Apes*, ed. A.E. Russon, K.A. Bard, and S.T. Parker, (Cambridge: Cambridge University Press), pp. 80–110.

de Waal, F.B.M. and Johanowicz, D.L. (1993), 'Modification of reconciliation behaviour through social experience: An experiment with two macaque species', *Child Development*, **64**, pp. 897–908.

de Waal, F.B.M. and Luttrell, L.M. (1989), 'Toward a comparative socioecology of the genus *Macaca*: Intergroup comparisons of rhesus and stumptail monkeys', *American Journal of Primatology*, **10**, pp. 83–109.

de Waal, F.B.M. and Luttrell, L.M. (1988), 'Mechanisms of social reciprocity in three primate species: Symmetrical relationship characteristics or cognition', *Ethology and Sociobiology*, **9**, pp. 101–18.

de Waal, F.B.M. and Luttrell, L M. (1985), 'The formal hierarchy of rhesus monkeys: An investigation of the bared-teeth display', *American Journal of Primatology*, **9**, pp. 73–85.

de Waal, F.B.M. and Ren, M.R. (1988), 'Comparison of the reconcilation behaviour of stumptail and rhesus macaques', *Ethology*, **78**, pp. 129–42.

de Waal, F.B.M., Uno, H., Luttrell, L.M., Meisner, L.F., and Jeannotte, L.A. (1996), 'Behavioural retardation in a macaque with autosomal trisomy and aging mother', *American Journal on Mental Retardation*, **100**, pp. 378–90.

de Waal, F.B.M.and van Hooff, J.A.R.A.M. (1981), 'Side-directed communication and agonistic interactions in chimpanzees', Behaviour, 77, pp. 164–98.

de Waal, F.B.M. and van Roosmalen, A. (1979), 'Reconciliation and consolation among chimpanzees', *Behavioural Ecology and Sociobiology*, **5**, pp. 55–66.

de Waal, F.B.M. and Yoshihara, D. (1983), 'Reconciliation and re-directed affection in rhesus monkeys', *Behaviour*, **85**, pp. 224–41.

Westermarck, E. (1912), *The Origin and Development of the Moral Ideas* (London: Macmillan).

Wickler, W. (1981 [1971]), *Die Biologie der Zehn Gebote: Warum die Natur für uns Kein Vorbid ist* (Munich: Piper).

Williams, G.C. (1988), 'Reply to comments on "Huxley's evolution and ethics in a sociobiological perspective" ', *Zygon*, **23**, pp. 383–407.

Williams, G.C. (1989), 'A sociobiological expansion of "Evolution and Ethics" ', *Evolution and Ethics*, (Princeton: Princeton University Press), pp. 179–214.

Wilson, D.S. and Sober, E. (1994), 'Reintroducing group selection to the human behavioural sciences', *Behaviour and Brain Sciences*, **17**, pp. 585–654.

Wrangham, R. (1975), *The Behavioural Ecology of Chimpanzees in Gombe National Park, Tanzania*. Unpublished Doctoral Dissertation, (Cambridge: Cambridge University).

Wrangham, R.W. (1979), 'On the evolution of ape social systems', *Social Science Information*, **18**, pp. 335–68.

York, A.D. and Rowell, T.E. (1988), 'Reconciliation following aggression in patas monkeys (*Erythrocebus patas)*', *Animal Behaviour*, **36**, pp. 502–9.

Part II
Mechanisms

[2]

PSYCHOLOGICAL ALTRUISM, EVOLUTIONARY ORIGINS, AND MORAL RULES[1]

PHILIP KITCHER

I

Biologists take an item of behavior to be altruistic just in case it raises the expected reproductive success of an organism other than the agent, while diminishing the expected reproductive success of the agent. In their usage, altruists are organisms who are disposed to perform altruistic behavior. Studies of kin selection and reciprocity have revealed that altruism of this sort might be favored by natural selection. But the mere fact that we, like other animals, might have dispositions to act in ways that promote the reproduction of those around us sheds little, if any, light on the connection between our moral precepts and practices and our evolutionary history, in part because the altruism that matters to us is not typically measured the Darwinian currency of reproduction, and in part because it has everything to do with the intentions of the agent.

That there is a connection between human morality and our evolutionary past, I have no doubt. That the connection can be stated simply is surely belied by the history of brave, but disastrous, ventures in evolutionary ethics. If the connection is to be illuminated then we must start by distinguishing a number of projects. The most ambitious would be to attempt to derive fundamental new moral principles or to revise old ones by considering the regimes of natural selection that shaped human social behavior. In my judgment, those endeavors come to grief on the difficulty of wringing moral conclusions out of biological premises. More promising is the thought that we can learn important things about the nature of morality by conceiving ourselves as part of the natural order, that parts of meta-ethics – particularly moral metaphysics and moral epistemology – should be reconfigured through the explicit recognition of the fact that our practices and precepts have long histories and that those histories eventually terminate on the savannah in the psychological dispositions of our remote ancestors and in the kinds of social arrangements that they fashioned. The central theme of this essay is that attention to our evolutionary past might offer us clues about human psychological tendencies, revealing us to be psychological altruists of a particular kind and providing us with a picture of moral agency and of the historical development of morality that leads us to rethink some main questions of moral philosophy.

I shall begin by trying to explain the distinction that has already figured in my discussion, the contrast between altruism in the biologist's sense and "psychological" altruism, the altruism that matters to morality.

II

To a first approximation, we think of people as altruistic when they adjust their preferences to their perceptions of the wants and needs of others. Here is a stylized situation: there is some divisible good that is valued by both A and B; when A is alone and comes across this good, then A prefers taking it all to any other available option; however, if A recognizes that B is present (and, I assume, that B also values the good), then A no longer prefers consuming the good entire but, instead, wants most to share the good with B. Moreover, this preference is generated by A's perception of B's valuation of the good, and it is not explained by A's having some other independent end that would be advanced by sharing with $B - A$ does not calculate that, in the long run, more good things will accrue from sharing with B.[2]

This stylized situation brings out what I take to be important features of our everyday notion of altruism, the one we link to morality and which we try to induce our children to follow. First, A's valuations differ between the solo situation and that in which B is present: it is not particularly unselfish for people to share things to which they attach no value. Second, A's preference when B is present is mediated by the perception of B's wants: altruism presupposes an empathetic ability to discern what others need or desire and to respond to it. Third, B's satisfaction is perceived as an end in itself, not as a means to the achievement of some of A's other goals, as if A were entering into a contract with B, or trying to impress B, or simply frightened of the consequences of not sharing with B.

Although the usual framework of decision theory may not be an adequate representation of the psychological states of agents, it does provide a way of making some of these points more precise, and eliciting some of the nuances of the notion of psychological altruism. Let's say that a valuational structure (or *valuation*) for an agent is a set of alternative actions that the agent takes to be available and a mapping from that set to the real numbers, so that the number assigned to an option represents the value the agent attaches to the expected outcome of performing that action; or, more generally, when the circumstances involve interactions, the valuation identifies sets of alternatives for all the agents involved in the interaction, and maps each combination of actions to expected values for all the agents.[3] We can now introduce a notion of psychological altruism as follows:

> The valuational structure that A forms in situation C is weakly altruistic with respect to B just in case:
>
> (1) in C, A ranks O_i ahead of O_j
> (2) in C^*, which is like C except only in the fact that the actions available to A have no impact on the well-being of B, A would rank the analogue of O_j, O_j^* ahead of the analogue of O_i, O_i^*
> (3) A's ranking of O_i ahead of O_j is explained by A's perception of B's preferences
> (4) A's ranking of O_i ahead of O_j is independent of A's expectations about B's future actions (and, in particular, about actions that would benefit A).

I take this to be a definition of *weak* altruism because it may be satisfied with only a minor reversal of preferences.[4] The most interesting notions are those in which the recognition of an impact on B produces a systematic change in A's assessment of all the options.

To generate a stronger conception, suppose that the values assigned to all of the options available in cases where B is present are functionally related to the values assigned when B is absent and to A's perception of B's valuations, in such a way that increases in the latter variable would yield increases in the value of the function: i.e. if V_{AB} is A's assignment when B is present, V_A is A's assignment when B is absent, and V_B^* A's perception of B's evaluation, then V_{AB} $= f(V_A, V_B^*)$ where $\partial f/\partial V_B^* > 0$. (The latter condition is intended to differentiate altruists who take others' values into account in a positive way, from the spiteful for whom others' wants cause a lowering of the associated value.) In general, the function f can take many forms, and can be responsive to other facets of the social situation, as when individuals attach a high value to an outcome if it is brought about in a particular way (think of cases in which the value of a joint course of action, for both parties, consists in the fact that it has been brought about through an attempt to respond to the needs of the other).[5] For the sake of simplicity in the present discussion, I am going to suppose that f takes a particular form, and that

$$V_{AB} = \theta V_A + (1 - \theta V_B^*) \qquad \text{where } 0 \leq \theta \leq 1.$$

This view of the altruistic adjustment of preferences fits very naturally those situations in which there is the possibility of sharing a divisible good, although it is by no means restricted to them.

When the relationship between the preferences in the presence and in the absence of the other takes this simple form, it is possible to identify dimensions along which an agent's altruism can be measured. First, we can gauge the *intensity* of A's altruism by the parameter θ. Those who are completely selfish give full weight to their own preferences in the absence of others, setting $\theta = 1$. At the opposite extreme are hyperaltruists whose preferences are completely determined by their perception of the other's evaluation, and for whom $\theta = 0$. Between these values, agents count as more altruistic the lower their value of θ, an important special case being $\theta = 0.5$, the point at which A treats B's evaluations as equal in importance with his own (we might think of this as "golden rule altruism").[6]

A second dimension of altruism concerns the *pervasiveness* of the altruistic response, measured by the range of contexts C in which the interests of the other affect the valuation. Combining these two independent measures, we can recognize different types of altruists – for example, those who make a more generous response (characterized by lower values of θ) across a narrower set of contexts C, can be distinguished from people who are prepared to set $\theta < 1$ across a broader range of contexts C but whose values of θ are always somewhat higher.

A third independent dimension of altruism, the *extent* of altruism, corresponds to the range of the individuals B whose perceived preferences A is willing to take into account. In a more extensive treatment than I shall attempt here it would be necessary to consider the ways in which A responds to the perceived needs of others in situations where several people might be affected by A's actions, differentiating those in which it is possible to form preferences that take all parties into account from those in which the responsive preference must give priority to some individuals over others, even though, had the privileged individuals not been present, A would have responded to the perceived evaluations of those who are currently slighted. (Consider, for example, a case in which both B and D want an indivisible good, in which A can help either to attain it, and in which A would want to help either B or D if the other were absent. It seems that there should be a requirement of altruistic coherence: if $V_B^* \geq V_D^*$, and if, in every context C, A is prepared to set the value of θ for B lower than the corresponding value for D, then when B and D are in conflict over the indivisible good, A's preference ought to respond to B rather than to D.)[7] For present purposes, I ignore multi-agent interactions and suggest that a third measure of altruism is the size of the set of individuals to whose perceived preferences an individual will respond.

In addition to intensity, pervasiveness, and extent, we should also consider the character of an altruist's *empathetic skill*. It is an all-too-familiar fact that some people who try to respond to the preferences of those around them are very bad at recognizing what values others would place on various outcomes (V_B^* is very different from V_B). Yet we should also appreciate the possibility that an empathetically skilled altruist may adjust the preference of the intended beneficiary, precisely because the altruist wishes either to factor out the beneficiary's reciprocal wish for the altruist's own good or because the altruist thinks that the evaluation assigned by the beneficiary would be revised in the light of further information (clearer thinking etc.).[8] So empathetic skill is compatible with various types of paternalism, including those that override the beneficiary's own altruistic

responses and those that identify a failure of the beneficiary's preferences to reflect what she "really" wants.[9] With respect to this fourth
dimension of altruism, I shall again simplify, supposing that A
correctly identifies B's preferences, and that considerations of paternalism are not pertinent.

A general theory of psychological altruism ought to investigate
the various ways in which A's preferences respond (the forms of the
function f), the character of the various dimensions and the requirements (if any) on altruistic coherence. But we don't need a general
theory to cope with the most elementary forms of altruism that are
relevant to morality, and, for the sake of exploring evolutionary
origins, we can start with cases in which one organism adjusts its
preferences by weighing its own independent valuations with the
valuations of others. It might seem that this simple conception is
rather thin, compared to the full range of human possibilities, but, as
we shall see, it will bring complexities enough.

III

I shall not defend at any length the idea that some nonhuman social
organisms are cognitively sophisticated, that they can recognize the
desires of organisms around them, and that, to the extent that our
preferences can be represented in terms of valuations, theirs can
too. [10] What is far more controversial is whether any organisms,
including ourselves, have dispositions to construct valuations that
are psychologically altruistic in my sense. There are two sources
of doubt, one based on the traditional concern that natural selection
cannot be expected to favor the emergence of such dispositions
nor to maintain them if they become prevalent, the other stemming
from the judgment, fashionable among primatologists, that the social
organization of primate life reveals "Machiavellian intelligence",
and even that increased powers of cognition in hominids reflect the
need to manipulate others and to avoid being manipulated.[11]

The first version is relatively easy to rebut. Consider a very simple
model of maternal care. Primates roaming on the savannah encounter
carcases that could serve as food. Imagine that a female, encountering a carcase in the absence of one of her offspring, would prefer
to devour it entire, but that, if one of her young is present (or, more
generally, can be summoned to the carcase before it spoils or is

taken by another animal) prefers to eat half and to give half to the young. If we suppose that food has decreasing marginal value, then the valuations might be as follows:

Mother's Valuation: Child Absent
Devour All	10
Eat Half	7

Mother' s Attribution to Child
Mother Devours All	0
Mother Shares Half	7

Suppose that the mother adjusts her valuation in the presence of the child using the parameter θ of Section II. Then we have:

Mother's Valuation: Child Present
Devour All	10θ
Share Half	7

Mothers for whom $\theta < 0.7$ will choose to share, and this action may be favored by kin selection. Hence, dispositions to give positive weight to the interests of relatives can be favored by natural selection. In similar fashion, one can show that natural selection can favor dispositions to reciprocate with non-relatives.[12]

Turn now to the second objection, which suggests that there will always be an alternative way to understand the proximal causes of behavior that benefits others, one that corresponds to the Hobbes-Machiavelli picture. How can we conceive the mother as calculating some future benefit to herself? One possibility, typically not made explicit but often, I think, underlying celebrations of the inevitability of selfishness, is that the very considerations that favor the evolutionary success of "altruism" undermine its genuineness. No doubt increased genetic representation in future generations is relevant to the evolution of psychological dispositions, but the important issue is whether organisms govern their behavior by recognizing such future "benefits". As a number of writers have pointed out, it is quite absurd to suppose that more than a few very select primates *could* calculate the genetic gains and losses (and those who do make their decisions in this way seem misguided, to say the least).[13] Natural selection does not need us to be clever, and it's enough that we have dispositions that generate fitness-enhancing actions even though we are quite oblivious as to why those dispositions enhance fitness.

Yet, even if we don't calculate in terms of future genetic "benefits", maybe we calculate nonetheless. A slightly more plausible suggestion is that maternal care proceeds from the expectation of future reciprocity – the child is expected to grow into a future ally, maybe eventually a caregiver. Here the consequences of the present action would be represented in terms we can imagine being within the mother's conceptual repertoire, but we are still supposing that animals have abilities to abstract from present conditions and to envisage a very different future, to overlook the weak juvenile and see a future strong ally. Even if we allow such amazing foresight, problems remain. If dispositions to share with young evolve under natural selection because of inclusive fitness considerations, then the expectations of future aid ought not to be an accurate guide to the kinds of behavior that selection would favor. Mothers ought to be providing some aid when there is little chance of reciprocity in the future, and should provide extra aid to offspring who can be expected to reciprocate. What this means is that, if the calculational story is to give values that correlate with the inclusive fitnesses, the perceived gains from future reciprocity have to be inflated. But why should mothers think that their care will be remembered, or, if remembered, that it will trigger a disposition to repay? Furthermore, if sharing is based on the expectation of returns, the young seem bad targets. Other, more mature, members of the group would appear to be better prospects for future aid.

There is a final version of the Hobbes-Machiavelli story. Perhaps there's an immediate benefit from sharing, deriving from the mothers' relief from the distress caused by the crying of their young. Instead of acting from their perception of children's needs, they try to promote their own ease by preventing wails, facial expressions, and bodily gestures that would upset them. It is easy to recall occasions on which mothers bribe howling children with food. But we should ask why mothers are so distressed by the cries of their offspring. If the answer is that they attach a high value to situations in which the needs of the young are met, then the Hobbesian story fails to provide any real alternative to the view that attributes altruistic dispositions:

mothers rank highly those outcomes in which they share with their young because they want to avoid the distressing cries and the cries are distressing precisely because of their disposition to value the well-being of the young. Even if the offspring are seen as manipulating their mothers, the susceptibility to manipulation is rooted in altruistic preferences.

This last version leads into a slightly different objection, often voiced by non-philosophers. Surely, on my account, altruists do what they want to do – and isn't doing what one wants to do the epitome of selfishness? This is an illuminating mistake which distorts a genuine insight. Psychological altruism, as many philosophers have recognized,[14] is not a matter of going against one's dominant desires but rather having dominant desires that are directed towards others. The insight that fuels the objection is that there is a connection between psychological altruism and self-denial, and my notion of psychological altruism makes that connection clear: the altruist prefers to do what she would not have wanted to do but for the benefit that the envisaged action would bring to the other. Altruism suppresses the wants one would have had in response to the perceived wishes of another individual.

I conclude that there is no basis for thinking that all the apparently other-directed dispositions that figure in primate social behavior can be reduced to self-interested calculation of some proxies for fitness. Once this point is accepted, we can make sense of a number of examples which those concerned to resist psychological altruism can only treat in a strained fashion. Jane Goodall recounts the story of a young female chimpanzee, Little Bee, who, when accompanied by her partially paralyzed mother, Madam Bee, would routinely slow her gait and would scramble up trees for fruit which she would then share. [15] Frans de Waal, in an extensive catalogue of examples of apparent animal kindliness, reports how a young male chimpanzee saw an older female who had often been friendly to him struggle to loosen a tire partly filled with water, and, after she gave up, solved the problem and presented her with the tire.[16] Finally, on at least one occasion, when zookeepers have tried to introduce a new male into a colony whose resident adult male has died, the remaining females have banded together to drive the newcomers out – until a seemingly ordinary male has been accepted, a male who is later discovered to have associated with one of the females when they were in the same group as juveniles.[17] In none of these instances is it easy to construct a self-interested explanation of the action – unless we assume that the agents are quite seriously confused – but, once we have acknowledged the general possibility of dispositions to form psychologically altruistic valuations, there is a natural account of what is occurring. Nonhuman primates, and, we may safely assume, hominids and humans too, have the capacity for sympathy.[18]

IV

How far, then, can we expect these altruistic dispositions to extend? My last two examples suggest that the boundaries of psychological altruism reach beyond the circle of immediate kin, but some may wonder whether charity inevitably ends at home. For, even though it is difficult to find a self-interested motivation for maternal care, the classic model of cooperative interactions among non-relatives, articulated in terms of repeated Prisoner's Dilemma, invites the rejoinder that the animals recognize the possibility of reciprocation and resolve to cooperate in the hope of future benefits. If that were so, condition (4) in the definition of weak altruism would not be satisfied. Thus it looks as though a regime of selection based on the recurrence of PD situations will always leave open the possibility that the psychological dispositions selected involve clever calculation and not genuine altruism. I shall approach this important concern somewhat obliquely.

Several years ago, I was struck by the implausibility of supposing that our primate ancestors constantly faced situations in which they were compelled to interact with a particular partner, situations whose structure took the form of PD. Much more likely, I believed, were occasions on which animals had the opportunity either to interact with a conspecific or to opt out and to act by themselves, and on which they would have some chance to choose their partners. Consider, for example, social grooming. If we conceive of this as directed at the removal of parasites from fur, then animals can either groom themselves or enter into a grooming partnership; pursuing the latter lays them open to the possibility of exploitation, and, making some plausible assumptions about the benefits of various kinds of grooming and the costs of spending the associated amounts of time, it was possible to show that this has the structure of an optional game.[19] Mathematical arguments supported the conclusion that high levels of cooperation were likely to develop and to be sustained in populations faced with a sufficiently large number of opportunities for playing optional PD with one another.[20] Indeed, John Batali and I were able to demonstrate that cooperation emerged more reliably and was maintained significantly more robustly under a regime of selection involving optional games than in the standard scenario (due to Robert Axelrod) in which the games are compulsory.[21] The next obvious task seemed to be to examine the social interactions of primates, and to see how much of it could be reconstructed as a series of optional games.

For reasons I can only summarize here, I am now convinced that emerging knowledge of primate social behavior belies the neat, mathematically tractable, models articulated by Axelrod, and modified by Batali and me. In brief, primate societies are typically inhomogeneous (not everyone is prepared to cooperate with everyone else, even when the others are not inclined to defect) and interactions are often asymmetrical (the beneficiaries do not always return the favor and are often not sanctioned for their failures to reciprocate). Thus, in cooperative hunting, those who engage in bringing down the prey are sometimes not rewarded and those who haven't taken part end up with pieces of the spoils.[22] Moreover, studies of the contents of chimpanzee feces suggest that meat does not play an important role in the diet of chimpanzees, thus undermining the idea that cooperative hunting is sufficiently significant for natural selection in these situations to shape altruistic dispositions.

As I have already noted, social grooming, understood as bringing relief from parasitic infestation, can be viewed as an optional PD, but any account restricted to considerations of hygiene would be hard-pressed to explain the enormous amounts of time that primates spend grooming one another. During some periods in the recorded histories of primate troops – particularly when social tensions are running high – the animals devote three to six hours per day to plucking and smoothing one another's fur.[23] Of course, as many primatologists have pointed out, grooming serves important social functions, but, until these are specified we have no basis for assigning the payoffs to

294   PHILIP KITCHER

various possible actions or for determining the structure of whatever game the animals are playing.[24]

The lack of fit between game-theoretic expectations and the character of primate social relations points to a more worrying problem. Although Axelrod's suggestion that animals are forced into partnerships that repeatedly play PD provided an analytic breakthrough, it should always have seemed implausible as a description of our evolutionary past. The modification I proposed, that the games are optional, not compulsory, was a step in the direction of greater realism, but it did not go far enough. My scenario took it for granted that ur-social animals could be regarded as forming a pool from which potential partners for interactions could be drawn, *but this is already to presuppose a minimal form of sociality*. It is crucial to ask how these animals could have become sufficiently tolerant of one another's presence to form any such pool. Posing that question makes it apparent that there is an earlier stage of interaction out of which some level of mutual tolerance emerges, and the dynamics of that phase of interactions have to be reflected in understanding the games animals play when they hunt together, divide up resources, groom one another, and so forth.[25]

The extent to which social relations, tolerance, cooperation and altruistic behavior extends beyond the family varies greatly even within the Great Apes. Male orang-utans are mostly solitary, and, until recently, it appeared that female-female interactions were relatively transitory and brief, that the largest social unit was a female travelling with one or two offspring.[26] Gibbons divide into small family groups (mother, father, and young) that are typically hostile to outsiders. Groups of gorillas typically contain several adult females, but often have only one adult male; to a first approximation, gorilla social life involves some cooperation among unrelated females and only aggressive interactions among adult males.[27] Only in chimpanzees and bonobos do we find larger social units with cooperation among unrelated adults of both sexes.

Chimpanzees live in bisexual groups (varying in size from about 20 to about 100), within which there are shifting patterns of alliance and dominance relations. Among bonobos, the groups are somewhat larger (roughly 50 to 150), with the same sorts of changing internal structures. A principal difference between the two groups is that the major associations in the wild seem to be among chimpanzee males and among bonobo females, although in both species, there are important social interactions among members of the other sex (and between members of opposite sexes). Study of hominid remains suggests that our ancestors lived in sexually mixed groups, and that their size was of the same order as that found in living chimpanzees and bonobos.[28] How did the chimp-bonobo-(hominid?) pattern of sociality evolve?

Any answer to this question must bring out the salient features that distinguish chimps and bonobos from the other great apes. I shall follow the approach developed by Richard Wrangham, who proposes that female behavior is shaped directly by ecological factors, particularly the distribution of food. Males have to adapt to this distribution, adjusting their behavior to maximize the chances of copulating with estrous females.[29] Crucial for our purposes in Wrangham's idea that mutually hostile communities of chimpanzees have "evolved from a hypothetical solitary-male system because males could afford to travel in small parties, even though the optimal foraging strategy was to travel alone; they were forced to do so because lone males therefore became vulnerable to attacks by pairs".[30] This remark identifies the source of sociality outside the immediate family as lying in the importance of coalitions and alliances. To put the point abstractly: *for understanding cooperative interactions among unrelated animals, PD (whether optional or compulsory) is not fundamental; the dynamics are set by the problem of forming coalitions and alliances.*[31]

We can develop this idea more generally than Wrangham does by not committing ourselves to his theses about the chief selection pressures on male and female behavior. Suppose, simply, that there is pressure on (sexually undifferentiated) organisms because there is competition for scarce resources, that the animals fight over these resources, and that stronger organisms typically win. We can envisage a five-step process that might have led from a situation in which there was no cooperation beyond the immediate family to the kind of social structure found in chimpanzees, bonobos (and hominids?).

I. Asociality – animals range in the fashion of organutans, finding some resources without contest ("scramble" competition) and competing directly for others ("contest" competition).

II. First coalitions – some animals arise that are disposed to act together in contest and to share the resources so obtained (not necessarily equally).

III. Escalation – because of the success of the early coalitions, larger coalitions form, sharing the benefits they earn in contests.

IV. Community stabilization – coalition size is ultimately limited by the difficulty of defending all the resources in a range, and the habitat becomes partitioned into ranges defended by stable communities, within which the resources are divided by the formation of subcoalitions.

V. Fitness enhancement through cooperation – by engaging in optional games (some of them, perhaps, PD), and behaving cooperatively, members of the stable communities increase their fitness.[32]

The game-theoretic structure here is much more intricate than the familiar PD. The following sketch will omit virtually all the technical details (some of them important).[33]

Suppose we have a population of N organisms in a habitat containing R resources. Any organism or coalition of organisms can visit v resources before those resources are renewed. Coalitions, however large, only visit the same number of resources as individuals. There are asymmetries in strength among members of the population, and contests among them are decided by a version of the standard Hawk-Dove game with assessment of strength.[34] The crucial point, for our purposes, is that the organisms can detect one another's strength, and, in any interaction, the strongest organism (or coalition) always wins at trivial cost. Assume that the strength of a coalition is the sum of strengths of the members. Finally, this world is *non-Rousseauian*, in that there are not enough resources for all to acquire the maximum that they could: $Nv > R$.

For almost all ways of developing these assumptions (all ways that do not assign special values to the crucial parameters), it is possible to show the following results:

(1) In an initial situation without coalitions, the population will contain at least one pair of organisms who can increase their fitness through coalition formation.

(2) Once the first coalition has formed, there will be other pairs of organisms who can increase their fitness through coalition-formation.

(3) As larger coalitions form, the fitness benefits are determined, in part, by sub-coalitional alliances.

(4) There is a point at which those excluded from the dominant coalition do better to scramble for resources than to form a rival coalition that would contest resources.

These results offer a non-constructive argument for the emergence under natural selection of the chimp-bonobo pattern of social inter-action among non-relatives: we expect a partitioning of the habitat into territories defended by groups that band together against rivals, and we expect that, within these groups, there will be subcoalitions that determine the fitness benefits.[35]

I suspect that it's not hard to see how the coalition-forming process begins and why it escalates, but the termination might appear more mysterious. Here's a quick version of the rationale. Suppose that there's a dominant coalition of size k that obtains v resources to split among its members. If $N-k > k$, there will typically be a way of permuting the members of the population to form a new dominant coalition, larger than the formerly dominant coalition but in which the payoffs to the continuing members of the former coalition are increased. (The other members of the coalition have to settle for less than the shares formerly given to those whom they displace.) But, if $(R-v)/(N-k + 1) > v/k$, the member of the k-membered coalition who receives the smallest share does better by simply avoiding the resources that the coalition visits, and scrambling for the remainder. Of course, this scramble competition is likely to give way to a situation in which those excluded from the k-membered coalition contest the $R-v$ resources that the k-membered coalition doesn't visit. Thus we obtain the partitioning of the habitat into territories controlled by sizeable coalitions.

So far my conclusions have been about animal social behavior, not directly about psychological altruism. Now, equipped with a sketch of the conditions under which cooperation among non-relatives might emerge, we can go back to the question with which this section began: can we expect natural selection to shape and maintain dispositions to form psychologically altruistic valuations towards non-relatives? I think that we can. If the selection regime is as I have taken it to be, then it is crucial for animals like chimpanzees, bonobos, and our hominid ancestors, to belong to a society, within which they have a place in subsocietal coalitions and alliances. Specifically, young animals, no longer under the constant protection of parents, will need allies if they are to gain anything at all in a competitive world. How are they to find their friends? Primatologists in the grip of the Hobbes-Machiavelli view of sociality-directed-by-manipulative-intelligence might propose that they are able to identify the fitness benefits and calculate their best option. That idea might have some plausibility if the animals could recognize themselves as engaging in something like iterated PD, but it is hopeless in the context of the many-person coalitional game. Game theorists are unable to identify best strategies for individuals playing that game, and I see no reason to think that our evolutionary ancestors and relatives had the cognitive resources to find a solution.

The strategies that chimpanzees and bonobos adopt seem not to be based on any tallying of costs and benefits. Instead, animals appear prepared to support particular members of their groups with whom they have a history of interactions, often dating back to periods early in their lives. I suggest that these friendships are based on dispositions to construct psychologically altruistic valuations, and that the ability to form such valuations was crucial in the transition to society with internal coalitions. Animals with such dispositions were able to serve as stable coalitional partners. Those who attempted to substitute the clever head for the kindly heart would have been likely to do far worse. This results partly from the fact that the complexity of the coalitional game makes it impossible to pick out detailed solutions. But it is also important to note that the merits of a strategy for allying oneself with others are very sensitive to details about the distribution of resources, the relative strengths of troop members, the reliability of other alliances, and a host of other facts that animals cannot be expected to know and use in their calculations.

Particularly in situations in which animals have little opportunity for winning contests by themselves, there will be a selective advantage to identifying with the interests of some of one's conspecifics (typically age-mates in the same predicament). The friendships of youth might thus be expected to be particularly deep and enduring.

Perhaps the most dramatic expression of this is the acceptance into a colony of animals whom some residents have known when they were juveniles together. Thus I suggest that the basic disposition to identify and sympathize with close kin was extended in our evolutionary past. We share with chimpanzees and bonobos the capacity for forming valuations that are altruistic in the sense of section II, and, if we had not had that capacity, we would not have been able to form the large mixed social units whose traces are found in the prehistoric record. Far from being anthropomorphic, sentimental, or self-deceiving, the hypothesis that we have tendencies to psychological altruism, towards kin and towards some of our non-relatives, seems the best explanation both of the behavior of living animals and of our having achieved the kind of sociality that we share with our evolutionary cousins.

V

The disposition to form altruistic valuations provides the weak with the opportunity to constrain the behavior of the strong, especially in situations where the consequences are complex and confusing. However this immediately raises the possibility that powerful animals, faced with clear-cut opportunities for gain, might benefit from having quite different tendencies. I now want to acknowledge the genuine insights of the popular primatological perspective that sees social intelligence as Machiavellian. There are occasions on which animals appear to calculate the selfish returns from various courses of action, most notably when they have a clear chance of becoming dominant within their social group.

Frans de Waal's original research on the Amhem Chimpanzee colony centered around the ways in which three high-status males – Yeroen, Luit, and Nikkie – related to one another and to the high-status females during times of transitions in power. Each of the males engaged in social interactions that are readily interpretable as directed towards improving their chances for retaining dominance, achieving dominance, or, at worst, serving as the principal lieutenant of the dominant male. In 1976, Luit aided the newly adult Nikkie in achieving dominance over the females in the colony, while Nikkie's diversionary tactics enabled Luit to dethrone Yeroen. Once he had

attained alpha rank, Luit's policy changed. He consolidated his position by siding with the females and with Yeroen against Nikkie. De
Waal interprets the switch as showing "just how much friendships
among chimpanzees are situation-liked".[36] De Waal elaborates the
point as follows:

> Coalitions based on personal affinities should be relatively stable; mutual trust
> and sympathy do not appear or disappear overnight. And yet there had been no
> evidence of stability so far, at least not in the coalitions of the adult males. . . . If
> friendship is so flexible that it can be adapted to a situation at will, a better name
> for it would be oppotunism.[37]

The subsequent twists and turns of the story underscore de Waal's
remarks. Yeroen deserted Luit to form a coalition with Nikkie, so that
Nikkie eventually became dominant with Yeroen as his lieutenant.
After a subsequent period of tension between the two allies, Luit
re-emerged at the top of the hierarchy, apparently in a weak coalition
with Nikkie. The uneasy situation was ended by a night fight, in
which Luit was fatally injured by the other two.[38]

To say that there are no stable friendships within chimpanzee
communities is too strong, for some coalitions endure for years,
even for virtually the entire lifetimes of the animals (before the power
struggle began, Yeroen and Luit had been long-time allies). What
de Waal's fascinating story reveals – and what is demonstrated by
similar instances, less fully documented, among wild chimpanzees –
is that the presence of a clear opportunity for self-advancement can
strain previous other-directed dispositions. In earlier sections, I've
suggested that studies of animal behavior and evolutionary considerations support the attribution to our evolutionary relatives and evolutionary ancestors of dispositions to form altruistic valuations, but it
should no more be supposed that those dispositions will be operative in every context than that they will be evenly directed towards
the surrounding individuals. The distinction of the dimensions of
psychological altruism is crucial here, as it is to the understanding of
the dynamics of coalitions and alliances. For the latter, it is important that altruistic dispositions differ in their extent. Now we should
recognize that the dispositions to form altruistic valuations are not
completely *pervasive*. Different degrees of altruism, characterized
by different values of the parameter θ can be triggered in different
situations, depending on the salience of features in the environment
and the nature of the background state of body and brain.

De Waal's characterization of the stable dispositions he expected to underlie friendship tacitly supposed that the dispositions would be completely pervasive, and it's clear that Luit and Yeroen had no such tendencies to respond to one another's preferences. However, de Waal's recognition of the "situation-linked" character of their mutual attitudes does seem correct. The stable dispositions of Luit, Yeroen, and other primates, seem best characterized by supposing them to be of the form:

> If the values of background variables are in S, then the valuation formed is characterized by $\theta = \ldots$

Animals do not have a fixed value of this parameter towards a conspecific, depending on the strength of the relationship. Of course, I don't know – and nor does anyone else – how to identify the important background variables or to explain how θ varies with them: all that can be said is that states in which there are clear possibilities for great advantage are likely to increase the value of θ, even setting it equal to 1.[39]

If indeed we, like chimpanzees and bonobos, have evolved to have dispositions of this form, then we will be vulnerable to two forms of inconsistency. One, *diachronic* inconsistency, exemplified by the vicissitudes of the relations among Yeroen, Luit, and Nikkie, occurs as different situations generate values of pertinent variables that trigger different dispositions, leading animals sometimes to form valuations directed towards others, sometimes to form selfish valuations. Even more interesting is the possibility of *synchronic* inconsistency. Suppose that there are two relevant background variables, x and y, and that A has both the following stable dispositions towards B: (a) when $x > r$, set $\theta = 0.5$; (b) when $y > s$, set $\theta = 1$. What happens to an animal that finds itself in a situation, where $x > r$ and $y > s$? Both dispositions are activated, two valuations are formed, but (at most) one can be expressed in action.

Possible, perhaps – but is there any reason to think that such situations ever occur in the psychological lives of animals. I think there is. Anyone who has spent some time watching social interactions among chimpanzees and bonobos has seen behavior that seems to express two conflicting attitudes. An animal hesitates, torn between different courses of action: we observe the chimpanzee holding a branch, rich in leaves, poised to strip them off and eat, and, at the same time, we see the body set to acknowledge the presence of a coalition partner and to share. The configuration of limbs and muscles is genuinely a mixture, and the animal's expression reveals the tension of the moment.[40]

Of course, we're familiar with similar phenomena in our own experience, and philosophers interested in *akrasia* and free will have found them especially significant. People who are trying to lose weight find themselves tempted by the aromas from the kitchen. They describe themselves as both wanting and not wanting the food, and, in support of their description, we can adduce the active salivary glands as well as the quickened pace. In many such instances, one such preference – one underlying valuation – will be seen as having priority, as being what the person "really" wants. But not in all. Thomas Schelling has been concerned to undermine what he takes to be a fiction of standard decision theory, to wit that there's a stable set of preferences that characterizes a person across a single period of time.[41] Schelling uses examples of programs of self-improvement to argue that we can't always identify one of the perspectives with the agent's "real" self. When we reflect on friends who struggle, more or less successfully, to master the Great Books, to play a musical instrument, or to set themselves a regular regime of exercise, we are often torn between thinking of someone as weak in resolve or as refusing to drive herself, interpretations that signal different ways of identifying which are the person's "real" values.

So I think that there is some reason to believe that we have evolved to have dispositions to form altruistic valuations that are incompletely pervasive, and that, on occasion, we experience genuine conflict among different attitudes to others, including our kin and close friends. I now want to suggest that this picture of our imperfect altruism has some interesting consequences.

VI

If animals' valuations were genuinely in flux, then their social lives would be very difficult. They are. Peace and mutual tolerance are typically hard won. Precisely because of this, we see so much close social interaction, including lengthy bouts of grooming, within primate troops. The dispositions to form other-directed valuations makes it possible for the animals to live together; the incomplete pervasiveness of the dispositions means that they frequently act in ways that strain their social relations. Day after day, the social fabric is torn and must be mended by hours of peace-making.[42]

Perhaps this was once the situation of our ancestors too. However, I conjecture that an important step in hominid evolution liberated us from the oscillations between quarrelling and making-up that seem to dominate chimpanzee and bonobo social life, opening up possibilities for forming larger, and more stable, societies. There are two ways in which the problem posed by synchronically inconsistent valuations can be resolved.

One rests on the internal melee. Given the state of background variables, a disposition to form one type of valuation may be relatively stronger or weaker, and the resultant action reflects differences in strength. Thus we may suppose that the salience of the opportunity for sole dominant status reinforced Luit's disposition to construct valuations that gave zero weight to the preferences of Nikkie, making that far stronger than the dispositions to form valuations accommodating them (valuations that seemed to direct his behavior on other occasions).

The other possibility for resolving valuational conflict supposes that agents are not simply passive in the face of competing dispositions but able to identify with some and to restrain others. Organisms that acquired a capacity for forming "higher-order volitions", coming to have desires about which of their valuations would be effective in action, or a "system of normative control", would be able to achieve a greater consistency in behavior, proving more reliable as allies, and avoiding some situations of social conflict.[43] Chimpanzees and bonobos are "wantons", whose dispositions are activated by the pressures of the moment, and whose social lives are thus only held together because of large investments of time in reactive peacemaking. At some point in hominid evolution, I suggest, our ancestors replaced this reactive strategy with an approach based on prevention. By regulating, however imperfectly, our clashing dispositions, we became capable of avoiding some situations in which our evolutionary cousins would have been swept into conflict.

This is a hypothesis about the evolutionary function of morality. It is guided by the obvious problem of two simple alternatives: if we are, by nature, selfish and unscrupulous, then it is unclear how we ever became social beings at all; if we are, by nature, sympathetic and altruistic, then it is not obvious why we need morality. The answer I have suggested proposes that we have a fragile capacity for altruism, enough to bring us together (as other social primates come together) to form enduring societies. Because of the fragility of the altruistic dispositions, social bonds are easily broken, requiring our cousins (and, I assume, our ancestors) to develop and deploy time-consuming techniques of social repair. Even a primitive ability to govern our first-order valuations by norms enabled us to solve the problem of social conflict differently, by pre-emption rather than by reaction. Our first efforts at morality thus reinforced a part of ourselves, the part that had made the transition to complex sociality possible in the first place, and responded to the genuine difficulties that our divided psychology posed for our social predicament.

I offer this as a hypothesis that identifies how the evolution of human morality might have gone in its earliest phases.[44] To develop evidence in its favor, we would need to look much more closely at facets of human prehistory that I have ignored here, at the details of chimpanzee and bonobo sociality, at the results of psychological studies of human beings and other primates, at the behavior of people who live in hunter-gatherer societies and at the ways in which they regulate internal conflicts. As a brief illustration, it is worth considering one aspect of those contemporary groups which seem most closely to resemble our hominid ancestors. From Westermarck on, [45] anthropologists have documented the pervasiveness and importance of systems of kinship, often apparently artificial. Being a member of a particular subgroup of the total community fixes the ways in which individuals are supposed to behave in situations of internal strife. A person is primarily identified as a member of the Bear Clan (say), imposing the obligation to show solidarity with another clan member who is in dispute with an outsider. Rules of the form "Always ally with the *X*s against others!" are by far the most prominent among the imperatives that govern the lives of hunter-gatherers. Why is this? In light of the approach I've adopted, there's an obvious explanation. Recall that the most fundamental problems in social interaction involve coalitions and alliances. The incomplete pervasiveness of our dispositions to form altruistic valuations towards erstwhile allies is the chief threat to the stability of the coalitions on which social life depends. Hence it is hardly surprising that the primary normative rules should tell us who our allies are and command us to stick with them.

The example I've just sketched foreshadows the way in which my speculations should be made more definite and how they should be defended. As with Darwin's "long argument", the defense must come in showing how a large body of phenomena can be coherently understood.[46] My uses of various pieces of primatological data in this essay should be viewed as first gestures in this direction.

I want to close this section by identifying, as clearly as possible, my view of the link between evolution and ethics. If my hypothesis is correct, then characteristics of moral agents, including incompletely pervasive dispositions to psychological altruism and the capacity for normative guidance have been shaped by natural selection. At some point in our prehistory, our ancestors developed a system of *proto-morality*, a set of rules that reinforced their fragile altruistic tendencies and that enabled them to live in social groups that were subject to less constant rupture and needed less frequent repair. The transition from proto-morality – consisting, perhaps, in a simple set of rules for identifying allies (apparently the core norms of hunter-gatherer societies) – to the complex moral systems and moral reflections of recorded history should not be seen, on my story, as a further elaboration of evolution under natural selection. I have made no explicit claims about the emergence of morality from proto-morality, but it seems to me overwhelmingly plausible that this history has been guided mainly, if not exclusively, by forces of cultural, rather than natural selection. The point of my conjecture is not to connect Darwinian evolution directly with contemporary moral norms, but to view it as yielding the preconditions for the historical development of morality, a process during which simple kinship rules have been profoundly enriched and, indeed, out of which the distinction of morality from other systems of rules has itself emerged. However, I do want to emphasize the idea that our moral demands have a long and complicated history, that, like other human practices, morality is not adopted by transcending one's society and one's training (and the long traditions that stand behind it). G.E. Moore may have found it evident that "[b]y far the most valuable things, which we know or can imagine, are certain states of consciousness, which may be roughly described as the pleasures of human intercourse and the enjoyment of beautiful objects",[47] but I think it far more likely that Moore's judgment depended on the intellectual ontogeny that led him out of late Victorian England and into Cambridge and Bloomsbury than that it rested on any sudden insight, independent of the long moral tradition in which he stood. Moore's refined reflections are at a far remove from the social lives of primates, but, on the perspective I commend, they have been shaped by the long history that leads from the inconsistent altruists on the savannah to the civilized discussions at King's and Trinity.

VII

Why should the admittedly speculative story I have told be taken to have any relevance at all to contemporary discussions in moral philosophy? Although I once believed that the project of uncovering the historical processes through which modern moral practices have emerged, tracing them to distant biological roots, was perfectly legitimate, I didn't view it as affecting either philosophical discussions in meta-ethics or substantive normative debates.[48] I now think that that view was mistaken on the first score: understanding the genealogy of morals has meta-ethical implications.

Consider some questions that are at the center of twentieth-century debates. Is there moral knowledge? Do moral statements have truth-values? Are moral reasons always overriding? Why should a rational egoist obey any moral rule? Such questions are typically treated as if the psychological character of moral agents could be ignored, as if the social contexts in which morality is learned are irrelevant, and as if the history of the tradition out of which contemporary moral practices come makes no difference. Moral metaphysics and moral epistemology look much like the analytic metaphysics and epistemology of the mid-century. I suggest that these areas of philosophy may benefit from the same kind of naturalism that a number of recent writers have urged in epistemology, philosophy of mind, and philosophy of science.[49]

Let me close by looking very briefly at one of the ways in which the story I have outlined might restructure debate. Persons, I've suggested, have the capacity for acquiring a store of rules and for deploying those rules to reinforce some of the valuations they generate, so that those valuations become effective in action. Call the set of stored rules the individual's *moral code*. The rudiments of moral codes are acquired during youth, largely as the result of the inculcation of standards to which the surrounding community professes allegiance, but moral codes can grow from there as a result of later experiences, including increasing awareness of other possibilities, frustrations born of attempting to honor demands that prove too taxing, efforts to achieve consistency and coherence, and so forth. Once a moral code is in place, it need not be confined to the task of identifying with a particular valuation already generated. The injunction to obey a particular rule may produce a new valuation, one that overrides valuations generated from the background psychological state.

Even at this schematic level of psychological detail, we should recognize four importantly different cases in which a person does what a moral rule commands.

(1)   Through a response to the situation which is quite independent of the rule, the agent constructs a single valuation that gives highest priority to the commended action, and that valuation leads directly to action. (A parent plunges into the water to rescue a drowning child, without having the thought that this is what duty requires.)[50]

(2)   Incompatible valuations are present as the result of rule-independent responses, and the guidance of the rule reinforces one. (Someone has a sympathetic response to the predicament of two friends whose wants cannot be simultaneously satisfied, and the rule reinforces one.)

(3)   The single valuation present as a result of rule-independent responses is not in accordance with the rule, and the rule generates a new, overriding valuation. (A parent feels sorry for a naughty child but recognizes that it is best to punish him.)

(4)   Continued performance of an initially unpleasant duty
        produces a new disposition to generate a valuation in
        accordance with what the rule commands, so that later
        actions are unmediated by the rule. (Someone begins work
        to aid people who originally seem repulsive, but ultimately
        come to appear lovely.)

Because contemporary moral philosophy has been beguiled by the
idea that people have a single, coherent, set of preferences (at least
at one time), two of these cases – (1) and (3) – have come to appear
central, so that our choices in moral psychology are between a warm-
hearted, but sloppy, Humean agent or a punctilious, but unfeeling,
Kantian taskmaster.[51]

This is a mistake. Moral agency depends fundamentally on both
the natural dispositions to sympathy and on the normative guidance
that makes up for their shortcomings – (2) and (4) are as central to
our moral lives as (1) and (3). Praiseworthy as sympathy can often
be, there is a lurking distrust of the sympathy of the wanton because
it is uncontrolled and unreliable. Likewise the worthiness of rising
above inclinations has to be understood against the background in
which normative control finds its functions, conditions in which we
have incompletely pervasive dispositions to sympathize with others,
dispositions that make sociality, and ultimately morality, possible.
A being who had no natural dispositions to regulate would not seem
fully human, and one who never had the appropriate sympathies
would appear perverse.

Our system of normative control is a capacity for adopting rules
that govern existing valuations and that bring new ones into being
(cases like (4)). We cannot think of the system as external to our
wishes, for it reinforces some and is responsible for the presence
of others, but neither can we view the operation of the system as
determining what the person "really" wants.[52]

According to my evolutionary account, the capacity for regulating the flux of our dispositions is a hominid solution for solving a
recurrent problem in the lives of social animals. If that problem is
to be solved, then the moral codes that individuals internalize must
reinforce valuations that promote stable patterns of interaction within the community, and should thus contain directives to take into
account the preferences of others – for it is precisely such valua-

tions that inspire actions maintaining social order. With this in mind, the traditional philosophical question "Why should a rational egoist obey the demands of morality?" should be reconceived.

A first response should be to protest that the rational egoist is a fiction, for (virtually) all of us have dispositions to form altruistic valuations. Although these dispositions may sometimes conflict with other dispositions, they cannot be dismissed. Moreover, our dispositions are shaped, and, in some instances, brought into being, by the socialization that we receive from birth in which we absorb the moral lore of our communities. Insofar as we can make sense of a person as an agent with a determinate, coherent, set of preferences, those preferences are shaped through the adoption of a moral code. Prior to the internalization of the code, there is only a flux of often-clashing dispositions.

The response invites a second version of the question – "Why should we not want to be wantons?". Surely the best approximation to the fictitious rational egoist is a character for whom the strongest valuation always wins out in the internal melee. Now this is not our predicament, for, on my hypothesis, our evolutionary history took a different direction, but, in any event, we should recognize that the social life of wantons would be very different from ours (perhaps like those of chimpanzees and bonobos). It would not have the unrestrained savagery Hobbes supposed to be the lot of people in the state of nature, but it would have the sporadic uneasiness found in the lives of our closest evolutionary relatives, and it is highly doubtful that it could accommodate most of the things that people have come to want.

The third, and most interesting, version of the question stems from the recognition that a system of morality may chafe. In extreme instances, a person may have a disposition to construct valuations opposed to the demands of the prevailing moral code, a disposition that cannot be overridden by his system of normative control, even though there is no defect in his capacity for normative control. How should situations of this kind be characterized, and when do they provide agents with justification for rejecting parts of the moral code? In this form, the question leads into the largest and most difficult questions in meta-ethics, questions about the justification of moral change, the possibility of seeing moral change as progressive, and about the feasibility of viewing the historical development of morality as a process that tracks moral truth. I believe that exploration of this area would reconfigure contemporary debates between moral cognitivists and non-cognitivists, but I hope to have said enough already to have defended the claim that my speculative evolutionary history may have philosophical implications.

NOTES

[1] The present essay is a fragment of a much longer work-in-progress. Parts of this work have been presented to audiences at the University of California at Irvine, University College London, Princeton University, Stanford University, Washington University, the University of California at San Diego, and the University of Illinois, as well as at the Pacific Division Meeting of the American Philosophical Association. Many people have given me helpful advice and constructive criticisms. I am particularly grateful to Richard Arneson and Brian Skyrms for their suggestions, and to David Brink for helpful written comments on the penultimate draft.

[2] I should note explicitly that it is this last condition that seems to pose most difficulties for an evolutionary explanation of psychological altruism. It is also worth pointing out that I want to adopt a liberal conception of B's "presence", taking B to be present when A recognizes that the available actions have an impact on B's preferences. The third condition may be easiest to satisfy when, although B is not physically present, A can perceive that B would benefit from certain options available to A: imagine cases in which A has the opportunity to summon B to share a good.

[3] Thus I intend the notion of valuational structure to subsume the concepts of utility matrix and of the representation of games in normal form.

[4] The minimal condition for A to be altruistic is that A should take B's preferences into some account in evaluating options – as my account of *weak* altruism requires. But we are typically interested in much stronger types of altruism, those which play a systematic role in the readjustment of preferences so that there is a difference to action. The ensuing discussion will show how these can be captured within the framework I am proposing.

[5] For a relatively simple version of a case like this, see O. Henry's short story *The Gift of the Magi*. A much more complex and interesting example occurs in the closing pages of Henry James' *The Golden Bowl*. Martha Nussbaum gives a brilliant discussion of the nuances of the Ververs' perceptions of their mutual accommodation in her essay " 'Finely Aware and Richly Responsible': Literature and the Moral Imagination" in *Love's Knowledge*, New York: Oxford University Press, 1990, pp. 148–167.

[6] I used this term in "The Evolution of Human Altruism", *Journal of Philosophy*, 90, 1993, 497–516. It is important to note that, in that essay, I had a rather different conception of the relationship among the psychological states of altruists, so that although the earlier account is similar to that presented here it is not the same.

[7] I mention this putative principle, because it reveals the kinds of questions that my account of psychological altruism ought to address. This essay concentrates only on simple cases, and the relations among dimensions of altruism will be explored in future work.

[8] In such instances, the altruist will think of the beneficiary's actual preferences as not according with the beneficiary's interests, either because the beneficiary intends to sacrifice her own interests to promote those of the altruist or because she fails to have a correct conception of what her interests are. If one subscribes to a view of a person's good on which it is not reducible to what the person would want under epistemically improved (or ideal) circumstances, then one may also think that there are further cases in which the altruist takes into account the beneficiary's real interests and not her preferences (or even her suitably refined preferences).

[9] In "The Evolution of Human Altruism", my notion of paternalistic altruism was only partial. I am grateful to Richard Arneson for helping me to see this.

[10] The recent literature in primatology supplies abundant evidence for the first two of these claims, and can, I think, be used to defend the third. See especially Dorothy Cheney and Robert Seyfarth *How Monkeys See The World* (Chicago: University of Chicago Press, 1990) Chapters 3 and 8; Jane Goodall *The Chimpanzees of Gombe* (Cambridge MA.: Harvard University Press, 1986); C. Bachmann and H. Kummer "Male Assessment of Female Choice in Hamadryas Baboons", *Behavioral Ecology and Sociobiology*, 6, 1980, pp. 315–321; and R. Byrne and A. Whiten (eds.) *Machiavellian Intelligence* (Oxford: Oxford University Press, 1988), particularly the essay by Nicholas Humphrey.

[11] Many, though not all, of the essays in *Machiavellian Intelligence* adopt this perspective, and, for a more pronounced articulation of the theme that intelligence is a tool for calculating egoists, see James Barkow, Leda Cosmides and John Tooby (eds.) *The Adapted Mind*, Oxford: Oxford University Press, 1992. Interestingly, some of the papers that inspired the tradition explicitly allow that the Hobbes-Machiavelli picture of ruthless, self-interested agents may be inadequate to account for the psychology of social animals. See, for example, Nicholas Humphrey "The Social Function of Intellect" (*Machiavellian Intelligence* pp. 13–26) especially p. 23.

[12] See "The Evolution of Human Altruism", where I offered both the kin selection example and a much more extensive argument for the possibility that golden-rule altruism can originate and be sustained under natural selection. For reasons that will become apparent later, the selection regime envisaged there, involving iterated optional Prisoner's Dilemma can only be a small part of the story of the evolution of psychological altruism. I should also note that my earlier discussion did not respond adequately to the second type of worry about altruistic dispositions, since if has seemed to some people that reciprocal altruism could be fuelled by explicit calculation.

[13] See Mary Midgley *Beast and Man*, Ithaca: Cornell University Press, 1978 and my *Vaulting Ambition*.

[14] For a particularly lucid formulation of this point, see Joel Feinberg "Psychological Egoism", reprinted in Feinberg (ed.) *Reason and Responsibility*, Belmont CA.: Wadsworth.

[15] Goodall *The Chimpanzees of Gombe*, Cambridge MA.: Harvard University Press, 1986.

[16] Frans de Waal, *Good Natured*, Cambridge MA.: Harvard University Press, 1996.

[17] De Waal, *Good Natured*, pp. 131–132.

[18] My usage of this term is intended to resonate with the suggestions of David Hume and Adam Smith, both of whom I take to have been broadly right about our sentiments. As we shall see, my picture of human agency diverges from theirs in some important respects.

[19] This was worked out in John Batali and Philip Kitcher "Evolution of Altruism in Optional and Compulsory Games", *Journal of Theoretical Biology*, 175, 1995, pp. 161–171.

[20] See "The Evolution of Human Altruism", and "Evolution of Altruism in Optional and Compulsory Games".

[21] Axelrod's pioneering work is summarized in *The Evolution of Cooperation*, New York: Basic Books, 1984, which served as the stimulus for the work that Batali and I undertook. Interestingly what we showed is not quite what we (and our predecessors) set out to prove, to wit the stability of cooperation. Rather, we demonstrated the instability of the absence of cooperation, and indeed, the absence of high levels of cooperation. (In our simulations, the frequency of cooperation in any generation is either extremely low or else very high). In "The Evolution of Human Altruism", I argued further that the iterated optional PD regime would favor the emergence of golden-rule altruism. I would now put the conclusion a little differently: the formal relationship among preferences corresponds to the golden-rule value, $\theta = 0.5$, but it is possible that this relationship is set up as the result of explicit calculation of expected benefits (in defiance of condition (4) on weak psychological altruism). This conclusion hasn't yet been tested on computer simulations.

[22] See, for example, Jane Goodall's detailed story of a hunt, *The Chimpanzees of Gombe* pp. 288–289.

[23] See De Waal *Chimpanzee Politics* and *Peacemaking Among Primates*. Of course, it is possible that the high frequency of grooming is an artefact of the captive situations in which primates have been most closely observed, although studies of wild populations do support the idea that grooming occurs far more often than is needed for purely hygienic reasons.

[24] Most obviously, if, in addition to the hygienic benefits, there's a sizeable social payoff *that can only be achieved through cooperative behavior*, then the game being played is no longer optional PD – indeed, the largest payoff by far may be associated with cooperation with animals who are longstanding allies.

[25] This is not to deny that the phenomena investigated by Axelrod and his successors (including Batali and me) don't have a place in the understanding of social relations, cooperative behavior, and altruism, but rather that these can only be properly understood when we have a clear view of the social matrix in which they are set. An analogous point applies to the kinds of interactions explored brilliantly by Brian Skyrms (*Evolution of the Social Contract*, Cambridge: Cambridge University Press, 1996). I develop this point in a commentary on Skyrms' book (to appear in a symposium in *Philosophy and Phenomenological Research*).

[26] See Peter Rodman and John Mitani "Orangutans: Sexual Dimorphism in a Solitary Species" in Barbara Smuts et al. (eds.) *Primate Societies*, Chicago: University of Chicago Press, 1987, pp. 146–154. I have heard from several people

that the "standard view" outlined in this article has been challenged by further studies of female associations, but I do not know of any published sources.

[27] See Donna Leighton "Gibbons: Territoriality and Monogamy", and Kelly Stewart and Alexander Harcourt "Gorillas: Variation in Female Relationships", both in Smuts *et. al. Primate Societies* (pp. 135–145, and 155–164, respectively).

[28] It may be pertinent to note that some human groups of hunter-gatherers tend to undergo fission when their size exceeds 150 to 300. One example is the Yanomamo; see Napoleon Chagnon "Fission in a Yanomamo Tribe", *The Sciences*, 16, 1976, pp. 14–18.

[29] See R. Wrangham "On the Evolution of Ape Social Systems", *Social Science Information*, 18, 1979, pp. 334–368; "An Ecological Model of Female-Bonded Primate Groups", *Behaviour*, 75, 1980, pp. 262–300; "Social Relationships in Comparative Perspective", in R. Hinde (ed.) *Primate Social Relationships: An Integrated Approach* (Oxford: Blackwell, 1983), and "Evolution of Social Structure" in Barbara Smuts et al. (eds.) *Primate Societies* (Chicago: University of Chicago Press), pp. 282–296. Wrangham bases his approach on the idea that the principal determinant of female reproductive success will be her access to food, and that the principal determinant of male reproductive success will be the ability to copulate as frequently as possible with an estrous female. On his account orang-utans pursue their relatively solitary lives because females can most efficiently forage for fruit by working alone, and males have the physical abilities to defend a home range including the smaller home ranges of several females. As will be clear below, I'm going to abstract from some of the particularities of Wrangham's discussion, offering a more general model of which his explanation would be a special case.

[30] "Evolution of Social Structure" p. 290. Compare Hobbes: "the weakest has strength enough to kill the strongest, either by secret machination, or by confederation with others that are in the same danger with himself" (*Leviathan* p. 183). However, Hobbes would not have thought that this could apply to the brutes, because, without speech "there had been amongst men neither Common-wealth, nor Society, nor Contract, nor Peace, no more than amongst Lyons, Bears, and Wolves" (*Leviathan* p. 100). Hobbes underrated the lions and the wolves, and knew nothing of the chimpanzees and bonobos.

[31] For similar suggestions to the effect that the most important issues about evolution of cooperation and sociality need to move beyond PD, see R. Noe "Alliance Formation among Male Baboons: Shopping for Profitable Partners", in A. Harcourt and F. de Waal (eds.) *Coalitions and Alliances in Humans an Other Animals*, Oxford: Oxford University Press, 1992, pp. 285–321. Several of Noe's points about baboons are analogues of claims I have made about chimpanzees. See, for example, his finding that coalitions and alliances can persist even when one of the parties never receives what is apparently the main prize.

[32] Note that the fitness values that occur in the payoff matrices for these games have to be computed by taking into account the consequences of the actions for the underlying alliances and coalitions to which the agent belongs. This articulates the point made earlier that the structure of animal interactions cannot be understood in isolation from the demands of the most fundamental game, here identified as the coalition-formation game.

[33] I hope to present these in a book that will work out the project begun in this essay.

[34] See, for example, John Maynard Smith, *Evolution and the Theory of Games*, Cambridge: Cambridge University Press, 1982, p. 108.

[35] The argument is non-constructive because it says that there will be coalitional structure without saying what it is. As will become clear, I don't know how to determine the structure, even given full information about the initial conditions. So the claim is roughly as useful as the mean-value theorem, which tells us that there is a maximum of a function within a particular interval, without specifying where it is. That sort of information was good enough for a celebrated application of Reaganomics – and, in the present context, we'll see that the absence of a constructive argument has philosophical implications.

[36] *Chimpanzee Politics*, Baltimore: Johns Hopkins, 1984, p. 126. My subsequent discussion will explain, and partially endorse, this reading of de Waal's. It's important to note, however, that the picture of cunning political agents, given in de Waal's first book is softened in his second (*Peace-Making Among Primates*, Cambridge M.: Harvard University Press, 1989), and replaced by an emphasis on animal sympathy and altruism in his most recent (*Good Natured*). Although I cannot fully document it here, I think that my analysis of the psychological altruism found in chimpanzees and bonobos explains the complex findings recorded by de Waal, offering a perspective from which all his observations fit together.

[37] *Chimpanzee Politics* p. 128.

[38] *Peacemaking Among Primates* Chapter 2. De Waal makes the important observation that Luit's desire to remain with his troop was so strong that it was difficult to remove him, even after he had been severely wounded.

[39] I should acknowledge that there are obvious concerns, given our ignorance of the nature of the dependence, about the testability of attributions of conflicting dispositions. A general response to these concerns would note that we should not be bound by any operational imperative; more specifically, I think that there are forms of behavior that provide evidence for the claim that an animal has a disposition to produce conflicting valuations (see the text below). The approach taken here has some kinship with Walter Mischel's emphasis on the failure of cross-situational consistency in people who have stable personality profiles (see W. Mischel and Y. Shoda "A Cognitive-Affective System Theory of Personality: Reconceptualizing Situations, Dispositions, Dynamics, and Invariance in Personality Structure", *Psychological Review*, 102, 1995, pp. 246–268; I am grateful to George Mandler for drawing my attention to the problem of trait consistency, and suggesting that I read Mischel's work). There are also connections between the approach I am adopting here and some of Anthony Damasio's work on decision-making and its neural disruption; see his *Descartes' Error*, New York: Putnam's 1994. For reasons of space, I can't elaborate the affinities here.

[40] This provides a start on an answer to worries about testability. We can sometimes suppose that conflicting valuations are reflected in different signals sent to various parts of the body – one valuation would be relevant to the state of one bodily system but not to another, and conversely, so that the overall state of he body is a mosaic. In general, we might test the presence of attitudes not by focusing on behavior but on readiness for behavior.

[41] T. Schelling "The Intimate Contest for Self-Command", Chapter 3 of *Choice and Consequence*, Cambridge MA.: Harvard University Press, 1984, pp. 57–82. The discussions on pp. 60–61 are especially pertinent.

PSYCHOLOGICAL ALTRUISM, EVOLUTIONARY ORIGINS, AND MORAL RULES 315

[42] For details of peace-making strategies, see de Waal *Peacemaking Among Primates*.

[43] I deliberately use terminology of Harry Frankfurt ("Freedom of the Will and the Concept of a Person", *Journal of Philosophy*, 68, 1970, pp. 5–20) and of Allan Gibbard (*Wise Choices, Apt Feelings* Cambridge MA.: Harvard University Press, 1990). It seems to me that Frankfurt's and Gibbard's insights can be connected with the evolutionary perspective I am developing. I suspect that Gibbard would approve of this, although Frankfurt might view it as misguided and unnecessary.

[44] A number of people (most forcefully Bas van Fraassen) have expressed surprise that the author of a book critical of human sociobiology might want to speculate in this area. I am sensitive to this concern. My best attempt to lay it to rest runs as follows: human morality is a natural phenomenon, that must fit, somehow, into our natural history; the last chapter of *Vaulting Ambition* identifies some ways in which I think that it does *not* fit, but my criticisms there will never be fully convincing unless I can devise a better account; any such account should not aspire to do more than show how human morality *could have* begun (in standard terminology, it is a "how possibly" explanation – see *Vaulting Ambition* p. 72 ff.); but finding any account that coheres with what we know about other primates and about ourselves is very hard, and it seems to me that standard philosophical accounts of our moral psychology are in constant danger of conjuring illusory faculties and abilities that are quite at odds with what we take ourselves to know about animal behavior; thus it seems worthwhile offering an avowedly conjectural picture and exploring how the problems of moral philosophy look from its perspective. (Although this is my favorite response to a common concern, it does not satisfy me on every day of the week.)

[45] Edward Westermarck's comparative survey of social rules in "primitive" societies, *The Origin and Development of the Moral Ideas*, London: MacMillan, 2 vols., 1906–1908, is a mine of comprehensive information.

[46] For an argument that this was Darwin's strategy see my essay "Darwin's Achievement" (in N. Rescher [ed.] *Reason and Rationality in Science*, University Press of America, 1985).

[47] Moore *Principia Ethica*, New York: Prometheus Books, 1988, p. 188 (section 113).

[48] See *Vaulting Ambition* Chapter 11, and "Four Ways of Biologicizing Ethics", in Elliott Sober (ed.) *Conceptual Issues in Evolutionary Biology* (second edition), Cambridge MA.: MIT Press, 1994.

[49] For representative versions of the kinds of naturalism I have in mind, see Alvin Goldman *Epistemology and Cognition* (Cambridge MA.: Harvard University Press, 1986), Fred Dretske *Naturalizing the Mind* (Cambridge MA.: MIT Press, 1994), and my own *The Nature of Mathematical Knowledge* (New York: Oxford University Press, 1983) and *The Advancement of Science* (New York: Oxford University Press, 1993).

[50] This, of course, is Bernard Williams' famous "one thought too many". See "Persons, Character and Morality", in Amelie Rorty (ed.) *The Identities of Persons*, Berkeley: University of California Press, 1976, pp. 197–216, especially pp. 213–215.

[51] Although (1) and (3) represent cartoon versions of what goes on in our moral lives, the cartoons are influential. A presently unpublished lecture by Barbara Herman has convinced me that Kant's actual views are much more nuanced

316   PHILIP KITCHER

and intricate, and Hume's moral psychology is also more sophisticated (as his discussions of artificial virtues make clear). But neither Hume nor Kant nor the philosophers they have influenced seem to escape the idea that we have a coherent set of preferences, and hence do not recognize the need to address the four types of moral agency as equally fundamental.

[52] Partly, this is for reasons offered by Schelling in *Choice and Consequence*, which show how systems of normative control can be inept or incomplete. Further problems result from the fact that we can recognize dispositions in the agent that produce valuations clashing with the system of normative control as genuine expressions of the agent's character and desires. The philosophical tradition is heavily inclined towards the view of the system of normative control as constituting the wishes of a "real" self, but psychodynamic conceptions stemming from Freud should remind us of the repressive possibilities of a system of normative control. See, for two very different examples, Freud's *The Ego and the Id*, and George Ainslie's *Picoeconomics* (Cambridge: Cambridge University Press, 1992). Samuel Scheffler has offered a penetrating account of the implications of some aspects of Freud's work for our understanding of central issues in moral psychology, and I think that the picture I sketch here could be articulated to capture his insights. (See Scheffler *Human Morality*, New York: Oxford University Press, 1992, Chapter 5.)

Department of Philosophy
University of California
San Diego, CA 92110-2492
U.S.A.

[3]

Summary of: 'Unto Others

The Evolution and Psychology of Unselfish Behavior'

Elliott Sober and David Sloan Wilson

The hypothesis of group selection fell victim to a seemingly devastating critique in 1960s evolutionary biology. In Unto Others *(1998), we argue to the contrary, that group selection is a conceptually coherent and empirically well documented cause of evolution. We suggest, in addition, that it has been especially important in human evolution. In the second part of* Unto Others, *we consider the issue of psychological egoism and altruism — do human beings have ultimate motives concerning the well-being of others? We argue that previous psychological and philosophical work on this question has been inconclusive. We propose an evolutionary argument for the claim that human beings have altruistic ultimate motives.*

I: Introduction

Part One of *Unto Others* (Sober & Wilson, 1998) addresses the biological question of whether evolutionary altruism exists in nature and, if so, how it should be explained. Part Two concerns the psychological question of whether any of our ultimate motives involves an irreducible concern for the welfare of others. Both questions are descriptive, not normative. And neither, on the surface, even mentions the topic of morality. How, then, do these evolutionary and psychological matters bear on issues about morality? And what relevance do these descriptive questions have for normative ethical questions? These are problems we'll postpone discussing until we have outlined the main points we develop in *Unto Others*.

A behaviour is said to be altruistic in the evolutionary sense of that term if it involves a fitness cost to the donor and confers a fitness benefit on the recipient. A mindless organism can be an evolutionary altruist. It is important to recognize that the costs and benefits that evolutionary altruism involves come in the currency of reproductive success. If we give you a package of contraceptives as a gift, this won't be evolutionarily altruistic if the gift fails to enhance your reproductive success. And parents who take care of their children are not evolutionarily altruistic if they rear more children to adulthood than do parents who neglect their children. Evolutionary altruism is not the same as helping.

The concept of psychological altruism is, in a sense, the mirror image of the evolutionary concept. Evolutionary altruism describes the fitness effects of a behaviour, not the thoughts or feelings, if any, that prompt individuals to produce those behaviours. In contrast, psychological altruism concerns the motives that cause a behavior, not its actual effects. If your treatment of others is prompted by your having an ultimate, noninstrumental concern for their welfare, this says nothing as to whether your actions will in fact be beneficial. Similarly, if you act only to benefit yourself, it is a further question what effect your actions will have on others. Psychological egoists who help because this makes them feel good may make the world a better place. And psychological altruists who are misguided, or whose efforts miscarry, can make the world worse.

Although the two concepts of altruism are distinct, they often are run together. People sometimes conclude that if genuine evolutionary altruism does not exist in nature, then it would be mere wishful thinking to hold that psychological altruism exists in human nature. The inference does not follow.

II: Evolutionary Altruism — Part One of *Unto Others*

1. The problem of evolutionary altruism and the critique of group selection in the 1960s

Evolutionary altruism poses a fundamental problem for the theory of natural selection. By definition, altruists have lower fitness than the selfish individuals with whom they interact. It therefore seems inevitable that natural selection should eliminate altruistic behaviour, just as it eliminates other traits that diminish an individual's fitness. Darwin saw this point, but he also thought that he saw genuinely altruistic characteristics in nature. The barbed stinger of a honey bee causes the bee to die when it stings an intruder to the nest. And numerous species of social insects include individual workers who are sterile. In both cases, the trait is good for the group though deleterious for the individuals who have it. In addition to these examples from nonhuman species, Darwin thought that human moralities exhibit striking examples of evolutionary altruism. In *The Descent of Man*, Darwin (1871) discusses the behaviour of courageous men who risk their lives to defend their tribes when a war occurs. Darwin hypothesized that these characteristics cannot be explained by the usual process of natural selection in which individuals compete with other individuals in the same group. This led him to advance the hypothesis of *group selection*. Barbed stingers, sterile castes, and human morality evolved because groups competed against other groups. Evolutionarily selfish traits evolve if selection occurs exclusively at the individual level. Group selection makes the evolution of altruism possible.

Although Darwin invoked the hypothesis of group selection only a few times, his successors were less abstemious. Group selection became an important hypothesis in the evolutionary biologist's toolkit during the heyday of the Modern Synthesis (c. 1930–1960). Biologists invoked individual selection to explain some traits, such as sharp teeth and immunity to disease; they invoked group selection to explain others, such as pecking order and the existence of genetic variation within species. Biologists simply used the concept that seemed appropriate. Discussion of putative group adaptations were not grounded in mathematical models of the group selection process, which hardly existed. Nor did naturalists usually feel the need to supply a mathematical model to support the claim that this or that phenotype evolved by individual selection.

All this changed in the 1960s when the hypothesis of group selection was vigorously criticized. It was attacked not just for making claims that are empirically false, but for being conceptually confused. The most influential of these critiques was George C. Williams' 1966 book, *Adaptation and Natural Selection*. Williams argued that traits don't evolve because they help groups; and even the idea that they evolve because they benefit individual organisms isn't quite right. Williams proposed that the right view is that traits evolve because they promote the replication of genes.

Williams' book, like much of the literature of that period, exhibits an ambivalent attitude towards the idea of group selection. Williams was consistently against the hypothesis; what he was ambivalent about was the grounds on which he thought the hypothesis should be rejected. Some of Williams' book deploys empirical arguments against group selection. For example, he argues that individual selection and group selection make different predictions about the sex ratio (the proportion of males and females) that should be found in a population; he claimed that the observations are squarely on the side of individual selection. But a substantial part of Williams' book advances somewhat *a priori* arguments against group selection. An example is his contention that the gene is the unit of selection because genes persist through many generations, whereas groups, organisms, and gene complexes are evanescent. Another example is his contention that group selection hypotheses are less parsimonious than hypotheses of individual selection, and so should be rejected on that basis.

The attack on group selection in the 1960s occurred at the same time that new mathematical models made it seem that the hypothesis of group selection was superfluous. W.D. Hamilton published an enormously influential paper in 1964, which begins with the claim that the classical notion of Darwinian fitness — an organism's prospects of reproductive success — can explain virtually none of the helping behaviour we see in nature. It can explain parental care, but when individuals help individuals who are not their offspring, a new concept of fitness is needed to explain why. This led Hamilton to introduce the mathematical concept of *inclusive fitness*. The point of this concept was to show how helping a relative and helping one's offspring can be brought under the same theoretical umbrella — both evolve because they enhance the donor's inclusive fitness. Many biologists concluded that helping behaviour directed at relatives is therefore an instance of selfishness, not altruism. Helping offspring and helping kin are both in one's genetic self-interest, because both allow copies of one's genes to make their way into the next generation. Behaviours that earlier seemed instances of altruism now seemed to be instances of genetic selfishness. The traits that Darwin invoked the hypothesis of group selection to explain apparently can be explained by 'kin selection' (the term that Maynard Smith, 1964, suggested for the process that Hamilton described), which was interpreted as an instance of individual selection. Group selection wasn't needed as a hypothesis; it was 'unparsimonious'.

Another mathematical development that pushed group selection further into the shadows was evolutionary game theory. Maynard Smith, one of the main architects of evolutionary game theory, wanted to provide a sane alternative to sloppy group selection thinking. Konrad Lorenz and others had suggested, for example, that animals restrain themselves in intraspecific combat because this is good for the species. Maynard Smith and Price (1973) developed their game of hawks versus doves to show how restraint in combat can result from purely individual selection. Each individual in the population competes with one other individual, chosen at random, to determine which will obtain some fitness benefit. Each plays either the hawk strategy of all-out fighting or the dove strategy of engaging in restrained and brief aggression. When a hawk fights a hawk, one of them gets the prize, but each stands a good chance of serious injury or death. When a hawk fights a dove, the hawk wins the prize and the dove beats a hasty retreat, thus avoiding serious injury. And when two doves fight, the battle is over quickly; there is a winner and a loser, but neither gets hurt. In this model, which trait does better depends on which trait is common and which is rare. If hawks are very common, a dove will do better than the average hawk — the average hawk gets injured a lot, but the dove does not. On the other hand, if doves are very common, a hawk will do better than the average dove. The evolutionary result is a polymorphism. Neither trait is driven to extinction; both are represented in the population. What Lorenz tried to explain by invoking the good of the species, Maynard Smith and Price proposed to explain purely in terms of individual advantage. Just as was true in the case of Hamilton's work on inclusive fitness, the hypothesis of group selection appeared superfluous. You don't *need* the hypothesis to explain what you observe. Altruism is only an appearance. Dovishness isn't present because it helps the group; the trait is maintained in the population because individual doves gain an advantage from not fighting to the death.

Another apparent nail in the coffin of group selection was Maynard Smith's (1964) 'haystack model' of group selection. Maynard Smith considered the hypothetical situation in which field mice live in haystacks. The process begins by fertilized females each finding their own haystacks. Each gives birth to a set of offspring who then reproduce among themselves, brothers and sisters mating with each other. After that, the haystack holds together for another generation, with first cousins mating with first cousins. Each haystack contains a group of mice founded by a single female that sticks together for some number of generations. At a certain point, all the mice come out of their haystacks, mate at random, and then individual fertilized females go off to found their own groups in new haystacks. Maynard Smith analyzed this process mathematically and concluded that altruism can't evolve by group selection. Group selection is an inherently weak force, unable to overcome the countervailing and stronger force of individual selection, which promotes the evolution of selfishness.

The net effect of the critique of group selection in the 1960s was that the existence of adaptations that evolve because they benefit the group was dismissed from serious consideration in biology. The lesson was that the hypothesis of group selection doesn't have to be considered as an empirical possibility when the question is raised as to why this or that trait evolved. You know *in advance* that group selection is not the explanation. Only those who cling to the illusion that nature is cuddly and hospitable could take the hypothesis of group adaptation seriously.

2. Conceptual arguments against group selection

In *Unto Others*, we argue that this seemingly devastating critique of group selection completely missed the mark. The purely conceptual arguments against group selection show nothing. And the more empirical arguments also are flawed.

Let us grant that genes — not organisms or groups of organisms — are the *units of replication*. By this we mean that they are the devices that insure heredity. Offspring resemble parents because genes are passed from the latter to the former. However, this establishes nothing about why the adaptations found in nature have evolved.

Presumably, even if the gene is the unit of replication, it still can be true that some genes evolve because they code for traits that benefit individuals — this is why sharp teeth and immunity from disease evolve. But the same point holds for groups: even if the gene is the unit of replication, it remains to be decided whether some genes evolve because they code for traits that benefit groups. The fact that genes are *replicators* is entirely irrelevant to the units of *selection* problem.

The idea that group selection should be rejected because it is unparsimonious also fails to pass muster. Here's an example of how the argument is deployed, in Williams (1966), in Dawkins (1976), and in many other places. Why do crows exhibit sentinel behaviour? Group selection was sometimes invoked to explain this as an instance of altruism. A crow that sights an approaching predator and issues a warning cry places itself at risk by attracting the predator's attention; in addition, the sentinel confers a benefit on the other crows in the group by alerting them to danger. Interpreted in this way, a group selection explanation may seem plausible. However, an alternative possibility is that the sentinel behaviour is not really altruistic at all. Perhaps the sentinel cry is difficult for the predator to locate, and maybe the cry sends the other crows in the group into a frenzy of activity, thus permitting the sentinel to beat a safe retreat. If the behaviour is selfish, no group selection explanation is needed. At this point, one might think that two empirical hypotheses have been presented and that observations are needed to test which is better supported. However, the style of parsimony argument advanced in the anti-group selection literature concludes without further ado that the group selection explanation should be rejected, just because an individual selection explanation has been *imagined*. Data aren't needed, because parsimony answers our question. In *Unto Others*, we argue that this is a spurious application of the principle of parsimony. Parsimony is a guide to how observations should be interpreted; it is not a substitute for performing observational tests.

There is another fallacy that has played a central role in the group selection debate. The fallacy involves defining 'individual selection' so that any trait that evolves because of selection is automatically said to be due to individual selection; the hypothesis that traits might evolve by group selection thus becomes a definitional impossibility. In *Unto Others*, we call this *the averaging fallacy*. To explain how the fallacy works, let's begin with the standard representation of fitness payoffs to altruistic (A) and selfish (S) individuals when they interact in groups of size two. The argument would not be different if we considered larger groups. When two individuals interact, the payoff to the row player depends on whether he is A or S and on whether the person he interacts with is A or S (b is the benefit to the recipient and c is the cost to the altruistic A-type's behaviour):

		the other player is	
		A	S
fitness of a	A	$x + b - c$	$x - c$
player who is	S	$x + b$	x

What is the average fitness of A individuals? It will be an average — an altruist has a certain probability (p) of being paired with another altruist, and the complementary probability $(1-p)$ of being paired with a selfish individual. Likewise, a selfish individual has a certain probability of being paired with an altruist (q) and the complementary probability $(1-q)$ of being paired with another selfish individual. Thus, the fitnesses of the two traits are

$$w(A) = p(x+b-c) + (1-p)(x-c) = pb + x - c$$
$$w(S) = (q)(x+b) + (1-q)(x) = qb + x.$$

By definition, the trait with the higher average fitness will increase in frequency, if natural selection governs the evolutionary process. The criterion for which trait evolves is therefore:

(1) $w(A) > w(S)$ if and only if $p-q > c/b$.

The quantity $(p-q)$, we emphasize, is the difference between two probabilities:

$(p-q)$ = the probability that an altruist has of interacting with another altruist
         minus the probability that a selfish individual has of interacting with
         an altruist.

This difference represents the *correlation* of the two traits.
    Two consequences of proposition (1) are worth noting:

(2)   When like interacts with like, $w(A) > w(S)$ if and only if $b > c$.

(3)   When individuals interact at random, $w(A) > w(S)$ if and only if $0 > c/b$.

Proposition (2) identifies the case most favourable for the evolution of A — as long as the benefit to the recipient is greater than the cost incurred by the donor, A will evolve. Proposition (3), on the other hand, describes a situation in which A cannot evolve, as long as c and b are both greater than zero.
    This analysis of the evolutionary consequences of the payoffs stipulated for traits A and S is not controversial. The fallacy arises when it is proposed that the selfish trait is the trait that has the higher average fitness, and that individual selection is the process that causes selfishness, so defined, to evolve. The effect of this proposal is that A is said to be selfish in situation (2) if $b > c$, while S is labelled selfish in situation (3), if $b,c > 0$. Selfishness is equated with 'what evolves', and individual selection is, by definition, the selection process that makes selfishness evolve. This framework entails that altruism cannot evolve by natural selection and that group selection cannot exist. We reject this definitional framework because it fails to do justice to the biological problem that Darwin and his successors were addressing. The question of what types of adaptations are found in nature is *empirical*. If altruism and group adaptations do not exist, this must be demonstrated by observation. The real question cannot be settled by this semantic sleight of hand.
    Our proposal is to define altruism and selfishness by the payoff matrix given above. What is true, by definition, is that altruists are less fit than selfish individuals *in the same group*. If b and c are both positive, then $x+b > x-c$. However, nothing follows from this as to whether altruists have lower fitness when one averages *across all groups*. This will be not be the case in the circumstance described in proposition (2), if $b > c$, but will be the case in the situation described in (3), if $b,c > 0$.

SUMMARY OF *UNTO OTHERS* 191

Given the payoffs described, groups vary in fitness; the average fitness in AA groups is *(x+b-c)*, the average fitness in AS groups is *(x + [b-c]/2)*, and the average fitness in SS groups is *x*. Group selection favours altruism; groups do better the more altruists they contain. Individual selection, on the other hand, favours selfishness. There is no individual selection within homogeneous groups; the only individual (i.e., within-group) selection that occurs is in groups that are AS. Within such groups, self-ishness outcompetes altruism. Here group and individual selection are opposing forces; which force is stronger determines whether altruism increases or declines in frequency in the ensemble of groups. Just as Darwin conjectured, it takes group selection for altruism to evolve.

Our proposal — that altruism and selfishness should be defined by the payoff matrix described above, and that group selection involves selection among groups, whereas individual selection involves selection within groups — is not something we invented, but reflects a long-standing set of practices in biology. Fitness averaged across groups is a criterion for which trait evolves. However, if one additionally wants to know whether group selection is part of the process, one must decompose this average by making within-group and between-group fitness comparisons.

This perspective on what altruism and group selection mean undermines the perva-sive opinion that kin selection and game-theoretic interactions are alternatives to group selection. It also allows us to re-evaluate Hamilton's claim that classical Dar-winian fitness cannot explain the evolution of helping behaviour (other than that of parental care) and that the concept of inclusive fitness is needed. The inclusive fitness of altruism reflects the cost to the donor and the benefit to the recipient, the latter weighed by the coefficient of relatedness (r) that donor bears to recipient:

$I(A) = x - c + br.$

The inclusive fitness of a selfish individual is

$I(S) = x.$

Notice that I(A) does not reflect the possibility that the altruist in question may receive a donation from another altruist, and the same is true of I(S) — it fails to reflect the possibility that a selfish individual may receive a donation from an altruist. The reason for these omissions is that we are assuming that altruism is *rare*. In any event, from these two inclusive fitnesses, we obtain 'Hamilton's rule' for the evolu-tion of altruism:

(4) $r > c/b.$

We hope the reader notices a resemblance between propositions (1) and (4). The coef-ficient of relatedness is a way of expressing the correlation of interactors. Contrary to Hamilton (1964), the concept of inclusive fitness is *not* needed to describe the cir-cumstances in which altruism will evolve.

The coefficient of relatedness 'r' is relevant to the evolution of altruism because related individuals tend to resemble each other. What is crucial for the evolution of altruism is that altruists tend to interact with altruists. This can occur because rela-tives tend to interact with each other, or because unrelated individuals who resemble each other tend to interact. The natural conclusion to draw is that kin selection is a kind of group selection, in which the groups are composed of relatives. When an altruistic individual helps a related individual who is selfish, the donor still has a lower fitness than the recipient. The fact that they are related does not cancel this fundamental fact. *Within* a group of relatives, altruists are less fit than selfish individuals. It is only because of selection *among* groups that altruism can evolve. This, by the way, is the interpretation that Hamilton (1975) himself embraced about his own work, but his changed interpretation apparently has not been heard by many of his disciples.

Similar conclusions need to be drawn about game theory. Perhaps the most famous study in evolutionary game theory is the set of simulations carried out by Axelrod (1984). Axelrod had various game theorists suggest strategies that individuals might follow in repeated interactions. Individuals pair up at random and then behave altruistically or selfishly towards each other on each of several interactions. The payoffs that come from each interaction are the ones described before. However, the situation is more complex because there are many strategies that individuals might follow. Some strategies are *unconditional* — for example, an individual might act selfishly on every move (ALLS) or it might act altruistically on every move. In addition, there are many *conditional* strategies, according to which a player's action at one time depends on what has happened earlier in his interactions with the other player. Axelrod found that the strategy suggested by Anatol Rappaport of Tit-for-Tat (TFT) did better than many more selfish strategies. TFT is a strategy of *reciprocity*. A TFT player begins by acting altruistically and thereafter does whatever the other player did on the previous move. Two TFT players act altruistically towards each other on every move; if there are n moves in the game, each obtains a total score of $n(x-b+c)$. When TFT plays ALLS, the TFT player acts altruistically on the first move and then shifts to selfishness thereafter; if there are n interactions, TFT receives $(x-c) + (n-1)x = (nx-c)$ in its interaction with ALLS, who receives $(x+b) + (n-1)x$. Finally, if two ALLS players interact, each receives nx.

It is perfectly true, as a biographical matter, that Maynard Smith developed evolutionary game theory as an alternative to the hypothesis of group selection. However, the theory he described in fact involves group selection. If TFT competes with ALLS, there is group selection in which groups are formed at random and the groups are of size 2. Groups do better the more TFTers they contain. There is individual selection within mixed groups, in which TFT does worse than ALLS. TFT is able to evolve only because group selection favouring TFT overcomes the opposing force of individual (within-group) selection, which favours ALLS.

3. Empirical arguments against group selection

Williams (1966) proposed that sex ratio provides an empirical test of group selection. If sex ratio evolves by individual selection, then a roughly 1:1 ratio should be present. On the other hand, if sex ratio evolves by group selection, a female-biased sex ratio will evolve if this ratio helps the group to maximize its productivity. Williams then claims that the sex ratios found in nature are almost all close to even. He concludes that the case against group selection, with respect to this trait at least, is closed.

A year later, Hamilton (1967) reported that female-biased sex ratios are abundant. One might expect that the evolution community would have greeted Hamilton's report as providing powerful evidence in favour of group selection. This is exactly what did not occur. Although Hamilton described his own explanation of the

SUMMARY OF *UNTO OTHERS* 193

evolution of 'extraordinary sex ratios' as involving group selection, this is not how most other biologists interpreted it. Williams' sound reasoning that individual selection should produce an even sex ratio traces back to a model first informally proposed by R.A. Fisher (1930). Fisher assumed that parents produce a generation of offspring; these offspring then mate with each other at random, thus producing the grandoffspring of the original parents. If the offspring generation is predominately male, then a parent does best by producing all daughters; if the offspring generation is predominately female, the parent does best by producing all sons. Selection favours parents who produce the minority sex, and the population evolves towards an even sex ratio as a result. Hamilton introduced a change in assumptions. He considered the example of parasitic wasps who lay their eggs in hosts. One or more fertilized females lays eggs in a host; the offspring of these original foundresses mate with each other, after which they disperse to find new hosts and the cycle starts anew. The important point about Hamilton's model is that offspring in different hosts don't mate with each other.

Williams observed, correctly, that the way for a group to maximize its productivity is for it to have the smallest number of males that is necessary to insure that all females are fertilized. Group selection therefore favours a female-biased sex ratio, and this in fact is what Hamilton's model explains. The wasps in a host form a group, and groups with a female-biased sex ratio are more productive than groups in which the sex ratio is even. This is how Hamilton (1967, footnote 43) interprets his model, but most of his readers apparently did not. Rather, they construed Hamilton's model as describing individual selection; the reason is that Hamilton analyzed his model by calculating what the 'unbeatable strategy' is — that is, the strategy whose fitness is greater than the alternatives. This is the sex ratio strategy that will evolve. To automatically equate the unbeatable strategy with 'what evolves by individual selection' is to commit the averaging fallacy. Instead of considering what goes on within hosts as an instance of individual selection and differences among hosts as reflecting the action of group selection, the mistake is to meld these two processes together to yield a single summary statistic, which reflects the fitnesses of strategies averaged across groups. There is nothing wrong with obtaining this average if one merely wishes to say what trait will evolve. However, if the goal, additionally, is to say whether group selection is in part responsible for the evolutionary outcome, one can't use a framework in which what evolves is automatically equated with pure individual selection.

The other empirical argument we mentioned before, which was thought to tell against the hypothesis of group selection, is Maynard Smith's (1964) haystack model. It is a little odd to call this argument 'empirical', since it did not involve the gathering of data. Rather, the argument was 'theoretical', based on the analysis of a hypothetical model. In any event, let's consider how Maynard Smith managed to reach the conclusion that group selection is a weak force, unequal to the task of overcoming the opposing force of individual selection. The answer is that Maynard Smith simply *stipulated* that the within-haystack, individual selection part of his process was as powerful as it could possibly be. He *assumes without argument* that altruism is driven to extinction in all haystacks in which it is mixed with selfishness; the only way that altruism can survive in a haystack is by being in a haystack that is 100 per cent altruistic. We do not dispute that, as a matter of definition, altruism must *decline* in frequency in all mixed haystacks. But the idea that it must *decline to zero* in all such haystacks is *not* a matter of definition. In effect, Maynard Smith explored a worse-case scenario for group selection. This tells us nothing as to whether altruism can evolve by group selection. Twenty years later, one of us (DSW) explored the question in a more general setting. The result is that altruism can evolve by group selection for a reasonable range of parameter values. The haystack model is not the stake through the heart of group selection that it was thought to be.

4. Multilevel selection theory is pluralistic

It is one thing to undermine fallacious arguments against group selection. It is something quite different to show that group selection has actually occurred and that it has been an important factor in the evolution of some traits. We attempt to do both in *Unto Others*. Sex ratio evolution is an especially well documented trait that has been influenced by group selection. But there are others — the evolution of reduced virulence in disease organisms, for example. Rather than discussing other examples, we want to make some general comments about the overall theory we are proposing.

First, our claim is not that *all* sex ratios in *all* populations are group adaptations. As Fisher argued, even sex ratios are plausibly regarded as individual adaptations. And as for the female-biased sex ratios found in nature, our claim is not that group selection was the *only* factor influencing their evolution. We do not claim that these groups have the smallest number of males consistent with all the females being fertilized. Rather, we claim that the biased sex ratios that evolve are *compromises* between the simultaneous and opposite influences of group and individual selection. Group selection rarely, if ever, occurs without individual selection occurring as well.

The more general point we want to emphasize is that hypotheses of group selection need to be evaluated on a trait-by-trait and a lineage-by-lineage basis. Group selection influenced sex ratio in some species, but not in others. And the fact that group selection did not influence sex ratio in human beings, for example, leaves open the question of whether group selection has been an important influence on other human traits. Unlike the monolithic theory of the selfish gene, which claims that *all* traits in *all* lineages evolved for the good of the genes, the theory we advocate, *multilevel selection theory,* is pluralistic. Different traits evolved because of different combinations of causes.

5. Group selection and human evolution

In *Unto Others*, we develop the conjecture that group selection was a strong force in human evolution. Group selection includes, but is not confined to, direct intergroup competition such as warfare. But, just as individual plants can compete with each other in virtue of the desert conditions in which they live (some being more drought-resistant than others), so groups can compete with each other without directly interacting (e.g., by some groups fostering co-operation more than others). In addition, cultural variation in addition to genetic variation can provide the mechanisms for phenotypic variation and heritability at the group level (see also Boyd and Richerson, 1985).

As noted earlier, the evolution of altruism depends on altruists interacting preferentially with each other. Kin selection is a powerful idea because interaction among kin is a pervasive pattern across many plant and animal groups. However, in many organisms, including especially human beings, individuals *choose* the individuals with whom they interact. If altruists seek out other altruists, this promotes the evolution of altruism. Although kin selection is a kind of group selection, there can be group selection that isn't kin selection; this, we suspect, is especially important in the case of human evolution. However, it isn't *uniquely* human — for example, even so-called lower vertebrates such as guppies can choose the social partners with which they interact.

An additional factor that helps altruism to evolve, which may be uniquely human, is the existence of cultural norms that impose social controls. Consider a very costly act, such as donating ten per cent of your food to the community. Since this act is very costly, a very strong degree of correlation among interactors will be needed to get it to evolve. However, suppose you live in a society in which individuals who make the donation are rewarded, and those who do not are punished. The act of donation has been transformed. It is no longer altruistic to make the donation, but selfish. Individuals in your group who donate do better than individuals who do not. However, it would be wrong to conclude from this that the existence of social controls make the hypothesis of group selection unnecessary. For where did the existence and enforcement of the social sanctions come from? Why do some individuals enforce the penalty for nondonation? This costs them something. A free-rider could enjoy the benefits without paying the costs of having a norm of donation enforced. Enforcing the requirement of donation is altruistic, even if donation is no longer altruistic. But notice that the cost of being an enforcer may be slight. It may not cost you anything like ten per cent of your food supply to help enforce the norm of donation. This means that the degree of correlation among interactors needed to get *this* altruistic behaviour to evolve is much less.

We believe that this argument may explain how altruistic behaviours were able to evolve in the genetically heterogeneous groups in which our ancestors lived. Human societies, both ancient and modern, are nowhere near as genetically uniform as bee hives and ant colonies. How, then, did co-operative behaviour manage to evolve in them? Human beings, we believe, did something that no other species was able to do. Social norms convert highly altruistic traits into traits that are selfish. And enforcing a social norm can involve a smaller cost than the required behaviour would have imposed if there were no norms. Social norms allow social organization to evolve by reducing its costs. Here again, it is important to recognize that culture allows a form of selection to occur whose elements may be found in the absence of culture. Bees 'police' the behaviour of other bees. What is uniquely human is the harnessing of socially shared values.

In addition to these rather 'theoretical' considerations, *Unto Others* also presents some observations that support the hypothesis that human beings are a group selected species. We randomly sampled twenty-five societies from the Human Relations Area File, an anthropological database, consulting what the files say about social norms. The actual contents of these norms vary enormously across our sample — for example, some societies encourage innovation in dress, while others demand uniformity. In spite of this diversity, cultural norms almost always require individuals to avoid conflict with each other and to behave benevolently towards fellow group members. Such constraints are rarely present with respect to outsiders, however. It also was striking how closely individuals can monitor the behaviour of group members in most traditional societies. Equally impressive is the emphasis on egalitarianism (among males — not, apparently, between males and females) found in many traditional societies; the norm was not that there should be complete equality, but that inequalities are permitted only when they enhance group functioning.

In addition to this survey data, we also describe a 'smoking gun' of cultural group selection — the conflict between the Nuer and Dinka tribes in East Africa. This conflict has been studied extensively by anthropologists for most of this century. The Nuer have gradually eroded the territory and resources of the Dinka, owing to the Nuer's superior group organization. The transformation was largely underwritten by people in Dinka villages defecting to the Nuers and being absorbed into their culture. We conjecture that this example has countless counterparts in the human past, and that the process of cultural group selection that it exemplifies has been an important influence on cultural change.

We think that Part I of *Unto Others* provides a solid foundation for the theory of group selection and that we have presented several well-documented cases of group selection in nonhuman species. Our discussion of human group selection is more tentative, but nonetheless we are prepared to claim that human beings have been strongly influenced by group selection processes.

III: Psychological Altruism — Part Two of *Unto Others*

Psychological egoism is a theory that claims that all of our ultimate desires are self-directed. Whenever we want others to do well (or badly), we have these other-directed desires only instrumentally; we care about what happens to others only because we think that the welfare of others has ramifications for ourselves. Egoism has exerted a powerful influence in the social sciences and has made large inroads in the thinking of ordinary people. In Part Two of *Unto Others*, we review the philosophical and psychological arguments that have been developed about egoism, both *pro* and *con*. We contend that these arguments are inconclusive. A new approach is needed; in Chapter 10, we present an evolutionary argument for thinking that some of our ultimate motives are altruistic.

It is easy to invent egoistic explanations for even the most harrowing acts of self-sacrifice. The soldier in a foxhole who throws himself on a grenade to save the lives of his comrades is a fixture in the literature on egoism. How could this act be a product of self-interest, if the soldier knows that it will end his life? The egoist may answer that the soldier realizes in an instant that he would rather die than suffer the guilt feelings that would haunt him if he saved himself and allowed his friends to perish. The soldier prefers to die and have no sensations at all rather than live and suffer the torments of the damned. This reply may sound *forced*, but this does not show that it must be *false*. And the fact that an egoistic explanation can be *invented* is no sure sign that egoism is *true*.

1. Clarifying egoism

When egoism claims that all our ultimate desires are self-directed, what do 'ultimate' and 'self-directed' mean?

There are some things that we want for their own sakes; other things we want only because we think they will get us something else. The crucial relation that we need to define is this:

S wants *m* solely as a means to acquiring *e* if and only if *S* wants *m*, *S* wants *e*, and *S* wants *m* only because she believes that obtaining *m* will help her obtain *e*.

An ultimate desire is a desire that someone has for reasons that go beyond its ability to contribute instrumentally to the attainment of something else. Consider pain. The most obvious reason that people want to avoid pain is simply that they dislike experiencing it. Avoiding pain is one of our ultimate goals. However, many people realize that being in pain reduces their ability to concentrate, so they may sometimes take an aspirin in part because they want to remove a source of distraction. This shows that the things we want as ends in themselves we also may want for instrumental reasons.

When psychological egoism seeks to explain why one person helped another, it isn't enough to show that *one* of the reasons for helping was self-benefit; this is quite consistent with there being another, purely altruistic, reason that the individual had for helping. Symmetrically, to refute egoism, one need not cite examples of helping in which *only* other-directed motives play a role. If people sometimes help for both egoistic and altruistic ultimate reasons, then psychological egoism is false.

Egoism and altruism both require the distinction between self-directed and other-directed desires, which should be understood in terms of a desire's propositional content. If Adam wants the apple, this is elliptical for saying that Adam wants it to be the case that *he has the apple*. This desire is purely self-directed, since its propositional content mentions Adam, but no other agent. In contrast, when Eve wants *Adam to have the apple*, this desire is purely other-directed; its propositional content mentions another person, Adam, but not Eve herself. Egoism claims that all of our ultimate desires are self-directed; altruism, that some are other-directed.

A special version of egoism is psychological hedonism. The hedonist says that the only ultimate desires that people have are attaining pleasure and avoiding pain. Hedonism is sometimes criticized for holding that pleasure is a single type of sensation — that the pleasure we get from the taste of a peach and the pleasure we get from seeing those we love prosper somehow boil down to the same thing (Lafollette, 1988). However, this criticism does not apply to hedonism as we have described it. The salient fact about hedonism is its claim that people are *motivational solipsists*; the only things they care about ultimately are states of their own consciousness. Although hedonists must be egoists, the reverse isn't true. For example, if people desire their own survival as an end in itself, they may be egoists, but they are not hedonists.

Some desires are neither purely self-directed nor purely other-directed. If Phyllis wants to be famous, this means that she wants others to know who she is. This desire's propositional content involves a relation between self and others. If Phyllis seeks fame solely because she thinks this will be pleasurable or profitable, then she may be an egoist. But what if she wants to be famous as an end in itself? There is no reason to cram this possibility into either egoism or altruism. So let us recognize *relationism* as a possibility distinct from both. Construed in this way, egoism avoids the difficulty of having to explain why the theory is compatible with the existence of some relational ultimate desires, but not with others (Kavka, 1986).

With egoism characterized as suggested, it obviously is not entailed by the truism that people act on the basis of their own desires, nor by the truism that they seek to have their desires satisfied. The fact that Joe acts on the basis of Joe's desires, not on the basis of Jim's, tells us *whose* desires are doing the work; it says nothing about whether the ultimate desires in Joe's head are *purely self-directed*. And the fact that

Joe wants his desires to be satisfied means merely that he wants their propositional contents to come true (Stampe, 1994). If Joe wants it to rain tomorrow, then his desire is satisfied if it rains, whether or not he notices the weather. To want one's desires satisfied is not the same as wanting the feeling of satisfaction that sometimes accompanies a satisfied desire.

Egoism is sometimes criticized for attributing too much calculation to spontaneous acts of helping. People who help in emergency situations often report doing so 'without thinking' (Clark and Word, 1974). However, it is hard to take such reports literally when the acts involve a precise series of complicated actions that are well-suited to an apparent end. A lifeguard who rescues a struggling swimmer is properly viewed as having a goal and as selecting actions that advance that goal. The fact that she engaged in no ponderous and self-conscious calculation does not show that no means/end reasoning occurred. In any case, actions that really do occur without the mediation of beliefs and desires fall outside the scope of both egoism and altruism.

A related criticism is that egoism assumes that people are more rational than they really are. However, recall that egoism is simply a claim about the ultimate desires that people have. As such, it says nothing about how people decide what to do on the basis of their beliefs and desires. The assumption of rationality is no more a part of psychological egoism than it is part of *motivational pluralism* — the view that people have both egoistic and altruistic ultimate desires.

2. Psychological arguments

It may strike some readers that deciding between egoism and motivational pluralism is easy. Individuals can merely gaze within their own minds and determine by introspection what their ultimate motives are. The problem with this easy solution is that there is no independent reason to think that the testimony of introspection is to be trusted in this instance. Introspection is misleading or incomplete in what it tells us about other facets of the mind; there is no reason to think that the mind is an open book with respect to the issue of ultimate motives.

In *Unto Others*, we devote most of Chapter 8 to the literature in social psychology that seeks to test egoism and motivational pluralism experimentally. The most systematic attempt in this regard is the work of Batson and co-workers, summarized in Batson (1991). Batson tests a hypothesis he calls the *empathy-altruism hypothesis* against a variety of egoistic explanations. The empathy-altruism hypothesis asserts that empathy causes people to have altruistic ultimate desires. We argue that Batson's experiments succeed in refuting some simple forms of egoism, but that the perennial problem of refuting egoism remains — when one version of egoism is refuted by a set of observations, another can be invented that fits the data. We also argue that even if Batson's experiments show that empathy causes helping, they don't settle whether empathy brings about this result by triggering an altruistic ultimate motive. We don't conclude from this that experimental social psychology will never be able to answer the question of whether psychological egoism is true. Our negative conclusion is more modest — empirical attempts to decide between egoism and motivational pluralism have not yet succeeded.

3. A bevy of philosophical arguments

Egoism has come under fire in philosophy from a number of angles. In Chapter 9 of *Unto Others*, we review these arguments and conclude that none of them succeeds. Here, briefly, is a sampling of the arguments we consider, and our replies:

— Egoism has been said to be *untestable*, and thus not a genuine scientific theory at all. We reply that if egoism is untestable, so is motivational pluralism. If it is true that when one egoistic explanation is discredited, another can be invented in its stead, then the same can be said of pluralism. The reason that egoism and pluralism have this sort of flexibility is that both make claims about the *kinds* of explanations that human behaviour has; they do not provide a detailed explanation of any particular behaviour. Egoism and pluralism are *isms*, which are notorious for the fact that they are not crisply falsifiable by a single set of observations.

— Joseph Butler (1692–1752) is widely regarded as having refuted psychological hedonism (Broad, 1965; Feinberg, 1984; Nagel, 1970). His argument can be outlined as follows:

1. People sometimes experience pleasure.

2. When people experience pleasure, this is because they had a desire for some external thing, and that desire was satisfied.

∴ Hedonism is false.

We think the second premise is false. It is overstated; although some pleasures are the result of a desire's being satisfied, others are not (Broad, 1965, p. 66). One can enjoy the smell of violets without having formed the desire to smell a flower, or something sweet. Since desires are propositional attitudes, forming a desire is a cognitive achievement. Pleasure and pain, on the other hand, are sometimes cognitively mediated, but sometimes they are not. This defect in the argument can be repaired; Butler does not need to say that desire satisfaction is the one and only road to pleasure. The main defect in the argument occurs in the transition from premises to conclusion. Consider the causal chain from a *desire* (the desire for food, say), to an *action* (eating), to a *result* — pleasure. Because the pleasure traces back to an antecedently existing desire, it will be false that the resulting pleasure caused the desire (on the assumption that cause must precede effect). However, this does not settle how two *desires* — the *desire for food* and the *desire for pleasure* — are related. Hedonism says that people desire food *because* they want pleasure (and think that food will bring them pleasure). Butler's argument concludes that this causal claim is false, but for no good reason. The crucial mistake in the argument comes from confusing two quite different items — the *pleasure* that results from a desire's being satisfied and the *desire for pleasure*. Even if the occurrence of pleasure presupposed that the agent desired something besides pleasure, nothing follows about the relationship between the *desire for pleasure* and the desire for something else (Sober, 1992; Stewart, 1992). Hedonism does not deny that people desire external things; rather, the theory tries to explain why that is so.

— We also consider the argument against egoism that Nozick (1974) presents by his example of an 'experience machine', the claim that hedonism is a paradoxical and irrational motivational theory, and the claim that egoism has the burden of proof. We conclude that none of these attacks on egoism is decisive.

There is one philosophical argument that attempts to support egoism, not refute it. This is the claim that egoism is preferable to pluralism because the former theory is more parsimonious. Egoism posits one type of ultimate desire whereas pluralism says there are two. We have two criticisms. First, this parsimony argument measures a theory's parsimony by counting the kinds of ultimate desires it postulates. The opposite conclusion would be obtained if one counted *causal beliefs*. The pluralist says that people want others to do well and that they also want to do well themselves. The egoist says that a person wants others to do well only because he or she *believes* that this will promote self-interest. Pluralism does not include this belief attribution. Our second objection is that parsimony is a reasonable tie-breaker when all other considerations are equal; it remains to be seen whether egoism and pluralism are equally plausible on all other grounds. In Chapter 10, we propose an argument to the effect that pluralism has greater evolutionary plausibility.

4. An evolutionary approach

Psychological motives are *proximate mechanisms* in the sense of that term used in evolutionary biology. When a sunflower turns towards the sun, there must be some mechanism inside the sunflower that causes it to do so. Hence, if phototropism evolved, a proximate mechanism that causes that behaviour also must have evolved. Similarly, if certain forms of helping behaviour in human beings are evolutionary adaptations, then the motives that cause those behaviours in individual human beings also must have evolved. Perhaps a general perspective on the evolution of proximate mechanisms can throw light on whether egoism or motivational pluralism was more likely to have evolved.

Pursuing this evolutionary approach does not presuppose that every detail of human behaviour, or every act of helping, can be explained completely by the hypothesis of evolution by natural selection. In Chapter 10, we consider a single fact about human behaviour, and our claim is that selection is relevant to explaining it. The phenomenon of interest is that human parents take care of their children; the average amount of parental care provided by human beings is strikingly greater than that provided by parents in many other species. We will assume that natural selection is at least part of the explanation of why parental care evolved in our lineage. This is not to deny that human parents vary; some take better care of their children than others, and some even abuse and kill their offspring. Another striking fact about individual variation is that mothers, on average, expend more time and effort on parental care than fathers. Perhaps there are evolutionary explanations for these individual differences as well; the question we want to address here, however, makes no assumption as to whether this is true.

In Chapter 10, we describe some general principles that govern how one might predict the proximate mechanism that will evolve to cause a particular behaviour. We develop these ideas by considering the example of a marine bacterium whose problem is to avoid environments in which there is oxygen. The organism has evolved a particular behaviour — it tends to swim away from greater oxygen concentrations and towards areas in which there is less. What proximate mechanism might have evolved that allows the organism to do this?

First, let's survey the range of possible design solutions that we need to consider. The most obvious solution is for the organism to have an oxygen detector. We call this the *direct solution* to the design problem; the organism needs to avoid oxygen and it solves that problem by detecting the very property that matters.

It isn't hard to imagine other solutions to the design problem that are less direct. Suppose that areas near the pond's surface contain more oxygen and areas deeper in the pond contain less. If so, the organism could use an up/down detector to make the requisite discrimination. This design solution is *indirect*; the organism needs to distinguish high oxygen from low and accomplishes this by detecting another property that happens to be correlated with the target. In general, there may be many indirect design solutions that the organism could exploit; there are as many indirect solutions as there are correlations between oxygen level and other properties found in the environment. Finally, we may add to our list the idea that there can be *pluralistic* solutions to a design problem. In addition to the monistic solution of having an oxygen detector and the monistic solution of having an up/down detector, an organism might deploy both.

Given this multitude of possibilities, how might one predict which of them will evolve? Three principles are relevant — *availability*, *reliability*, and *efficiency*.

Natural selection acts only on the range of variation that exists ancestrally. An oxygen detector might be a good thing for the organism to have, but if that device was never present as an ancestral variant, natural selection cannot cause it to evolve. So the first sort of information we'd like to have concerns which proximate mechanisms were *available* ancestrally.

Let's suppose for the sake of argument that both an oxygen detector and an up/down detector are available ancestrally. Which of them is more likely to evolve? Here we need to address the issue of *reliability*. Which device does the more reliable job of indicating where oxygen is? Without further information, not much can be said. An oxygen detector may have any degree of reliability, and the same is true of an up/down detector. There is no *a priori* reason why the direct strategy should be more or less reliable than the indirect strategy. However, there is a special circumstance in which they will differ. It is illustrated by the following diagram:

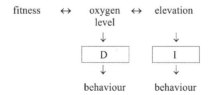

The double arrows indicate correlation; avoiding oxygen is correlated with fitness, and elevation is correlated with oxygen level. In the diagram, there is no arrow from elevation to fitness except the one that passes through oxygen level. This means that elevation is correlated with fitness *only because* elevation is correlated with oxygen, and oxygen is correlated with fitness. There is no *a priori* reason why this should be true. For example, if there were more predators at the bottom of ponds than at the top, then elevation would have two sorts of relevance for fitness. However, if oxygen level 'screens off' fitness from elevation in the way indicated, we can state the following principle about the reliability of the direct device D and the indirect device I:

(D/I) If oxygen level and elevation are less than perfectly correlated, and if D detects
       oxygen level at least as well as I detects elevation, then D will be more reliable
       than I.

This is the Direct/Indirect Asymmetry Principle. Direct solutions to a design problem
aren't always more reliable, but they are more reliable in this circumstance.

A second principle about reliability also can be extracted from this diagram. Just as
scientists do a better job of discriminating between hypotheses if they have more evidence rather than less, so it will be true that the marine bacterium we are considering
will make more reliable discriminations about where to swim if it has two sources of
information rather than just one:

(TBO) If oxygen level and elevation are less than perfectly correlated, and if D and I
      are each reliable, though fallible, detectors of oxygen concentration, then D
      and I working together will be more reliable than either of them working alone.

This is the Two-is-Better-than-One Principle. It requires an assumption — that the
two devices do not interfere with each other when both are present in an organism.

The D/I Asymmetry and the TBO Principle pertain to the issue of reliability. Let us
now turn to the third consideration that is relevant to predicting which proximate
mechanism will evolve, namely *efficiency*. Even if an oxygen detector and an elevation detector are both available, and even if the oxygen detector is more reliable, it
doesn't follow that natural selection will favour the oxygen detector. It may be that an
oxygen detector requires more energy to build and maintain than an elevation detector. Organisms run on energy no less than automobiles do. Efficiency is relevant to a
trait's overall fitness just as much as its reliability is.

With these three considerations in hand, let's return to the problem of predicting
which motivational mechanism for providing parental care is likely to have evolved
in the lineage leading to human beings. The three motivational mechanisms we need
to consider correspond to three different rules for selecting a behaviour in the light of
what one believes:

(HED) Provide parental care if, and only if, doing so will maximize pleasure and
      minimize pain.

(ALT) Provide parental care if, and only if, doing so will advance the welfare of one's
      children.

(PLUR) Provide parental care if, and only if, doing so will either maximize pleasure
       and minimize pain, or will advance the welfare of one's children.

(ALT) is a relatively direct, and (HED) is a relatively indirect, solution to the design
problem of getting an organism to take care of its offspring. Just as our marine bacterium can avoid oxygen by detecting elevation, so it is possible in principle for a
hedonistic organism to provide parental care; what is required is that the organism be
so constituted that providing parental care is the thing that usually maximizes its
pleasure and minimizes its pain (or that the organism believes that this is so).

Let's consider how reliable these three mechanisms will be in a certain situation.
Suppose that a parent learns that its child is in danger. Imagine that your neighbour
tells you that your child has just fallen through the ice on a frozen lake. Here is how
(HED) and (ALT) will do their work:

child needs help → parent believes child needs help → parent feels anxiety and fear

The altruistic parent will be moved to action just by virtue of believing that its child needs help. The hedonistic parent will not; rather, what moves the hedonistic parent to action are the feelings of anxiety and fear that are caused by the news. It should be clear from this diagram that the (D/I) Asymmetry Principle applies; (ALT) will be more reliable than (HED). And by the (TBO) Principle, (PLUR) will do better than both. In this example, hedonism comes in last in the three-way competition, at least as far as reliability is concerned.

The important thing about this example is that the feelings that the parent has are *belief mediated*. The only reason the parent *feels* anxiety and fear is that the parent *believes* that its child is in trouble. This is true of many of the situations that egoism and hedonism are called upon to explain, but it is not true of all. For example, consider the following situation in which pain is a direct effect, and belief a relatively indirect effect, of bodily injury:

fingers are burned → pain → belief that one's fingers have been injured

In this case, hedonism is a direct solution to the design problem; it would be a poor engineering solution to have the organism be unresponsive to pain and to have it withdraw its fingers from the flame only after it forms a belief about bodily injury. In this situation, *belief is pain-mediated* and the (D/I) Asymmetry Principle explains why a hedonistic focus on pain makes sense. However, the same principle indicates what is misguided about hedonism as a design solution when *pain is belief-mediated*, which is what occurs so often in the context of parental care.

If hedonism is less reliable than both pure altruism and motivational pluralism, how do these three mechanisms compare when we consider their availability and efficiency? With respect to availability, we make the following claim: *if hedonism was available ancestrally, so was altruism*. The reason is that the two motivational mechanisms differ in only a modest way. Both require a belief/desire psychology. And both the hedonistic and the altruistic parent want their children to do well; the only difference is that the hedonist has this propositional content as an instrumental desire while the altruist has it as an ultimate desire. If altruism and pluralism did not evolve, this was not because they were unavailable as variants for selection to act upon.

What about efficiency? Does it cost more calories to build and maintain an altruistic or a pluralistic organism than it does to build and maintain a hedonist? We don't see why. What requires energy is building the hardware that implements a belief/desire psychology. However, we doubt that it makes an energetic difference whether the organism has one ultimate desire rather than two. People with more beliefs apparently don't need to eat more than people with fewer. The same point seems to apply to the issue of how many, or which, ultimate desires one has.

In summary, pure altruism and pluralism are both more reliable than hedonism as devices for delivering parental care. And, with respect to the issues of availability and efficiency, we find no difference among these three motivational mechanisms. This suggests that natural selection is more likely to have made us motivational pluralists than to have made us hedonists.

From an evolutionary point of view, hedonism is a bizarre motivational mechanism. What matters in the process of natural selection is an organism's ability to survive and be reproductively successful. Reproductive success involves not just the production of offspring, but the survival of those offspring to reproductive age. So what matters is the survival of one's own body and the bodies of one's children. Hedonism, on the other hand, says that organisms care ultimately about the states of their own consciousness, and about that alone. Why would natural selection have led organisms to care about something that is peripheral to fitness, rather than have them set their eyes on the prize? If organisms were unable to conceptualize propositions about their own bodies and the bodies of their offspring, that might be a reason. After all, it might make sense for an organism to exploit the indirect strategy of deciding where to swim on the basis of elevation rather than on the basis of oxygen concentration, if the organism cannot detect oxygen. But if an organism is smart enough to form representations about itself and its offspring, this justification of the indirect strategy will not be plausible. The fact that we evolved from ancestors who were cognitively less sophisticated makes it unsurprising that avoiding pain and attaining pleasure are two of our ultimate goals. But the fact that human beings are able to form representations with so many different propositional contents suggests that evolution supplemented this list of what we care about as ends in themselves.

IV: Evolutionary Altruism, Psychological Altruism, and Ethics

The study of ethics has a *normative* and a *descriptive* component. Normative ethics seeks to say what is good and what is right; it seeks to identify what we are obliged to do and what we are permitted to do. Descriptive ethics, on the other hand, is neutral on these normative questions; it attempts to *describe* and *explain* morality as a cultural phenomenon, not *justify* it. How does morality vary within and across cultures, and through time? Are there moral ideas that constitute cultural universals? And how is one to explain this pattern of variation?

Although we think our work on evolutionary and psychological altruism bears on these questions, we also think that it is important not to blur the problems. Psychological altruism is not the same as morality. And an explanation of why human beings hold a moral principle is not, in itself, a justification (or a refutation) of that principle.

We say that psychological altruism is not the same as morality because individuals can have concerns about the welfare of specific others without their formulating those concerns in terms of ethical principles. A mother chimp may want her offspring to have some food, but this does not mean that she thinks that all chimps should be well-fed, or that all mothers should take care of their offspring. Egoistic and altruistic desires are both desires about specific individuals. Having self-directed preferences is not sufficient for having a morality; the same goes for other-directed preferences.

Why, then, did morality evolve? People can have specific likes and dislikes without this producing a socially shared moral code. And if everyone dislikes certain things, what is the point of there being a moral code that says that those things should be shunned? If everyone hates sticking pins in their toes, what is the point of an ethic that tells people that it is wrong to stick pins in their toes? And if parents invariably love their children, what would be the point of having a moral principle that tells parents that they ought to love their children? Behaviours that people do spontaneously by virtue of their own desires don't need to have a moral code laid on top of them. The obvious suggestion is that the social function of morality is to get people to do things that they would not otherwise be disposed to do, or to strengthen dispositions that people already have in weaker forms. Morality is not a mere redundant overlay on the psychologically altruistic motives we may have.

Functionalism went out of style in anthropology and other social sciences in part because it was hard to see what feedback mechanism might make institutions persist or disappear. Even if religion promotes group solidarity, how would that explain the persistence of religion? The idea of selection makes this question tractable. We hope that *Unto Others* will allow social scientists to explore the hypothesis that morality is a group adaptation. We do not deny that moral principles have functioned as ideological weapons, allowing some individuals to prosper at the expense of others in the same group. However, the hypothesis that moralities sometimes persist and spread because they benefit the group is not mere wishful thinking. Darwin's idea that features of morality can be explained by group selection needs to be explored.

What, if anything, do the evolutionary and psychological issues we discuss in *Unto Others* contribute to normative theory? Every normative theory relies on a conception of human nature. Sometimes this is expressed by invoking the *ought implies can principle*. If people ought to do something, then it must be possible for them to do it. Human nature circumscribes what is possible. We do not regard human nature as unchangeable. In part, this is because evolution isn't over. Genetic and cultural evolution will continue to modify the capacities that people have. But if we want to understand the capacities that people *now* have, surely an understanding of our evolutionary past is crucial. One lesson that may flow from the evolutionary and psychological study of altruism is that prisoners' dilemmas are in fact rarer than many researchers suppose. Decision theory says that it is irrational to co-operate (to act altruistically) in one-shot prisoners' dilemmas. However, perhaps some situations that appear to third parties to be prisoners' dilemmas really are not. Payoffs are usually measured in dollars, or in other tangible commodities. But if people sometimes care about each other, and not just about money, they are not irrational when they choose to co-operate in such interactions. Narrow forms of egoism make such behaviours appear irrational. Perhaps the conclusion to draw is not that people *are* irrational, but that the assumption of egoism needs to be rethought.

References

Axelrod, Robert (1984), *The Evolution of Co-operation* (New York: Basic Books).
Batson, C. Daniel (1991), *The Altruism Question: Toward A Social-Psychological Answer* (Hillsdale, NJ: Lawrence Erlbaum Associates).

Boyd, Robert and Richerson, Peter (1985), *Culture and the Evolutionary Process* (Chicago: University of Chicago Press).

Broad, C.D (1965), *Five Types of Ethical Theory* (Totowa, NJ: Littlefield, Adams).

Butler, Joseph (1726), *Fifteen Sermons preached at the Rolls Chapel*, reprinted in part in *British Moralists*, Volume 1, ed. L.A. Selby-Bigge (New York: Dover Books, 1965; originally published Oxford: The Clarendon Press, 1897).

Clark, R.D. and Word, L.E. (1974), 'Where is the apathetic bystander? Situational characteristics of the emergency', *Journal of Personality and Social Psychology*, **29**, pp. 279–87.

Darwin, Charles (1871), *The Descent of Man and Evolution in Relation to Sex* (London: Murray).

Dawkins, Richard (1976), *The Selfish Gene* (New York: Oxford University Press).

Feinberg, J. (1984), 'Psychological egoism', in *Reason at Work*, ed. S. Cahn, P. Kitcher and G. Sher (San Diego, Calif.: Harcourt Brace and Jovanovich), pp. 25–35.

Fisher, Ronald (1930), *The Genetical Theory of Natural Selection* (New York: Dover, 1958).

Hamilton, W.D. (1964), 'The Genetical evolution of social behaviour I and II', *Journal of Theoretical Biology*, **7**, pp. 1–16, pp. 17–52.

Hamilton, W.D. (1967), 'Extraordinary sex ratios', *Science*, **156**, pp. 477–88.

Hamilton, W.D. (1975), 'Innate social aptitudes of man — an approach from evolutionary genetics', in *Biosocial Anthropology*, ed. R. Fox (New York: John Wiley) pp. 133–15 .

Kavka, Gregory (1986), *Hobbesian Moral and Political Theory* (Princeton, NJ: Princeton University Press).

Lafollette, Hugh (1988), 'The truth in psychological egoism', in *Reason and Responsibility*, 7th edition, ed. J. Feinberg (Belmont, Calif.: Wadsworth), pp. 500–7.

Maynard Smith, John (1964), 'Group selection and kin selection', *Nature*, **201**, pp. 1145–6.

Maynard Smith, John and Price, George (1973), 'The logic of animal conflict', *Nature*, **246**, pp.15–18.

Nagel, Thomas (1970), *The Possibility of Altruism* (Oxford: Oxford University Press).

Nozick, Robert (1974), *Anarchy, State, and Utopia* (New York: Basic Books).

Sober, Elliott (1992), 'Hedonism and Butler's stone', *Ethics*, **103**, pp. 97–103.

Sober, Elliott and Wilson, David Sloan (1998), *Unto Others: The Evolution and Psychology of Unselfish Behavior* (Cambridge, MA: Harvard University Press).

Stampe, Dennis (1994), 'Desire', in *A Companion to the Philosophy of Mind*, ed. S. Guttenplan (Cambridge, Mass.: Basil Blackwell), pp. 244–50.

Stewart, R.M. (1992), 'Butler's argument against psychological hedonism', *Canadian Journal of Philosophy*, **22**, pp. 211–21.

Williams, George C. (1966), *Adaptation and Natural Selection* (Princeton, NJ: Princeton University Press).

[4]

Game Theory, Rationality and Evolution of the Social Contract

Brian Skyrms

Game theory based on rational choice is compared with game theory based on evolutionary, or other adaptive, dynamics. The Nash equilibrium concept has a central role to play in both theories, even though one makes extremely strong assumptions about cognitive capacities and common knowledge of the players, and the other does not. Nevertheless, there are also important differences between the two theories. These differences are illustrated in a number of games that model types of interaction that are key ingredients for any theory of the social contract.

Introduction

The Theory of Games was conceived by von Neumann and Morgenstern as a theory of interactive decisions for ideally rational agents. So was the theory of the social contract, from the Sophists to Thomas Hobbes. Hobbes wanted to bring the rigor and certainty of Euclidean geometry to social philosophy. If he fell somewhat short of this goal, even by the standards of his own time, perhaps the theory of games could be utilized to complete the project. This idea has been pursued in different ways by John Harsanyi, John Rawls, David Gauthier and Ken Binmore.

Rational Choice

But the foundation of the classical theory of games on the theory of rational decision has itself proved more complicated than it seemed at the time. Von Neumann and Morgenstern thought of the theory of rationality as work in progress which, when complete, would specify a unique correct rational act for any decision maker in any decision situation. If such a theory of rationality were in hand, then it appears that Nash equilibrium would be the unique outcome of the decisions of rational agents in an interactive decision situation. A *Nash equilibrium* is defined as a specification of an act for each agent, such that no agent can gain by unilateral change of her act. Now, the argument goes, each decision maker can figure out what the other decision makers will do from the theory of rationality. Thus, the only state consistent with rationality is that each player maximizes payoff given knowledge of what the other players do —

a Nash equilibrium. You can find this argument applied within the theory of two-person zero-sum games — within which it makes a good deal of sense — in *The Theory of Games and Economic Behavior* (von Neumann & Morgenstern, 1947), p. 148.

There are some tacit assumptions to the argument. In the first place, it is not only assumed that each agent is rational, but also that every agent knows so — in order to deduce the actions of the others from the theory. But, if the theory is predicated on all the agents being able to deduce the actions of the others, then agents must not only know that others are rational, but also know that others know that they are — otherwise others might not use the theory, and so forth. That is, we assume not only that all the agents are rational, but also that rationality is *common knowledge*. Likewise we must assume that the agents correctly identify the decision situation to which they are applying the theory, and that they know that all others do so as well, and that they know that all others know this, and so forth.

In the second place, it is assumed that the theory of rationality singles out a unique correct act. This is 'almost' true in the theory of zero-sum games. Where multiple alternatives are allowed, they are interchangeable in an appropriate sense. But it is wildly false when we move into the territory of non-zero sum games. For instance, consider the following co-ordination game. Two players each independently pick from a list of three colours. If they pick the same, they win a prize; otherwise they lose. The *symmetry* of the situation precludes any reasonable unique prescription by any theory of rational choice.

Since the payoffs are invariant under permutations of colours and the colours themselves are assumed to have no significance other than as labels, the strategy that is uniquely recommended must also be invariant under permutations of colours. The only Nash equilibrium invariant in this way consists in each player independently randomizing between the three colours with equal probabilities. This is a unique prescription, but hardly a reasonable one since it leads the players to mis-coordinate two-thids of the time. This is the Curse of Symmetry that plagues theories that, like Harsanyi and Selten (1988), pursue the goal of a uniquely selected rational act for each player.

Once the hope of a theory of rationality that always singles out a unique rational act is given up, the von Neumann-Morgenstern justification of Nash equilibrium unravels. Perhaps I choose Red, thinking that you will choose Red and you choose Green, thinking that I will choose Green. We are acting rationally, given our beliefs, but we mis-coordinated and are not at a Nash equilibrium.

It is time to back up and see what sort of results we can get from assumptions of rational choice. To do so, we cannot continue to be vague about the term 'rationality'. We say that, for a given player, act B is *weakly dominated* by act A, if, no matter what the other players do, act A leads a payoff at least as great as act B does, and for *some* combination of acts by other players, act A leads to a greater payoff than act B. We say that act A *strongly* dominates act B if, for *all* combinations of acts by other players, act A leads to a greater payoff than act B. We say that a player is *Bayes-rational* if she acts so as to maximize her expected payoff, given her own degrees of belief and her own evaluation of consequences. If a player is *Bayes-rational*, she will not choose a *strongly dominated* act because the act that strongly dominates it must have greater expected payoff, no matter what her degrees-of-belief about what other players will do. If a player is *Bayes rational* and, in addition, gives every combination of other players' actions non-zero probability, then she will not choose a *weakly dominated* act, because in this case the act that weakly dominates will have greater expected payoff.

Suppose that the setting for rational choice based game theory is one where not only are all the players Bayes-rational, but furthermore it is common knowledge among the players that they are all Bayes-rational. Assume also that the game being played is also common knowledge. What does this tell us about the outcomes? Each player's play must maximize expected payoff given her beliefs about others players' play, and each player's beliefs about other players' beliefs and play must be consistent with them maximizing expected payoff, and each player's beliefs about other players' beliefs about her beliefs and her play must be consistent with her maximizing expected payoff, and so forth to arbitrary high levels. Bernheim (1984) and Pearce (1984) investigate strategies that are consistent with common knowledge of rationality, which they call *rationalizable* (or 'correlated rationalizable' in the general case). They show that such strategies are those that remain after *iterated deletion of strongly dominated strategies*. (You identify all the dominated strategies for all the players and delete them. After they are deleted, other strategies may become strongly dominated. Then you delete those, and so forth until you come to a stage where there are no strongly dominated strategies to delete.)

Common knowledge of Bayesian rationality imposes much weaker constraints on play than Nash equilibrium. In our co-ordination game, *any* profile of play is consistent with common knowledge of Bayesian rationality. But in some situations, it leads to a unique outcome. Consider the following 'Beauty Contest' game (Moulin, 1986; Nagel, 1995). A finite number of people play. Each names an integer from 0 to 100. The person who is closest to half the mean wins a prize, others get nothing. (In case of a tie, the prize is shared.) In the beauty contest game, the prize is independent of the actions of the players.[1] The unique outcome consistent with common knowledge of Bayesian rationality is that each person chooses zero.

First, notice that the highest that half of the mean could be (if everyone chose 100) is fifty, so it makes no sense choosing a number greater than fifty. But everyone knows this, so no one will choose more than fifty, in which case it makes no sense choosing a number greater than twenty-five. Iterate this argument, and any number greater than zero gets eliminated as a rational choice. This shows that iterated deletion of weakly dominated strategies eliminates everything but zero. A slightly more complex argument using mixed strategies can be given to show that iterated elimination of strongly dominated strategies eliminates everything but zero. Everyone choosing zero is the unique outcome of this game consistent with common knowledge of Bayesian rationality. If you try this experiment in a room full of people, no matter how sophisticated, you will not get the result predicted by common knowledge of rationality. This is well-documented in the experimental literature (Nagel, 1995; Ho, Weigelt and Camerer, 1996; Stahl, 1996; Camerer, 1997; Duffy and Nagel, 1997).

These are called 'Beauty Contest' games because they use the kind of iterated reasoning that Keynes described in a famous analogy to the stock market:

> . . . professional investment may be likened to those newspaper competitions in which the competitors have to pick out the six prettiest faces from a hundred photographs, the prize being awarded to the competitor whose choice most nearly corresponds to the

---

[1] This sentence was added during peer commentary in response to a query from Herbert Gintis.

average preferences of the competitors as a whole . . . . It is not a case of choosing which, to the best of one's judgement, are really the prettiest, nor even those which average opinion thinks are the prettiest. We have reached the third degree where we devote our intelligences to anticipating what the average opinion expects the average opinion to be. There are some, I believe, who practice the fourth, fifth and higher degrees (pp. 155–6).

Common knowledge of rationality requires arbitrarily high degrees of this reasoning.

Is it so surprising that people typically fail to satisfy the assumption of common knowledge of Bayesian rationality? Even to state the assumption requires an infinite hierarchy of degrees of belief, degrees of belief over degrees of belief, and so on. The model can be shown to be mathematically consistent, but it is an abstraction far removed from reality. One might retreat from the assumption of common knowledge of Bayesian rationality and try to base game theory on the simple assumption that the players *are* Bayesian rational. The resulting theory is so weak as to be hardly worth pursuing. It says that players will not choose strongly dominated acts — acts with an alternative that carries a better payoff better no matter what the other players do. (This assumption itself is difficult to square with some experimentally observed behaviour.) One is reminded of Spinoza's characterization of his predecessors: ' . . . they conceive of men, not as they are, but as they themselves would like them to be . . . '.

Adaptive Dynamics

Hobbes wanted a theory of the social contract based in self-interested rational choice. David Hume represents a different tradition. For Hume, the social contract is a tissue of conventions which have grown up over time. I cannot resist reproducing in full this marvellously insightful passage from his *Treatise:*

> Two men who pull on the oars of a boat do it by an agreement or convention, tho' they have never given promises to each other. Nor is the rule concerning the stability of possession the less deriv'd from human conventions, that it arises gradually, and acquires force by a slow progression, and by our repeated experience of the inconveniences of transgressing it. On the contrary, this experience assures us still more, that the sense of interest has become common to all our fellows, and gives us a confidence of the future regularity of their conduct: And 'tis only on the expectation of this, that our moderation and abstinence are founded. In like manner are languages gradually establish'd by human conventions without any promise. In like manner do gold and silver become the common measures of exchange, and are esteem'd sufficient payment for what is of a hundred times their value (p. 490).

Hume is interested in how we actually got the contract we now have. He believes that we should study the processes that lead to a gradual establishment of social norms and conventions. Modern Humeans, such as Sugden, Binmore, Gibbard, take inspiration as well from Darwinian dynamics. The social contract has evolved, and will continue to evolve. Different cultures, with their alternative social conventions, may be instances of different equilibria, each with its own basin of attraction. The proper way to pursue modern Humean social philosophy is via dynamic modelling of cultural evolution and social learning.

In Evolutionary theory, as in classical game theory, there is strategic interaction in the form of frequency-dependent selection, but there is no presumption of rationality — let alone common knowledge of rationality — operative. Indeed the organisms that are evolving may not even be making decisions at all. Nevertheless, game theoretic ideas have been fruitfully applied to strategic interaction in evolution. And the key equilibrium concept is almost the same.

The most striking fact about the relationship between evolutionary game theory and economic game theory is that, at the most basic level, a theory built of hyper-rational actors and a theory built of possibly non-rational actors are in fundamental agreement. This fact has been widely noticed, and its importance can hardly be over-estimated. Criticism of game theory based on the failure of rationality assumptions must be reconsidered from the viewpoint of adaptive processes. There are many roads to the Nash equilibrium concept, only one of which is based on highly idealized rationality assumptions.

However, as we look more closely at the theory in the evolutionary and rational choice settings, differences begin to emerge. At the onset, a single population evolutionary setting imposes a symmetry requirement which selects Nash equilibria which might appear implausible in other settings. Furthermore, refinements of the Nash equilibrium are handled differently. Standard evolutionary dynamics, the replicator dynamics, does not guarantee elimination of weakly dominated strategies. In a closely related phenomenon, when we consider extensive form games, evolutionary dynamics need not eliminate strategies which fail the test of sequential rationality. Going further, we shall see that if we generalize the theories to allow for correlation, we find that the two theories can diverge dramatically. Correlated evolutionary game theory can even allow for the fixation of strongly dominated strategies. These are strategies which fail under even the weakest theory of rational choice — the theory that players are in fact Bayes rational.

The situation is therefore more complicated than it might at first appear. There are aspects of accord between evolutionary game theory and rational game theory as well as areas of difference. This is as true for cultural evolution as for biological evolution. The phenomena in question thus have considerable interest for social and political philosophy, and touch some recurrent themes in Hobbes and Hume.

Evolutionary Game Theory

Let us consider the case of individuals who are paired at random from a large population to engage in a strategic interaction. Reproduction is asexual and individuals breed true. We measure the payoffs in terms of evolutionary fitness — expected number of offspring. Payoff to a strategy depends on what strategy it is paired against, so we have frequency dependent selection in the population. We write the payoff of strategy A when played against strategy B as $U(A|B)$. Under these assumptions, the expected Fitness for a strategy is an average of its payoffs against alternative strategies weighted by the population proportions of the other strategies:

$$U(A) = SUM_i \, U(A|B_i) \, P(B_i)$$

The expected fitness of the population is the average of the fitnesses of the strategies with the weights of the average being the population proportions:

$$UBAR = SUM_j \, U(A_j) \, P(A_j)$$

We postulate that the population is large enough that we may safely assume that you get what you expect.

274   B. SKYRMS

The population dynamics is then deterministic. Assuming discrete generations with one contest per individual per generation, we get:

DARWIN MAP: $P'(A) = P(A) U(A)/UBAR$

where $P(A)$ is the proportion of the population today using strategy A and P' is the proportion tomorrow. Letting the time between generations become small, we find the corresponding continuous dynamics:

DARWIN FLOW: $dP(A)/dt = P(A) [U(A)-UBAR]/UBAR$

The Darwin flow has the same orbits as the simpler dynamics obtained by discarding the denominator, although the velocity along the orbits may differ:

REPLICATOR FLOW: $dP(A)/dt = P(A) [U(A)-UBAR]$

In the following discussion, we will concentrate on this replicator dynamics. It is of some interest that the replicator dynamics emerges naturally from a number of different models of cultural evolution based on imitation (Binmore, Gale and Samuelson, 1995; Bjornerstedt and Weibull, 1995; Sacco, 1995; Samuelson, 1997; Schlag, 1994; 1996).

The replicator flow was introduced by Taylor and Jonker (1978) to build a foundation for the concept of evolutionarily stable strategy introduced by Maynard-Smith and Price (1973). The leading idea of Maynard-Smith and Price was that of a strategy such that if the whole population uses that strategy, it cannot be successfully invaded. That is to say, that if the population were invaded by a very small proportion of individuals playing a different strategy, the invaders would have a smaller average payoff than that of the natives. The definition offered is that A is evolutionarily stable just in case both:

(i) $U(A|A) \geq U(B|A)$ (for all B different from A) and

(ii) If $U(A|A) = U(B|A)$ then $U(A|B)>U(B|B)$

The leading idea of Maynard-Smith and Price only makes sense for mixed strategies if individuals play randomized strategies. But the replicator dynamics is appropriate for a model where individuals play pure strategies, and the counterparts of the mixed strategies of game theory are polymorphic states of the population. We retain this model, but introduce the notion of an evolutionarily stable state of the population. It is a polymorphic state of the population, which would satisfy the foregoing inequalities if the corresponding randomized strategies were used. An evolutionarily stable *strategy* corresponds to an evolutionarily stable *state* (ESS) in which the whole population uses that strategy.

Nash from Nature

Condition (i) in the definition of ESS looks a lot like the definition of Nash equilibrium. It is, in fact, the condition that $<A, A>$ is a Nash equilibrium of the associated two person game in which both players have the payoffs specified in by the fitness matrix, $U(\_|\_)$. the second condition adds a kind of stability requirement. The requirement is sufficient to guarantee strong dynamical stability in the replicator dynamics:

EVERY ESS is a STRONGLY DYNAMICALLY STABLE (or ATTRACTING) EQUILIBRIUM in the REPLICATOR DYNAMICS

This is more than sufficient to guarantee Nash equilibrium in the corresponding game:

> IF A is a DYNAMICALLY STABLE EQUILIBRIUM in the REPLICATOR DYNAMICS then <A,A> is a NASH EQUILIBRIUM of the corresponding TWO-PERSON GAME.

(The converse of each of the foregoing propositions fails. For more details see Hofbauer and Sigmund, 1988; van Damme, 1987; Weibull 1997).

Evidently, Nash equilibrium has an important role to play here, in the absence of common knowledge of rationality or even rationality itself. The reason is quite clear, but nevertheless deserves to be emphasized. The underlying dynamics is adaptive: it has a tendency towards maximal fitness. Many alternative dynamics — of learning as well as of evolution — share this property. (See Borgers and Sarin, 1997, for a connection between reinforcement learning and replicator dynamics, but compare the discussion of Fudenberg and Levine, 1998, Ch. 3). There is a moral here for philosophers and political theorists who have attacked the theory of games on the basis of its rationality assumptions. Game theory has a far broader domain of application than that suggested by its classical foundations.

Symmetry

For every evolutionary game, there is a corresponding symmetric two-person game, and for every ESS in the Evolutionary game, there is a corresponding symmetric Nash equilibrium in the two-person game. Symmetry is imposed on the Nash equilibrium of the two-person game because the players have no identity in the evolutionary game. Different individuals play the game. The things which have enduring identity are the strategies. Evolutionary games are played under what I call 'The Darwinian Veil of Ignorance' in Chapter I of my book, *Evolution of the Social Contract*. Evolution ignores individual idiosyncratic concerns simply because individuals do not persist through evolutionary time.

For example, consider the game of Chicken. There are two strategies: Swerve; Don't. The fitnesses are:

$$U(S|S) = 20$$
$$U(S|D) = 15$$
$$U(D|S) = 25$$
$$U(D|D) = 10$$

In the two-person game, there are two Nash equilibria in pure strategies: player one swerves and player two doesn't, player two swerves and player one doesn't. There is also a mixed Nash equilibrium with each player having equal chances of swerving. In the evolutionary setting, there are just swervers and non-swervers. The only equilibrium of the two-person game that corresponds to an ESS of the evolutionary game is the mixed strategy. It corresponds to an evolutionary stable polymorphic state where the population is equally split between swervers and non-swervers. Any departure from the state is rectified by the replicator dynamics, for it is better to swerve when the majority don't and better not to swerve when the majority do.

The evolutionary setting has radically changed the dynamical picture. If we were considering learning dynamics for two fixed individuals, the relevant state space would be the unit square with the x-axis representing the probability that player one would swerve and the y-axis representing the probabilities that player two would swerve. With any reasonable learning dynamics, the mixed equilibrium of the two-person game would be highly unstable and the two pure equilibria would be strongly stable. The move to the evolutionary setting in effect restricts the dynamics to the diagonal of the unit square. The probability that player one will encounter a given strategy must be the same as the probability that player two will. It is just the proportion of the population using that strategy. On the diagonal, the mixed equilibrium is now strongly stable.

For another example which is of considerable importance for social philosophy, consider the simplest version of the Nash bargaining game. Two individuals have a resource to divide. We assume that payoff just equals the amount of the resource. They each must independently state the minimal fraction of the resource that they will accept. If these amounts add up to more than the total resource, there is no bargain struck and each players gets nothing. Otherwise, each gets what she asks for.

This two person game has an infinite number of Nash equilibria of the form: Player one demands x of the resource and player two demands (1-x) of the resource (0<x<1). Each of these Nash equilibria is strict — which is to say that a unilateral deviation from the equilibrium not only produces no gain; it produces a positive loss. Here we have the problem of multiple Nash equilibria in especially difficult form. There are an infinite number of equilibria and, being strict, they satisfy all sorts of refinements of the Nash equilibrium concept (see van Damme, 1987).

Suppose we now put this game in the evolution context that we have developed. What pure strategies are evolutionarily stable? There is exactly one: Demand half! First, it is evolutionarily stable. In a population in which all demand half, all get half. A mutant who demanded more of the natives would get nothing; a mutant who demanded less would get less. Next, no other pure strategy is evolutionarily stable. Assume a population composed of players who demand x, where x<1/2. Mutants who demand 1/2 of the natives will get 1/2 and can invade. Next consider a population of players who demand x, where x>1/2. They get nothing. Mutants who demand y, where 0<y<(1-x) of the natives will get y and can invade. So can mutants who demand 1/2, for although they get nothing in encounters with natives, they get 1/2 in encounters with each other. Likewise, they can invade a population of natives who all demand 1. Here the symmetry requirement imposed by the evolutionary setting by itself selects a unique equilibrium from the infinite number of strict Nash equilibria of the two-person game. The 'Darwinian Veil of Ignorance' gives an egalitarian solution.

This is only the beginning of the story of the evolutionary dynamics of bargaining games. Even in the game I described, there are evolutionarily stable polymorphic states of the population which may be of considerable interest. (But see Alexander and Skyrms, 1999, and Alexander, 1999, for local interaction models where these polymorphisms almost never occur.) And in more complicated evolutionary games we can consider individuals who can occupy different roles, with the payoff function of the resource being role-dependent. However, at the most basic level, we have a powerful illustration of my point. The evolutionary setting for game theory here makes a dramatic difference in equilibrium selection, even while it supports a selection of a Nash equilibrium.

Weakly Dominated Strategies

The strategic situation can be radically changed in bargaining situations by introducing sequential structure. Consider the Ultimatum game of Güth, Schmittberger and Schwarze (1982). One player — the Proposer — demands some fraction of the resource. The second player — the Responder — is informed of the Proposer's proposal and either takes it or leaves it. If he takes it the first player gets what she demanded and he gets the rest. If he leaves it, neither player gets anything.

The are again an infinite number of Nash equilibria in this game, but from the point of view of rational decision theory they are definitely not created equal. For example, there is a Nash equilibrium where the Proposer has the strategy 'Demand half' and the Responder has the strategy 'Accept half or more but reject less'. Given each player's strategy, the other could not do better by altering her strategy. But there is nevertheless something odd about the Responder's strategy. This can be brought out in various ways. In the first place, the Responder's strategy is weakly dominated. That is to say that there are alternative strategies which do as well against all possible Proposer's strategies, and better against some. For example, it is weakly dominated by the strategy 'Accept all offers'. A closely related point is that the equilibrium at issue is not subgame perfect in the sense of Selten (1965). If the Proposer were to demand 60 per cent, this would put the responder into a subgame, which in this case would be a simple decision problem: 40 per cent or nothing. The Nash equilibrium of the subgame is the optimal act: Accept 40 per cent. So the conjectured equilibrium induces play on the subgame which is not an equilibrium of the subgame.

For simplicity, let's modify the game. There are ten lumps of resource to divide; lumps are equally valuable and can't be split; the proposer can't demand all ten lumps. Now there is only one subgame perfect equilibrium — the Proposer demands nine lumps and the Responder has the strategy of accepting all offers. If there is to be a Bayesian rational response to any possible offer, the Responder must have the strategy of accepting all offers. And if the Proposer knows that the Responder will respond by optimizing no matter what she does, she will demand nine lumps.

It is worth noting that the subgame perfect equilibrium predicted by the foregoing rationality assumptions does not seem to be what is found in experimental trials of the ultimatum game — although the interpretation of the experimental evidence is a matter of considerable controversy in the literature. Proposers usually demand less than 9 and often demand five. Responders often reject low offers, in effect choosing zero over one or two. I do not want to discuss the interpretation of these results here. I only note their existence. What happens when we transpose the ultimatum game into an evolutionary setting?

Here the game itself is not symmetric — the proposer and responder have different strategy sets. There are two ways to fit this asymmetry into an evolutionary framework. One is to model the game as an interaction between two disjoint populations. A proposer is drawn at random from the proposer population and a responder is drawn at random from the responder population and they play the game with the payoff being expected number of offspring. The alternative is a single population model with roles. Individuals from a single population sometimes are in the role of Proposer and sometimes in the role of Responder. In this model, individuals are required to have the more complex strategies appropriate to the symmetrized game, for example:

> If in the role of Proposer demand x;
> If in the role of Responder and proposer demands z, take it;
> Else if in the role of Responder and proposer demands z', take it;
> Else if in the role of responder and the proposer demands z'',leave it;
> etc.

The evolutionary dynamics of the two population model was investigated by Binmore, Gale and Samuelson (1995) and that of the one population symmetrized model by myself (1996; 1998b) for the replicator flow and by Harms (1994; 1997) for the Darwin map.

Despite some differences in modelling, all these studies confirm one central fact. Evolutionary dynamics need not eliminate weakly dominated strategies; evolutionary dynamics need not lead to subgame perfect equilibrium. Let me describe my results for a small game of Divide Ten, where proposers are restricted to two possible strategies: Demand Nine; Demand Five. Responders now have four possible strategies depending on how they respond to a demand of nine and how they respond to a demand of five. The typical result of evolution starting from a population in which all strategies are represented is a polymorphic population which includes weakly dominated strategies.

One sort of polymorphism includes Fairmen types who demand five and accept five but reject greedy proposals; together with Easy Riders who demand five and accept all. The Fairman strategy is weakly dominated by the Easy Rider strategy, but nevertheless some proportion of Fairmen can persist in the final polymorphism. Another sort of polymorphism consists of Gamesmen who demand nine and accept all, together with Mad Dogs who accede to a demand of nine but reject a Fairman's demand of five. Mad Dog is weakly dominated by Gamesman but nevertheless some proportion of Mad Dogs can persist in the final polymorphism. Which polymorphism one ends up with depends on what population proportions one starts with. However, in either case one ends up with populations which include weakly dominated strategies.

How is this possible? If we start with a completely mixed population — in which all strategies are represented — the weakly dominated strategies must have a smaller average fitness than the strategies which weakly dominate them. The weakly dominating strategies by definition do at least as well against all opponents and better against some. Call the latter the Discriminating Opponents. As long as the Discriminating Opponents are present in the population, the weakly dominating do better than the weakly dominated, but the Discriminating Opponents may go extinct more rapidly than the weakly dominated ones. This leaves a polymorphism of types which do equally well in the absence of Discriminating Opponents. This theoretical possibility is, in fact, the typical case in the ultimatum game. This conclusion is not changed, but rather reinforced, if we enrich our model by permitting the Proposer to demand 1, 2, 3, 4, 5, 6, 7, 8, 9. Then we get more complicated polymorphisms which typically include a number of weakly dominated strategies.

It might be natural to expect that adding a little mutation to the model would get rid of the weakly dominated strategies. The surprising fact is that such is not the case.

The persistence of weakly dominated strategies here is quite robust to mutation. It is true that adding a little mutation may keep the population completely mixed, so that weakly dominated strategies get a strictly smaller average fitness than those strategies which weakly dominate them, although the differential may be very small. But mutation also has a dynamical effect. Other strategies mutate into the weakly dominated ones. This effect is also very small. In polymorphic equilibria under mutation these two very small effects come into balance.

That is not to say that the effects of mutation are negligible. These effects depend on the mutation matrix, Mij, of probabilities that one given type will mutate into another given type. We model the combined effects of differential reproduction and mutation by a modification of the Darwin Flow:

$$dP(A_i)/dt = (1-e)[P(A_i) (U(A_i)-UBAR)/UBAR] + e[SUM_j P(A_j) M_{ij} - P(A_i)]$$

(See Hofbauer and Sigmund, 1988, p. 252)

There is no principled reason why these probabilities should all be equal. But let us take the case of a uniform mutation matrix as an example. Then, in the ultimatum game, introduction of mutation can undermine a polymorphism of Fairmen and Easy Riders and lead to a polymorphism dominated by Gamesmen and Mad Dogs. With a uniform mutation matrix and a mutation rate of 0.001, there is a polymorphism of about 80 per cent gamesmen and 20 per cent Mad Dogs — with other types maintained at very small levels by mutation. Notice that the weakly dominated strategy Mad Dog persists here at quite substantial levels. Other mutation matrices can support polymorphisms consisting mainly of Fairmen and Free Riders. In either case, we have equilibria that involve weakly dominated strategies.

A crack has appeared between game theory based on rationality and game theory based on evolutionary dynamics. There are rationality-based arguments against weakly dominated strategies and for sub-game perfect equilibrium. We have seen that evolutionary dynamics does not respect these arguments. This is true for both one-population and two-population models. It is true for both continuous and discrete versions of the dynamics. It remains true if mutation is added to the dynamics. I will not go into the matter here, but it also remains true if variation due to recombination — as in the genetic algorithm — is added to the dynamics (see Skyrms, 1996, Ch. 2). We shall see in the next section that the crack widens into a gulf if we relax the assumption of random pairing from the population.

Strongly Dominated Strategies

Consider a two-person game in which we assume that the players are Bayesian rational and know the structure of the game, but nothing more. We do not assume that the players know each other's strategies. We do not assume that Bayesian rationality is Common Knowledge. In general, this assumption will not suffice for Nash equilibrium or even Rationalizability. But one thing that it does is guarantee that players will not play strongly dominated strategies.

In certain special games, this is enough to single out a unique Nash equilibrium. The most well known game of this class is the Prisoner's Dilemma. In this game both players must choose between co-operation [C] and defection [D]. For each player:

$U(C|C) \;=\; 10$
$U(D|C) \;=\; 15$
$U(C|D) \;=\; 0$
$U(D|D) \;=\; 5$

Defection strongly dominates co-operation and if players optimize they both defect.

If we simply transpose our game to an evolutionary setting, nothing is changed. The game is symmetric; players have the same strategy sets and the payoff for one strategy against another does not depend on which player plays which. Evolutionary dynamics drives the co-operators to extinction. Defection is the unique evolutionarily stable strategy and a population composed exclusively of defectors is the unique evolutionarily stable state. So far evolution and rational decision theory agree.

Let us recall, however, that our evolutionary model was based on a number of simplifying assumptions. Among those was the assumption that individuals are paired *at random* from the population to play the game. This may or may not be plausible. There is a rich biological literature discussing situations where it is not plausible for biological evolution (see Hamilton, 1964; Sober and Wilson, 1998). With regard to cultural evolution I believe that many social institutions exist in order to facilitate non-random pairing (Milgrom *et al.*, 1990). A more general evolutionary game theory will allow for correlated pairing (Skyrms, 1996, Ch. 3).

Here the fundamental objects are conditional pairing probabilities, $P(A|B)$, specifying the probability that someone meets an A-player given that she is a B-player. These conditional pairing proportions may depend on various factors, depending on the particular biological or social context being modelled. The expected fitness of a strategy is now calculated using these conditional probabilities:

$$U(A) = \mathrm{SUM}_i\, U(A|B_i)\, P(B_i|A)$$

Now suppose that nature has somehow — I don't care how — arranged high correlation between like strategies among individuals playing the Prisoner's Dilemma. For instance, suppose:

$$P(C|C) = P(D|D) = 0.9 \text{ and } P(C|D) = P(D|C) = 0.1$$

Then the fitness of co-operation exceeds that of defection and the strategy of co-operation takes over the population. Correlated evolutionary game theory permits models in which a strongly dominated strategy is selected.

We can, of course, consider correlation in two person games between rational decision makers. This was done by Aumann (1974; 1987) in his seminal work on correlated equilibrium. Aumann takes the natural step of letting the 'coin flips' of players playing mixed strategies be correlated. The resulting profile is a joint correlated strategy. Players know the joint probability distribution and find out the results of their own 'coin flips'. If, whatever the results of those flips, they do not gain in expected utility by unilateral deviation, they are at a correlated equilibrium. However, this natural generalization is quite different than the generalization of mixed strategies as polymorphic populations that I have sketched. In particular, there is only one Aumann correlated equilibrium in Prisoner's Dilemma. It has each player defecting. More generally, Aumann correlated equilibria do not use strongly dominated strategies.

In fact, evolutionary game theory can deal with two kinds of mixed strategy. The first kind arises when individuals themselves use randomized strategies. The second kind interprets a population polymorphism as a mixed strategy. I have focused on the second kind in this paper. We have assumed that individuals play pure strategies in order to preserve the rationale for the replicator dynamics. If one drops the independence assumption from the first kind of mixed strategy, one gets Aumann correlated equilibrium. If one drops the independence assumption from the second kind, one gets the kind of correlated evolutionary game theory I am discussing here. In this setting, new phenomena are possible — the most dramatic of which include the fixation of strongly dominated strategies.

Prisoner's Dilemma is so widely discussed because it is a simple paradigm of the possibility of conflict between efficiency and strict dominance. Everyone would be better off if everyone co-operated. Co-operation is the efficient strategy. Whatever the other player does, you are better off defecting. Defection is the dominant strategy. In the game theory of rational choice, dominance wins. But in correlated evolutionary game theory, under favourable conditions of correlation, efficiency can win.

The point is general. If there is a strategy such that it is best if everyone uses it, then sufficiently high auto-correlation will favour that strategy. Perfect correlation imposes a kind of Darwinian categorical imperative under these conditions. Others do unto you as you do unto them. Then a strategy, A, such that $U(A|A)>U(B|B)$ for all B different from A, will be carried to fixation by the replicator dynamics from any initial state in which it is represented in the population.

On the other hand, in some strategic situations, anti-correlation may promote social welfare. Suppose that there are two types of a species that do not do well when interacting with themselves, but each of which gains a large payoff when interacting with the other type. Then the payoff to each type and the average payoff to the species might be maximized if the types could somehow anti-correlate. The situation hypothesized is one where mechanisms for detection might well evolve to support pairing with the other type. Consider species that reproduce sexually.

The introduction of correlation, in the manner indicated, takes evolutionary game theory into largely uncharted waters — unexplored by the traditional game theory based on rational choice. The basic structure of the theory is relaxed, opening up novel possibilities for the achievement of efficiency. One of the simplest and most dramatic examples is the possibility of the evolution of the strongly dominated strategy of co-operation in the one-shot Prisoner's Dilemma. (For various types of correlation generated by learning dynamics see Vanderschraaf and Skyrms, 1994. For correlation generated by spatial interaction, see Alexander, 1999, and Alexander and Skyrms, 1999. For the role of correlation in a general treatment of convention see Vanderschraaf, 1995; 1998.)

Utility and Rationality

From the fundamental insight that, in the random pairing model Darwin supports Nash, we have moved to an appreciation of ways in which game theory based on rational choice and game theory based on evolution may disagree. In the Ultimatum game, evolution does not respect weak dominance or subgame perfection. In the Prisoner's Dilemma with correlation evolution may not respect strong dominance.

Is this an argument for the evolution of irrationality? It would be a mistake to leap to this conclusion. Irrationality in Bayesian terms does not consist in failing to maximize Darwinian fitness, but rather in failing to maximize expected utility. One can conjecture utility functions which save the Bayes rationality of *prima facie* irrational behaviour in putative Ultimatum or Prisoner's Dilemma games. Whether this is the best way to explain observed behaviour remains to be seen. But if it were, we could talk about the evolution of utility-functions that disagree with Darwinian fitness rather than the evolution of irrationality.

However that may be, when we look at the *structure* of game theory based on rational choice and that of game theory based on evolutionary dynamics, we find that beyond the areas of agreement there are also areas of radical difference.

Avoiding the Curse of Symmetry

Let me close by returning to the example with which I started. That is the co-ordination game where players get a positive payoff if, and only if, they choose the same colour. Suppose it is a population that is evolving a custom and the possibilities are 'Choose red' and 'Choose green'. Corresponding to the mixed strategy delivered up by hyper-rational equilibrium selection, there is a population state where half the population chooses red and half chooses green. If we assume random encounters and the replicator dynamics, this population state is indeed a dynamic equilibrium, but it is dynamically unstable. A population in this state is like a ball rolling along a knife-edge. If the population has a little greater proportion on the red side, the replicator dynamics will carry it to a state where all choose red; if it has a little greater proportion on the green side, the dynamics will carry it to a state where all choose green. Variant adaptive dynamics will do the same.

This is a simplified picture, which encapsulates essential features of the evolution of conventions in general. One case of special interest is the evolution of the meanings of symbols in the kind of sender-receiver games introduced by David Lewis (1969). Rational-choice equilibrium selection theory would have to lead players to randomized 'babbling equilibria' where no meaning at all is generated. Evolutionary dynamics leads to the fixation of meaning in signalling system equilibria. You can read about it in Skyrms (1996, Ch.5; 1999).

Conclusion

As an explanatory theory of human behaviour, dynamical models of cultural evolution and social learning hold more promise of success than models based on rational choice. Under the right conditions, evolutionary models supply a rationale for Nash equilibrium that rational choice theory is hard-pressed to deliver. Furthermore, in cases with multiple symmetrical Nash equilibria, the dynamic models offer a plausible, historically path-dependent model of equilibrium selection. In conditions, such as those of correlated encounters, where the evolutionary dynamic theory is structurally at odds with the rational choice theory, the evolutionary theory provides the best account of human behaviour.

References

Alexander, J. (1999), 'The (spatial) evolution of the equal split', Working Paper Institute for Mathematical Behavioural Sciences U.C.Irvine.

Alexander, J. and Skyrms, B. (1999), 'Bargaining with neighbors: is justice contagious?', Working Paper Logic and Philosophy of Science U.C.Irvine.

Aumann, R. J. (1974), 'Subjectivity and correlation in randomized strategies', *Journal of Mathematical Economics*, **1**, pp. 67–96.

Aumann, R. J. (1987), 'Correlated equilibrium as an expression of Bayesian rationality', *Econometrica* **55**, pp. 1–18.

Binmore, K. (1993), 'Game theory and the social contract Vol. 1', *Playing Fair* (Cambridge, MA: MIT Press).

Binmore, K. (1998), 'Game theory and the social contract Vol. 2', *Just Playing* (Cambridge,MA: MIT Press).

Binmore, K., Gale, J. and Samuelson, L. (1995), 'Learning to be imperfect:The ultimatum game', *Games and Economic Behaviour*, **8**, pp. 56–90.

Bernheim, B. D. (1984), 'Rationalizable strategic behaviour', *Econometrica* **52**, pp. 1007–28.

Bjornerstedt, J. and Weibull, J. (1995), 'Nash equilibrium and evolution by imitation', in *The Rational Foundations of Economic Behavior*, ed. K. Arrow *et al.* (New York: MacMillan), pp. 155–71.

Bomze, I. (1986), 'Non-cooperative two-person games in biology: A classification', *International Journal of Game Theory*, **15**, pp. 31–57.

Borgers, T. and Sarin, R. (1997), 'Learning through reinforcement and the replicator dynamics', *Journal of Economic Theory*, **77**, pp. 1–14.

Camerer, V. (1997), 'Progress in behavioural game theory', *Journal of Economic Perspectives*, **11**, pp. 167–88.

Duffy, J. and Nagel, R. (1997), 'On the robustness of behaviour in experimental "Beauty Contest" games', *The Economic Journal*, **107**, pp. 1684–700.

Fudenberg, D. and Levine, D. (1998), *The Theory of Learning in Games* (MIT :Cambridge, MA).

Gauthier, D. (1969), *The Logic of the Leviathan* (Oxford: Oxford University Press).

Gauthier, D. (1986), *Morals by Agreement* (Oxford: Clarendon Press).

Gibbard, A. (1990), *Wise Choices, Apt Feelings:A Theory of Normative Judgement* (Oxford: Clarendon Press).

Güth, W., Schmittberger, R. and Schwarze, B. (1982), 'An experimental analysis of ultimatum bargainin',g *Journal of Economic Behaviour and Organization*, **3**, pp. 367–88.

Güth, W., and Tietz, R. (1990), 'Ultimatum bargaining behaviour: A survey and comparison of experimental results', *Journal of Economic Psychology*, **11**, pp. 417–49.

Hamilton, W. D. (1964), 'The genetical evolution of social behaviour', *Journal of Theoretical Biology*, **7**, pp. 1–52.

Harms, W. (1994), 'Discrete replicator dynamics for the ultimatum game with mutation and recombination', Technical Report, (University of California, Irvine).

Harms, W. (1997), 'Evolution and ultimatum bargaining', *Theory and Decision*, **42**, pp. 147–75.

Harsanyi, J. and Selten, R. (1988) *A General Theory of Equilibrium Selection in Games* (Cambridge, MA: MIT Press).

Ho, T. H., Weigelt, K. and Camerer, C. (1996), 'Iterated dominance and iterated best-response in experimental *p*-beauty contests', Social Science Working Paper 974, California Institute of Technology.

Hofbauer, J. and Sigmund, K. (1988), *The Theory of Evolution and Dynamical Systems* (Cambridge: Cambridge University Press).

Hume, D. (1975), *Enquiries Concerning Human Understanding and Concerning the Principles of Morals*, reprinted from the posthumous edition of 1777 with text revised and notes by P. H. Nidditch (Oxford:Clarendon Press).

Keynes, J.M. (1936), *The General Theory of Employment, Interest and Money* (New York: Harcourt Brace).

Lewis, D. (1969), *Convention* (Cambridge, MA:Harvard University Press).

Maynard-Smith, J. and Price, G.R. (1973), 'The logic of animal conflict', *Nature*, **146**, pp. 15–18.

Maynard-Smith, J. and Parker, G.R. (1976), 'The logic of asymmetric contests', *Animal Behaviour*, **24**, pp. 159–75.

Milgrom, P., North, D. and Weingast, B. (1990), 'The role of institutions in the revival of trade: The law merchant, private judges, and the champagne fairs', *Economics and Politics*, **2**, pp. 1–23.

Moulin, H. (1986), *Game Theory for the Social Sciences* (New York: New York University Press).

Nagel, R. (1995), 'Unravelling in guessing games:An experimental study', *American Economic Review*, **85**, pp. 1313–26.

Pearce, D.G. (1984), 'Rationalizable strategic behaviour and the problem of perfectio',n *Econometrica*, **52**, pp. 1029–50.

Sacco, P.L. (1995), 'Comment', in *The Rational Foundations of Economic Behavior*, ed. K. Arrow *et al.* (New York: Macmillan).

Samuelson, L. (1997), *Evolutionary Games and Equilibrium Selection* (Cambridge,MA:MIT Press).

Samuelson, L. (1988), 'Evolutionary foundations of solution concepts for finite two-player normal form games', in *Proceedings of the Second Conference on Theoretical Aspects of Reasoning About Knowledge*, ed. M. Vardi (Los Altos, CA: Morgan Kaufmann).

Schlag, K. (1994), 'Why imitate and if so how? Exploring a model of social evolution', *Discussion Paper B-296* (Department of Economics, University of Bonn).

Schlag, K. (1996), 'Why imitate and if so, how? A bounded rational approach to many armed bandits', *Discussion Paper B-361* (Department of Economics, University of Bonn).

Selten, R. (1975), 'Re-examination of the perfectness concept of equilibrium in extensive games', *International Journal of Game Theory*, **4**, pp. 25–55.

Selten, R. (1965), 'Spieltheoretische Behandlung eines Oligopolmodells mit Nachfragetragheit', *Zeitschrift fur die gesamte Staatswissenschaft*, **121**, pp. 301–24, 667–89.

Skyrms, B. (1990), *The Dynamics of Rational Deliberation* (Cambridge,MA: Harvard University Press).

Skyrms, B. (1994a), 'Darwin meets 'The logic of decision': Correlation in evolutionary game theory', *Philosophy of Science*, **61**, pp. 503–28.

Skyrms, B. (1994b), 'Sex and justice', *The Journal of Philosophy*, **91**, pp. 305–20.

Skyrms, B. (1995), 'Introduction to the Nobel symposium on game theory', *Games and Economic Behaviour*, **8**, pp. 3–5.

Skyrms, B. (1996), *Evolution of the Social Contract* (New York: Cambridge University Press).

Skyrms, B. (1998a), 'Mutual aid' in *Modelling Rationality, Morality and Evolution*, ed. Peter Danielson (Oxford: Oxford University Press).

Skyrms, B. (1998b), 'Evolution of an anomaly', *Protosoziologie*, **12**, pp. 192–211.

Skyrms, B. (1999), 'Evolution of inference', in *Dynamics in Human and Primate Societies*, ed. T. Kohler and G. Gumerman. SFI Studies in the Sciences of Complexity. (New York: Oxford University Press).

Sober, E. (1992), 'The evolution of altruism: Correlation, cost and benefit', *Biology and Philosophy*, **7**, pp. 177–87.

Sober, E. and Wilson, D.S. (1998), *Unto Others:The Evolution and Psychology of Unselfish behaviour* (Cambridge, MA: Harvard University Press).

Spinoza, B. (1985), *Ethics:The Collected Works of Spinoza, Volume I*, trans. E. Curley (Princeton: Princeton University Press).

Stahl, D.O. (1996), 'Boundedly rational rule learning in a guessing game', *Games and Economic behaviour*, **16**, pp. 303–30.

Sugden, R. (1986), *The Economics of Rights, Co-operation and Welfare* (Oxford: Blackwell).

Taylor, P. and Jonker, L. (1978), 'Evolutionarily stable strategies and game dynamics', *Mathematical Biosciences*, **16**, pp.76–83.

van Damme, E. (1987), *Stability and Perfection of Nash Equilibria* (Berlin: Springer).

Vanderschraaf, P. (1998), 'Knowledge, equilibrium and convention', *Erkenntnis*, **49**, pp. 337–69.

Vanderschraaf, P. (1995), 'Convention as correlated equilibrium', *Erkenntnis*, **42**, pp. 65–87.

Vanderschraaf, P. and Skyrms, B. (1994), 'Deliberational correlated equilibrium', *Philosophical Topics*, **21**, pp. 191–227.

von Neumann, J. and Morgenstern, O. (1947), *Theory of Games and Economic Behaviour, 2nd. ed.* (Princeton: Princeton University Press).

Weibull, J. (1997), *Evolutionary Game Theory* (Cambridge, MA: MIT Press).

[5]

If *Homo Economicus* Could Choose His Own Utility Function, Would He Want One with a Conscience?

By ROBERT H. FRANK*

A blush may reveal a lie and cause great embarrassment at the moment. But in situations that require trust, there can be great advantage in being known to be a blusher. This paper develops a model in which tastes are determined endogenously for their capacity to help solve the so-called precommitment problem. The tastes that emerge are very different from those assumed in conventional models of rational choice.

The rational choice model takes tastes as given,[1] and assumes that people pursue self-interest. The model performs well much of the time, yet apparent contradictions abound. Travelers on interstate highways leave tips for waitresses they will never see again. Participants in bloody family feuds seek revenge even at ruinous cost to themselves. People walk away from profitable transactions whose terms they believe to be "unfair." The British spend vast sums to defend the desolate Falklands, even though they have little empire left against which to deter future aggression. In these and countless other ways, people do not seem to be maximizing utility functions of the usual sort.

In this paper, I investigate the familiar theme that seemingly irrational behavior can sometimes be explained without departing from the utility-maximization framework.

Efforts along these lines have generally focused on a more careful assessment of the constraints people face. The alternative I explore here is to suggest that tastes differ in systematic ways from those portrayed in standard economic models. Instead of treating tastes as a datum, I retreat a step and ask, "What kind of tastes would maximize the attainment of selfish objectives?" This is essentially the behavioral biologist's approach (W. D. Hamilton, 1964; E. O. Wilson, 1975, 1978; Richard Dawkins, 1976; Robert Trivers, 1985). It treats tastes not as ends in themselves, but as means for attaining important material objectives.[2]

My focus is a specific example in which people have opportunities to cheat without possibility of reprisal. This example illustrates that it will sometimes be in a selfish person's interest to have a utility function that predisposes him not to cheat, even when he is *certain* he would not be caught. The example also suggests other specific elements of the utility function that might help people resolve important market failures and bargaining problems. The common feature of

*Professor of Economics, Department of Economics, Cornell University, Ithaca, NY 14853. I thank Robert Trivers, Jack Hirshleifer, Thomas Schelling, Robert Axelrod, Henry Hansmann, Thomas Eisner, Simon Levin, Stephen Emlen, Elizabeth Adkins Regan, and an anonymous referee for encouragement and helpful comments on an earlier draft. This research was supported by National Science Foundation Grant No. SES-8605829.
[1] Gary Becker and George Stigler (1977) provide the clearest statement of the modern economists' hands-off approach to tastes. For three recent exceptions that are closely related to the theme of this paper, see Thomas Schelling (1978), George Akerlof (1983), and Jack Hirshleifer (1984). See also Hirshleifer (1977; 1978a,b), Philip Coelho (1985), Paul Rubin and Chris Paul (1979), Rubin (1982), and myself (1985a,b; 1986, 1988).

[2] In the biologist's view, tastes are no different from other characteristics. All are selected for their capacity to promote survival and reproduction. A wolf who "likes" to stand up to intruders runs the risk of serious injury. But because the animal is known to have that preference, it may do much better, on the average, than one who is known to prefer passive retreat. If so, the success conferred by this taste will lead to an increase in its frequency in the gene pool.

these problems is that their solution requires people to make *ex ante* commitments to behave in ways that will not be self-serving *ex post*.

I. The Commitment Problem

Thomas Schelling (1960) provides a vivid illustration of this class of problems. He describes a kidnapper who suddenly gets cold feet. He wants to set his victim free, but is afraid he will go to the police. In return for his freedom, the victim gladly promises not to do so. The problem, of course, is that both realize that it will no longer be in his interest to keep this promise once he is free. And so the kidnapper reluctantly concludes he must kill his victim.

Schelling suggests the following way out of the dilemma: "If the victim has commited an offense whose disclosure would lead to blackmail, he may confess it; if not, he might commit one in the presence of his captor, to create a bond that will ensure his silence" (1960, pp. 43–44). The blackmailable act serves here as a "commitment device," something that provides the victim with an incentive to keep his promise. The need for such devices arises frequently in the course of economic interaction.

II. The Costs of Monitoring and the Breakdown of Cooperation

Consider, for example, the case of a joint venture. Numerous opportunities exist in which it is possible to increase output through cooperation. Unfortunately, however, performance in many such ventures is either impossible or prohibitively costly to monitor.

It follows that a tendency to cheat when there is no possibility of punishment will often eliminate opportunities for mutual gain. Consider a joint venture in which *A* and *B* may either behave honestly or dishonestly. Suppose the payoffs to the four possible combinations of strategies are as given in Table 1, where *H* denotes honest, *D* denotes dishonest, and where $x_4 > x_3 > x_2 > x_1$. Suppose also that the parties interact only once, so that cheating cannot be pen-

TABLE 1—THE PAYOFFS TO *A* FROM INTERACTION WITH *B*

		B	
		H	*D*
A	*H*	x_3	x_1
	D	x_4	x_2

alized. These payoffs create a prisoner's dilemma. *D* is the dominant strategy, even though the resultant payoff, x_2, is lower than if each had chosen H.[3] Parties *A* and *B* need a commitment device of the sort described by Schelling.

III. The Conscience as a Commitment Device

Schelling's device worked by altering the relevant material incentives, an approach that will not always be practical. Fortunately, commitment devices can also work in other ways. Consider, for example, a person with the capacity to experience strong feelings of guilt upon breaking a promise—that is, a person with a conscience. This person will often honor his promises even when material incentives favor breaking them. It is precisely this capacity of emotional forces to override rational calculations that makes them candidates for commitment devices.

Of course, merely *having* a conscience does not solve the commitment problem; one's potential trading partners must also know about it. But how can they? Surely it is insufficient for a person merely to *declare* how he or she feels. ("I have a conscience. Trust me.") A strategically important emotion can be communicated credibly only if it is accompanied by a signal that is at least partially insulated from direct control.

---

[3] Robert Axelrod (1984), Trivers (1971), and others have demonstrated the tendency for cooperation to emerge in repeated prisoner's dilemmas. My focus here is on whether cooperation can be self-serving when the game is played only once, or, if played more than once, when cheating cannot be detected. For an excellent survey of biological models of cooperation and altruism, see Hirshleifer (1977).

Many observable physical symptoms of emotional arousal satisfy this requirement. Posture, the rate of respiration, the pitch and timbre of the voice, perspiration, facial muscle tone and expression, movement of the eyes, and a host of other readily observable physical symptoms vary systematically with a person's affective condition (Melvin Konner, 1982; Paul Ekman, 1985). In most people, at least some of these symptoms are almost completely insulated from voluntary control (Ekman; Charles Darwin, 1872). If natural selection caused (Darwin), or even permitted, these symptoms to convey such feelings as fear, joy, and grief, there is no reason why it could not have allowed these same symptoms to be related in a characteristic way to feelings of guilt as well.[4]

If there are gains from appearing to be trustworthy, opportunists will inevitably attempt to mimic the symptoms of trustworthiness. Needless to say, character traits would be useless as commitment devices if people who lacked them could *perfectly* mimic their symptoms. But mimicry usually entails costs of its own, and is thus seldom perfect.[5] Granted, lie detection, even the

highly technical sort practiced by experienced professionals, is subject to error (David Lykken, 1981). Even so, clues to deceit are reliable in a probabilistic sense and, as will become clear, this is all that is needed for the conscience to serve as a commitment device.

For the sake of discussion, suppose that a mutant did in fact appear with a conscience accompanied by a statistically reliable signal. How might such a mutant have fared? In models of natural selection, a gene for a given trait will increase in frequency relative to its alleles (alternative genes on the same locus) if the trait leads, on average, to a higher material payoff. We must thus compare the mutant's expected payoff to that of persons who lack a conscience. The next section develops a formal model for this purpose.[6]

IV. An Illustrative Model

Consider a population each of whose members bears one of two traits, *H* or *D*. Those bearing *H* are honest, those bearing *D* are not.[7] To be honest here means to refrain from cheating one's partner in a cooperative venture, even when cheating cannot be punished. To be dishonest means always to cheat under the same circumstances.

Suppose people face one of two options:

1) They may pair with someone in a joint venture whose payoffs are as given above in Table 1. This venture will occur only once and dishonest behavior cannot be punished; or

---

[4] Trivers argues that guilt could have evolved within the framework of his reciprocal altruism model as a means of protecting people against the possibility of being caught cheating: "...it is possible that the common psychological assumption that one feels guilt even when one behaves badly in private is based on the fact that many transgressions are *likely* to become public knowledge" (1971, p. 50).

[5] There are abundant cases in nature where mimics coexist with the genuine bearers of a trait for extended periods. For example, there are butterflies, such as the monarch, whose strategy for defending themselves is to have developed a foul taste. This taste itself would be useless unless predators had some way of telling which butterflies to avoid. And so the monarch has developed a conspicuous pattern of wing markings that its predators have learned to interpret for this purpose. The safety afforded by these wing markings has created a profitable opportunity for mutant butterflies who bear similar markings but lack the bad taste that normally accompanies them. Merely by looking like the unpalatable monarchs, viceroy butterflies have escaped predation without having had to expend the bodily resources required to produce the objectionable taste itself. Coexistence of the two types implies that mimicry is either incomplete or else imposes important costs on the viceroy. For an illuminating discussion of the phenome-

non of deceptive signaling in animals, see John Krebs and Dawkins (1984).

[6] For a more reader-friendly version of this model, see my forthcoming book (1988, ch. 3).

[7] More precisely, imagine a population with two alleles, a_1 and a_2, at a given genetic locus. Individuals homozygous in a_1 are "honest," those homozygous in a_2 are not. Regardless of what effect honest behavior may have on fitness, each allele in heterozygous individuals will be affected equally by the interactions of these individuals with others. We may thus disregard these individuals when analyzing the competition between a_1 and a_2. In the text, *H* refers to individuals homozygous in a_1, *D* to those homozygous in a_2.

2) They may pair with someone in an alternative venture in which behavior can be perfectly monitored. Call this the "work-alone" option, because it does not involve activities that require trust. It too will occur only once, and its payoff is x_2 for each person, irrespective of the combinations of types who interact. Thus, the work-alone option offers the same payoff as when two dishonest persons interact in venture 1.

Many readers will find it more plausible to assume that persons who cheat one another in venture 1 do *worse* than those who participate in ventures that do not require trust (i.e., to assume a payoff larger than x_2 for the work-alone option). But since the goal of this exercise is to see whether honest persons can prosper in the material world, I retain the more conservative assumption that the work-alone option pays only x_2. If honest persons can survive with that payoff, they will certainly do so with an even higher one.

Since working alone pays only x_2, it will never appeal to a dishonest person, who can always do at least as well in venture 1. Its obvious attraction for an honest person is that there is no chance of being cheated; the downside, however, is that its payoff is less than x_3, the gain from successful cooperation in venture 1. (More below on how honest persons choose between the two options.)

In addition to their behavioral differences, let the H's and D's differ with respect to the genes that influence some observable characteristic S, which is also influenced by random environmental forces. More specifically, suppose that the observable characteristic for person i takes the value,

$$S_i = \mu_i + \varepsilon_i,$$

where $\mu_i = \mu_H$ if i is honest, $= \mu_D$ if i is dishonest, $\mu_H > \mu_D$, and ε_i is an independently, identically distributed random variable with zero mean. The μ_i component of S_i is heritable, the ε_i component is not.

Case 1: A Perfectly Reliable Signal. If the variance of ε_i were zero, everyone could tell with certainty whether any given person was an H or a D. In that case, H's would pair only with other H's in venture 1, and would

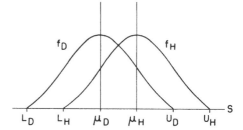

FIGURE 1. PROBABILITY DENSITIES FOR AN OBSERVABLE SIGNAL OF TRUSTWORTHINESS

receive a payoff of x_3. The D's would be left to pair with one another, for which they would receive only x_2. Thus, if the signal that accompanied each trait were *perfectly* reliable, the H's would soon drive the D's to extinction.

Case 2: Imperfect Signals. By contrast, when ε_i has sufficiently large variance, S_i provides merely a measure of the probability that i is trustworthy. As an illustration of this more interesting case, suppose people draw their S values independently from the probability densities f_D and f_H shown in Figure 1.[8]

In terms of the earlier discussion, the relative position of the two densities reflects imperfect mimicry by the D's of the trait used by the H's to identify themselves as trustworthy. For the densities shown, whose ranges do not overlap completely, we are sure that individuals with $S > U_D$ are honest, and that those with $S < L_H$ are dishonest. Individuals whose S values lie in the region

---

[8] Each density shown in Figure 1 is derived from the standard normal density, $f(S)$, in the following manner: First the tails are truncated at two standard deviations from the mean by subtracting $f(2)$ from every point. Let I denote the area under the resulting functions. The functions are then normalized by dividing each by I. Letting 2 and 3 be the respective means, we thus have

$$f_H = (1/\sqrt{2}\pi)(1/I)\left[\exp(-(x-3)^2/2) - \exp(-2)\right],$$

and

$$f_D = (1/\sqrt{2}\pi)(1/I)\left[\exp(-(x-2)^2/2) - \exp(-2)\right].$$

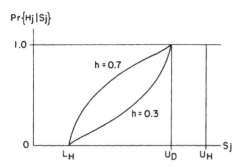

FIGURE 2. THE PROBABILITY OF BEING HONEST, CONDITIONAL ON *S*

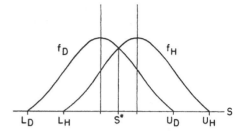

FIGURE 3. THE THRESHOLD SIGNAL VALUE WHEN $h = 1/2$ AND $x_2 = (x_1 + x_3)/2$

where the two densities overlap may be either *H* or *D*. If *h* denotes the proportion of the population bearing *H*, the probability that an individual *j* with $S = S_j$ is honest is then given by

$$(1) \quad \Pr\{H_j|S_j\} = \frac{h f_H(S_j)}{h f_H(S_j) + (1-h) f_D(S_j)}.$$

For the two densities shown in Figure 1, $\Pr\{H_j|S_j\}$ is plotted in Figure 2 for two different values of *h*.

A. *The Threshold Signal Value*

Given the payoffs in Table 1, both *H*'s and *D*'s will clearly do best by interacting (if they choose venture 1) with other individuals whom they believe to be *H*'s. Consider the problem confronting individual *i*, who is *H* and must decide whether to interact with individual *j*, who has $S = S_j$. Let $E(X_{ij}|S_j)$ denote the expected payoff to *i* from pairing with this individual. If the alternative is the work-alone option (which has payoff x_2), the condition that makes it worth *i*'s while to work with *j* is given by

$$(2) \quad E(X_{ij}|S_j) = x_3 \Pr\{H_j|S_j\}$$

$$+ x_1[1 - \Pr\{H_j|S_j\}] \geq x_2.$$

Call S^* the value of S_j that satisfies $E(X_{ij}|S_j) = x_2$. The value S^* is the smallest

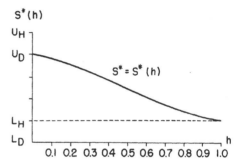

FIGURE 4. THE THRESHOLD SIGNAL VALUE AS A FUNCTION OF *h*

value of S_j for which it would be as attractive for *i* to work with *j* as to work alone. Using the densities from Figure 1, this threshold signal value is shown in Figure 3 for the particular case of $h = 1/2$ and $x_2 = (x_1 + x_3)/2$.

Using equations (1) and (2), it can easily be shown that S^* is a decreasing function of *h*: As the proportion of the population that is honest grows, it becomes ever more probable that an individual whose *S* lies between L_H and U_D is honest, and ever more likely, therefore, that the expected payoff from interacting with that individual will exceed x_2. It is also easy to show that S^* decreases with increases in $(x_3 - x_2)$ and increases with increases in $(x_2 - x_1)$. The greater the payoff to successful cooperation, and the smaller the penalty for being cheated, the lower will be the threshold signal for cooperation. For the particular case of $(x_4 - x_3) = (x_3 - x_2)$

$=(x_2 - x_1) = x_1$, the schedule $S^*(h)$ that corresponds to the densities from Figure 1 is plotted in Figure 4.

B. *The Sorting Process Whereby Interacting Pairs Form*

It is not strictly correct to say that an individual who is honest faces only the alternatives of working alone or working with someone with $S = S_j$. There are other members of the population besides j; and, of those with $S > S^*$, i will want to choose a partner with the highest possible S value. The problem is that *everyone* prefers to interact with such a partner, and there are only so many of them to go around.

The bearer of a high S value, whether he is honest or not, owns a valuable asset. The natural way to exploit it will be to use it to attract a partner whose own S value is also high. Without going into technical details, it can be seen that the outcome will be for the two with the highest S values to pair up, then the next two, and so on, until all those with $S \geq S^*$ are paired. Needless to say, the expected total payoff to any pair increases with the S values of its members. If the population is sufficiently dense, as assumed here, the S values within each pair will be virtually the same.[9]

C. *The Expected Payoff Functions*

With the foregoing discussion in mind, let us now investigate what happens if a mutation provides a small toehold for the H's in a population that initially consisted entirely of D's. Will the H's make headway or be driven to extinction? To answer this question, we must compare the average payoffs

[9]In less dense populations, S values could differ within pairs, suggesting that the member in any pair with the lower S value might have to pay his partner in order to induce him to participate. Any such transfer payments will obviously not affect the average payoff to members of a cooperating pair. Even if the assumption of dense populations did not hold, therefore, complications of this sort would create no difficulties: We are concerned only with average payoffs to the two genotypes here.

for H's and D's, denoted $E(X|H)$ and $E(X|D)$, respectively, when h is near zero. To calculate $E(X|H)$, note from equations (1) and (2) that $S^*(h)$ approaches U_D as h approaches zero (see Figure 4). As the share of the population that is honest approaches zero, the rational strategy for an honest person, in the limit, is to interact only with people whose S values lie to the right of U_D —that is, with people who are certain to be honest. As h approaches zero, the expected payoff to the H's thus approaches the limiting value

$$(3) \quad \lim_{h \to 0} E(X|H) = x_3 \int_{U_D}^{U_H} f_H(S)\, dS$$

$$+ x_2 \int_{L_H}^{U_D} f_H(S)\, dS.$$

The second term on the right-hand side of equation (3) reflects the fact that an H with $S < U_D$ does best by not interacting—that is, by choosing the work-alone option. (He would be delighted to interact with an H with $S > U_D$, but that person would not agree to interact with *him*.)

The corresponding limiting expected payoff for the D's is simply x_2. So when the proportion of the population that is honest is very small, we have $E(X|H) > E(X|D)$, which means h will grow.

What if we had started at the opposite extreme—in a population in which the proportion of D's was near zero? In the limit, the decision rule for the H's would be "interact with anyone whose S value exceeds L_H." As h approaches one, the expected payoff for the D's would thus approach

$$(4) \quad \lim_{h \to 1} E(X|D) = x_2 \int_{L_D}^{L_H} f_D(S)\, dS$$

$$+ x_4 \int_{L_H}^{U_D} f_D(S)\, dS.$$

The corresponding expected payoff for H will approach x_3 as h approaches one. Comparing the right-hand side of equation (4) with x_3, we see that the former will be larger for sufficiently large values of x_2 or x_4, or for sufficiently extensive overlap of the ranges of f_H and f_D. Assuming that $E(X|D)$ does exceed x_3 for values of h close to one, h will then begin to fall.

To investigate the behavior of this population as it evolves away from either of the two pure states, we need general expressions for $E(X|H)$ and $E(X|D)$.

The first step in finding $E(X|H)$ is to use equations (1) and (2) to find S^* as a function of h (as was done to generate Figure 4). Given $S^*(h)$, we then calculate P_H and P_D, the respective shares of the honest and dishonest populations with $S > S^*$:

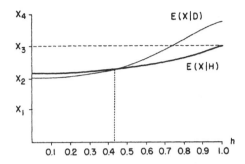

FIGURE 5. EXPECTED PAYOFFS FOR H AND D

(5) $$P_H = \int_{S^*}^{U_H} f_H(S) \, dS;$$

(6) $$P_D = \int_{S^*}^{U_D} f_D(S) \, dS.$$

Given equations (5) and (6), $E(X|H)$ can be computed as a weighted sum of the payoffs from (1) working alone, (2) interacting with someone who is honest, and (3) interacting with someone who is dishonest. Noting that $(1 - P_H)$ is the probability that an honest individual works alone, and using $\lambda = hP_H/(hP_H + (1-h)P_D)$ to denote the share of the population with $S > S^*$ that is honest, we have

(7) $E(X|H) = (1 - P_H)x_2$

$$+ P_H[\lambda x_3 + (1-\lambda)x_1].$$

In like fashion, $E(X|D)$ can be calculated as a weighted sum of the payoffs to working alone, interacting with someone who is honest, and interacting with someone who is dishonest:

(8) $E(X|D) = (1 - P_D)x_2$

$$+ P_D[\lambda x_4 + (1-\lambda)x_2].$$

For my illustrative densities from Figure 1, $E(X|H)$ and $E(X|D)$ are plotted as functions of h in Figure 5. Both $E(X|H)$ and $E(X|D)$ will in general be increasing functions of h. For the particular densities

from Figure 1, the two curves cross exactly once.

The equilibrium value of h is the one for which $E(X|D)$ and $E(X|H)$ intersect. For an equilibrium to be stable, $E(X|H)$ must cross $E(X|D)$ from above. Since $E(X|H)$ will always be greater than $E(X|D)$ at $h = 0$, the intersection of the two curves (assuming they intersect at all) will thus correspond to a stable equilibrium value of h, like the one depicted in Figure 5.

Case 3: Costs of Scrutiny. The model can be made more complete, and the existence of an equilibrium assured, if we add the assumption that resources must be expended in order to inspect the S values of others. Let us assume specifically that these values can be observed only by people who have borne the cost C of becoming sensitized.

First consider the limiting case of a population in which h approaches zero. A necessary condition for any interaction to occur at all is that $x_3 - C > x_2$. Assuming that condition is met, only those honest persons whose own S values exceed U_D will find it worthwhile to spend C to become sensitized. H's with $S < U_D$ will work alone (because there is no prospect of their finding an H with whom to interact). Thus the latter H's will receive payoff x_2, the same as $E(X|D)$ when h is zero. No D's will pay to become sensitized because they too have no chance of finding an unclaimed H, so the expected payoff for D's here is x_2. For very small values of h, then, some H's will fare the

600   THE AMERICAN ECONOMIC REVIEW   SEPTEMBER 1987

same as the D's, others better. Thus, as before, h will grow when it starts near zero.

For values of h near one, no one will find it worthwhile to pay C to become sensitized. With everyone's S thus effectively out of sight, even D's with $S < L_H$ will find honest partners with whom to interact. $E(X|D)$ will thus approach x_4 as h approaches one. So, unlike the case where S could be observed for free, it is no longer possible for the H's to take over the entire population, regardless of how close x_3 may be to x_4.

A key step in understanding the nature of the equilibrium that results when h starts near zero is to recognize that it will always be most strongly in the interests of those with the highest S values to bear the expense of becoming sensitized. To see this, recall from the earlier discussion that the bearer of a high S value owns a valuable asset he can use to attract a partner whose own S value is also high. It is in the interests of bearers of high S values to interact with others who are H. And, having become sensitized, they are then able to identify others with high S values (the people who are most likely to be H) and interact with them. Persons with relatively low S values, who can only hope to claim a partner whose own S value is also low, therefore face a lower expected payoff to the investment of becoming sensitized.

Knowing this enables one to conclude that the equilibrium h must be sufficiently low that it pays at least *some* of the population to become sensitized. This can be shown by noting that, if the contrary is assumed, we get a contradiction. That is, suppose it is assumed that, in equilibrium, the expected payoff to the person with the highest S value is higher if he does *not* become sensitized. Recall that his expected payoff if he had become sensitized, call it $E(X, C)$, would be

$$(9) \qquad E(X, C) = x_3 - C,$$

which is already assumed to be greater than x_2. We know that if it does not pay the H with the highest S value to become sensitized, it does not pay any other H to do so either. And when nobody pays C, the odds

of H's with low S values pairing successfully are the same as for those with high S values. All of the H's must thus expect a payoff higher than x_2. This means it pays all of the H's to interact, and all will expect the same payoff, call it $E(X, 0)$:

$$(10) \qquad E(X, 0) = hx_3 + (1 - h)x_1.$$

But the expected payoff for each D when everyone interacts is $hx_4 + (1 - h)x_2$, which is clearly larger than $E(X, 0)$.

Thus, if it is assumed that the equilibrium h is so large that it does not pay even those with the highest S values to become sensitized, we are forced to conclude that the expected payoff to D exceeds the expected payoff to H at that value of h. This means that it could not have been an equilibrium value after all. The equilibrium h must therefore be small enough that it is in the interests of at least those with the highest S values to become sensitized.

To gain further insight into the nature of the equilibrium here, let us again consider a population in which h started near zero and has grown to some positive value, say, h_0. At h_0, who will interact with whom, and in whose interests will it be to become sensitized? We can show it will pay those persons with S greater than some threshold value, $S^*(h_0) < U_D$, to become sensitized and interact with others with like S values. And we can also show that it may or may not pay H's with $S < S^*(h_0)$—that is, those who do not find it profitable to become sensitized— to interact at random with other members of the truncated population (people with $S < S^*(h_0)$).

The threshold value $S^*(h_0)$ is determined as follows. Again making use of the notation defined in equations (5) and (6), note that

$$(11) \qquad 1 - P_H = \int_{L_H}^{S^*(h_0)} f_H(S)\, dS;$$

$$(12) \qquad 1 - P_D = \int_{L_D}^{S^*(h_0)} f_D(S)\, dS,$$

are the respective shares of the honest and dishonest populations for which $S < S^*(h_0)$. Let $\delta = (1 - P_H)h_0 / [(1 - P_H)h_0 + (1 -$

$P_D)(1 - h_0)]$ denote the share of this truncated population that is honest. $S^*(h_0)$ is the value of S that solves the following equation:

$$(13) \quad \Pr\{ H|S^*(h_0)\} x_3$$

$$+ (1 - \Pr\{ H|S^*(h_0)\}) x_1 - C$$

$$= \max\{ x_2, \delta x_3 + (1 - \delta) x_1\}.$$

The left-hand side of equation (13) is the expected payoff to an H with $S = S^*(h_0)$ who pays C to become sensitized and then interacts with someone with the same S value (see equation (1)). The second expression within the brackets on the right-hand side of equation (13) is the expected payoff to that same H (or, for that matter, to any other H) if he does not become sensitized and interacts with a randomly chosen person from the truncated population (again, those D's and H's with $S < S^*(h_0)$). If x_2 is larger than that figure, as it is likely to be for small values of h_0, all H's with $S < S^*(h_0)$ will work alone. Otherwise all H's will interact —those with $S > S^*(h_0)$ interacting with others with $S > S^*(h_0)$, those with $S < S^*(h_0)$ interacting with randomly chosen members of the remaining population. Depending on the particular densities used for f_H and f_D, either outcome is possible.

Once $S^*(h_0)$ is determined, it is straightforward, albeit tedious, to calculate $E(X|H)$ and $E(X|D)$. Using the rule that h grows when $E(X|H) > E(X|D)$, the equilibrium value of h is the one for which the two expected payoff curves cross. If we again employ the densities from Figure 1, and use $C = x_1/4$, the expected payoff curves and the equilibrium h are as depicted in Figure 6.

Two features of Figure 6 merit discussion. Note first the discontinuities in the two payoff curves. To see why these occur, recall that it would not be in the interests of anyone to pay the costs of scrutiny if the population initially consisted almost entirely of H's. All of the H's, even those with the highest S values, expect a higher payoff interacting at random with people from the population at large. For values of h close to

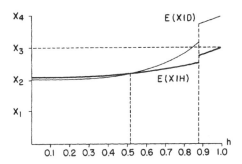

FIGURE 6. EQUILIBRIUM WHEN THERE ARE COSTS OF SCRUTINY

one, the expected payoff to the D's exceeds that to the H's, which, as noted earlier, causes h to fall. Once h reaches the value for which

$$(14) \quad E(X,0) = hx_3 + (1 - h)x_1 = x_3 - C,$$

however, it abruptly becomes in the interests of those H's with $S > U_D$ to become sensitized, thereby to interact selectively with one another, rather than with individuals chosen at random. At the value of h that solves equation (14), the H's who incur the costs of becoming sensitized are exactly compensated for those costs by the higher gains from selective interaction. But their sudden unavailability for interaction with the remainder of the population produces an abrupt decline in everyone else's expected payoff. This decline is reflected in the discontinuities in the expected payoff curves in Figure 6.

Note also that the equilibrium value of h in Figure 6 is larger than the corresponding value in Figure 5. Although the costs of scrutiny reduce the expected payoffs to both groups in absolute terms, the reduction is more pronounced for the D's. This happens because the costs of scrutiny cause the schedule $S^*(h)$ to shift to the right. And given that f_D lies to the left of f_H, any such shift in $S^*(h)$ will cause a proportionally greater reduction to the D's payoff than to the H's. It is perhaps fitting that, in their struggle with the D's, the H's are abetted by the fact that it requires some effort to observe the symptoms of trustworthiness.

This model treats people as if they were either honest or dishonest. In reality, however, honesty is not a binary trait. A more fully articulated theory would likely predict the coexistence of honest and dishonest impulses within each individual. The resulting environment would be one in which people search hopefully for others they can trust, while remaining ever vigilant against the possibility of being cheated—in short, an environment very much like the one that confronts us.

This outcome is at once in harmony with the economists' view that people pursue self-interest and with their critics' view that people often transcend selfish tendencies. The clear irony is that the gains from cooperation cannot be fully exploited here unless one adopts at the outset what everyone would call a genuinely unselfish point of view.[10]

The model thus suggests that people will often refrain from cheating not because they fear being caught, but because cheating simply makes them feel bad. This view is consistent with an extensive body of empirical evidence assembled by Jerome Kagan; as he summarizes his interpretation of that evidence:

> Construction of a persuasive rational basis for behaving morally has been the problem on which most moral philosophers have stubbed their toes. I believe they will continue to do so until they recognize what Chinese philosophers have known for a long time: namely, feeling, not logic, sustains the superego.   [1984, p. *xiv*]

Feelings may indeed sustain the superego. But in the model discussed here, it is logic that ultimately sustains those feelings.

---

[10] This is not to say, of course, that we never encounter the more narrowly selfish utility function generally assumed by economists. People of the sort who inhabit economic models surely do exist; but most of us (economists included!) make every effort to steer clear of them.

V. Other Aspects of the Utility Function

A. *Concerns About Fairness*

Like feelings of guilt, concerns about fairness may also play a useful role in the utility function.[11] People often walk away from profitable opportunities when they perceive the terms of a transaction to be "unfair." To make sense of such behavior in the framework of conventional economic models, we must imagine that they are investing in the development of reputations for being tough bargainers. That may sometimes be so. But much of the time people walk away just because they would feel even worse if they did not. And even the strategy of developing a tough reputation would be much easier to execute if people actually *disliked* participating in profitable, but one-sided, transactions. People who felt that way would not be lying when they threatened to abandon negotiations. Their preference structure would serve, in effect, as a commitment device, one that makes it in their interests to reject an unfair bargain.

B. *Anger and Vengeance*

Anger, too, often motivates behavior that does not appear self-serving. Customers who are wronged by a supplier may give up several days of their time and suffer much inconvenience in the process of tracking down a $50 refund. People often behave this way even when they are visiting distant locations, where there is no reasonable prospect that their behavior will help them develop reputations for toughness. The proximate cause is that behaving any other way would leave them feeling angry and frustrated.

Viewing tastes as commitment devices suggests a clear rationale for such behavior. If a capacity for anger is observable in a person, suppliers will know he is commited to seek redress, even if it does not "pay" him to do so. This, in turn, makes it rational for the buyer to trust the supplier, even

---

[11] See my book (1985b, chs. 2, 5, 6).

when some dimensions of product quality cannot be observed before purchase. The buyer's capacity for anger thus benefits both himself *and* the supplier. For, as George Akerlof's lemons model demonstrates (1970), the alternative here would be for no transaction to take place at all.

VI. Concluding Remarks

Tastes are important. Seemingly minor alterations in the utility function can produce large changes in the conclusions that emerge from economic models. As I have emphasized elsewhere (1984a,b; 1985a,b), the presence of concerns about relative position will affect the wages people earn and also how they spend them. Similarly, the presence of a capacity for anger makes it easier to understand a variety of behaviors that are difficult to rationalize by reference to utility functions of the usual sort.[12] Finally, conscience and other moral sentiments also play a powerful role in the choices people make.

Modern economists regard behaviors arising from these emotions as lying outside the scope of our standard model. But it is neither necessary nor productive to view them this way. Our utility-maximization framework has proven its usefulness for understanding and predicting human behavior. With more careful attention to the specification of the utility function, the territory to which this model applies can be greatly expanded.

[12] Hirshleifer (1978b) makes a similar point.

REFERENCES

Akerlof, George, "The Market for 'Lemons'," *Quarterly Journal of Economics*, August 1970, *84*, 488–500.

_____, "Loyalty Filters," *American Economic Review*, March 1983, *73*, 54–63.

Axelrod, Robert, *The Evolution of Cooperation*, New York: Basic Books, 1984.

Becker, Gary and Stigler, George, "De Gustibus Non Est Disputandum," *American Economic Review*, March 1977, *67*, 76–90.

Coelho, Philip, "An Examination into the Causes of Economic Growth: Status as an Economic Good," *Research in Law and Economics*, 1985, *7*, 89–116.

Darwin, Charles, *The Expression of Emotions in Man and Animals* (1872), Chicago: University of Chicago Press, 1965.

Dawkins, Richard, *The Selfish Gene*, New York: Oxford University Press, 1976.

Ekman, Paul, *Telling Lies*, New York: W. W. Norton, 1985.

Frank, Robert H., (1984a) "Interdependent Preferences and the Competitive Wage Structure," *Rand Journal of Economics*, Winter 1984, *15*, 510–20.

_____, (1984b) "Are Workers Paid Their Marginal Products?," *American Economic Review*, September 1984, *74*, 549–71.

_____, (1985a) "The Demand for Unobservable and Other Nonpositional Goods," *American Economic Review*, March 1985, *75*, 101–16.

_____, (1985b) *Choosing the Right Pond*, New York: Oxford University Press, 1985.

_____, "The Nature of the Utility Function," in A. MacFadyen and H. MacFadyen, eds., *Economic Psychology*, Amsterdam: North-Holland, 1986.

_____, *Beyond Self-Interest: Prisoner's Dilemmas and the Strategic Role of the Emotions*, New York: W. W. Norton, 1988.

Hamilton, W. D., "The Genetical Theory of Social Behavior," *Journal of Theoretical Biology*, 1964, *7*, 1–32.

Hirshleifer, Jack, "Economics from a Biological Viewpoint," *Journal of Law and Economics*, April 1977, 20:1–52.

_____, (1978a) "Competition, Cooperation, and Conflict in Economics and Biology," *American Economic Review Proceedings*, May 1978, *68*, 238–43.

_____, (1978b) "Natural Economy Versus Political Economy," *Journal of Social and Biological Structures*, October 1978, *1*, 319–37.

_____, "The Emotions as Guarantors of Threats and Promises," Department of Economics Working Paper No. 337, UCLA, August 1984.

Kagan, Jerome, *The Nature of the Child*, New York: Basic Books, 1984.

Konner, Melvin, *The Tangled Wing*, New York: Holt, Rinehart, and Winston, 1982.

Krebs, John R. and Dawkins, Richard, "Animal Signals: Mind Reading and Manipulation," in J. R. Krebs and N. B. Davies, eds., *Behavioral Ecology: An Evolutionary Approach*, 2nd ed., Sunderland: Sinauer Associates, 1984.

Lykken, David T., *A Tremor in the Blood*, New York: McGraw-Hill, 1981.

Rubin, Paul, "Evolved Ethics and Efficient Ethics," *Journal of Economic Behavior and Organization*, June/Sept. 1982, *3*, 161–74.

_____, **and Paul, Chris,** "An Evolutionary Model of Taste for Risk," *Economic In-quiry*, October 1979, *17*, 585–96.

Schelling, Thomas C., *The Strategy of Conflict*, New York: Oxford University Press, 1960.

_____, "Altruism, Meanness, and Other Potentially Strategic Behaviors," *American Economic Review Proceedings*, May 1978, *68*, 229–30.

Trivers, Robert L., "The Evolution of Reciprocal Altruism," *Quarterly Review of Biology*, March 1971, *46*, 35–57.

_____, *Social Evolution*, Menlo Park: Benjamin/Cummings, 1985.

Wilson, Edward O., *Sociobiology*, Cambridge: Belknap Press, Harvard University Press, 1975.

_____, *On Human Nature*, Cambridge: Harvard University Press, 1978.

Part III
Altruism

[6]

Recent Work on Human Altruism and Evolution*

Neven Sesardic

Altruism and evolution do not mix well. Paraphrasing Quine (1969, p. 126) one might say that "inveterately altruistic creatures have a pathetic tendency to die before reproducing their kind." Such a view that rather simple Darwinian forces work strongly against the preservation of altruistic traits is actually the background against which various explanations of the genesis of human altruism are being defended and discussed today. Indeed, if we call "paradoxical" any situation where we have seemingly convincing evidence in favor of each of two (or more) propositions that are seemingly mutually irreconcilable, then participants in the current debate about the emergence of human altruism are also haunted by a paradox. Moreover, since the debate is so conspicuously and so persistently revolving around this basic difficulty it seems methodologically appropriate to make the "paradox of altruism" the cornerstone of my presentation of the recent work on the topic.

INTRODUCTION: THE PARADOX OF ALTRUISM

To begin with, here is a crude version of the paradox of altruism: on one hand it seems that the existence of human altruism is an undeniable psychological fact, but on the other hand it seems, on evolutionary grounds, that altruism cannot exist, because species with this trait are expected to have gone extinct through the process of natural selection. (Selfishness increases biological fitness, and only the fittest survive.)

The alleged incompatibility between these two propositions is easily resolved. A standard strategy is to remove the sting of the para-

* I would like to thank Elliott Sober and an anonymous referee for *Ethics* for their very helpful critical comments on the first draft of this article. I have also benefited from discussions after presenting the paper to audiences in Zagreb, Dubrovnik, at Notre Dame, Purdue, Rutgers, and at the University of Minnesota.

dox by distinguishing two meanings of 'altruism', psychological and evolutionary.[1] For our purposes, psychological altruism (altruism$_p$) and evolutionary altruism (altruism$_e$) will be defined as follows:

A is behaving altruistically$_p$ = $_{df}$ A is acting with an intention to advance the interest of others at the expense of his own interests.

A is behaving altruistically$_e$ = $_{df}$ The effect of A's behavior is an increase of fitness of some other organisms at the expense of its own fitness.

Invoking this conceptual distinction we can perfectly consistently state both that altruism$_p$ is a (psychological) fact and that altruism$_e$ is an (evolutionary) impossibility. However, a deeper and subtler difficulty remains. Namely, even after avoiding a direct contradiction by separating the two senses of 'altruism' one can argue that these two senses are not completely unconnected, and that for this reason we may still be left with an epistemic tension between believing in the reality of psychological altruism and at the same time doubting the existence of evolutionary altruism.

Indeed, I shall try to show that we cannot get rid of the paradox of altruism by simply making the clashing statements speak about different things (i.e., different altruisms). With this purpose in mind I shall replace the initial crude version of the paradox of altruism with a more sophisticated form. The new version will consist of four propositions (instead of two), each of them again being seemingly very plausible despite all of them appearing to be mutually incoherent. This proposed reconstruction could do more than just help to exhibit the logical skeleton of our puzzle. Adding more structure and precision to the formulation of the paradox leads almost by itself to a novel, neat classification of alternative approaches, and (I hope) to a more fruitful comparison of these rival views. By making transparent the basic points of disagreement it could perhaps also lead to a deeper understanding of the genesis of human altruism.

Here is the "incongruous tetrad," the four conflicting propositions that I offer as a reconstruction of the paradox of altruism:

(1) Altruism$_e$ is a selectively disadvantageous trait.
(2) Altruism$_p$ tends to lead to altruism$_e$.
(3) Altruism$_p$ exists.
(4) Altruism$_p$ is a product of natural selection.

1. For good discussions of this important distinction see Kavka 1986; Sober 1988, 1993a; and Wilson 1992.

Our predicament is in short this: how is it possible (3) that altruism$_p$ exists and (4) that it is a product of natural selection if (2) it tends to lead to altruism$_e$ which itself (1) is a selectively disadvantageous trait? Proposition (1) is a statement about evolutionary altruism, (3) and (4) are statements about psychological altruism, and (2) supplies a link between them that creates a logical strain in the tetrad.

I have to explain why I have introduced (3) as a separate claim about psychological altruism, although its truth is obviously presupposed by (4). (Altruism$_p$ cannot be a product of natural selection unless it exists.) The reason is that (4) has two components: it presupposes that altruism$_p$ exists, and it states that altruism$_p$ is a product of natural selection. For the sake of clarity these two components ought to be considered and evaluated separately. The supposition—expressed by (3)—can be attacked by insisting that the behavior satisfying the definition of psychological altruism simply does not exist; or, alternatively, conceding the existence of psychological altruism, one can deny (4) by asserting that this kind of behavior is actually not a product of natural selection. Both of these arguments have been defended in the literature, and it seemed to me that it would only invite unnecessary confusion to fuse both controversial points into one sentence.

Note also that on both definitions of altruism the so-called reciprocal altruism is a misnomer: it is not altruism at all. Evolutionarily altruistic acts imply the net loss of fitness of the actor, and psychologically altruistic acts imply the intention of the actor to genuinely sacrifice his own interests. This terminological decision to exclude reciprocal altruism from the scope of altruism proper accords well with a widespread biological and philosophical usage. For instance, Peter Singer says that "reciprocal altruism is not really altruism at all; it could more accurately be described as enlightened self-interest" (Singer 1981, p. 42). Rawls prefers not to talk about reciprocal altruism but to call it simply reciprocity (Rawls 1971, p. 503), and Robert Trivers in his classical paper "The Evolution of Reciprocal Altruism" claims that "models that attempt to explain altruistic behavior in terms of natural selection are models designed to take the altruism out of altruism" (Trivers 1978, p. 213). The feeling that reciprocity is not altruism *sensu stricto* is best expressed in one of La Rochefoucauld's maxims: "When we help others in order to commit them to help us under similar circumstances, [the] services we render them are, properly speaking, services we render to ourselves in advance."[2]

There is, however, an additional, substantive reason for shutting the door on debating reciprocal altruism in the present context. Speak-

2. Sober is also arguing against classifying reciprocal altruism as altruism (1988, p. 84).

ing in evolutionary terms, reciprocal altruism has no puzzling features. Its being easily explainable as a selectively advantageous trait (to individual organisms) robs it of any biological "queerness." Hence the crucial difficulty reflected in (1) does not apply to it. On the psychological side, similarly, reciprocal altruism is unproblematic: it can arise through natural selection simply by riding on its biological counterpart (evolutionary reciprocal altruism), whose evolutionary credentials are impeccable, as we have just seen. Therefore, one who wants to focus on the paradox of altruism as here formulated is well advised to get reciprocal altruism out of the way as a red herring.

Our paradox can be resolved by choosing between the following two strategies. Either one can reject (at least) one of the propositions of the incongruous tetrad or else one can attempt to prove that contrary to the appearances the whole set is in reality perfectly coherent. The first (eliminativist) strategy comes in four possible variants (i.e., each one of the four assumptions can be dropped); interestingly enough, all these variants had their advocates in the continuing debate about altruism. On the other hand, the second (reconciliationist) strategy splits upon analysis into three possible versions that, again happily, represent the currently most important theoretical standpoints. Moreover all these enumerated options exhaust the logical space of possible solutions. (Of course, I do not want to say that there will be no novel approaches to the problem, only that any of them will have to fall into some place in my scheme.) Let us therefore follow this emerging order and discuss in turn the two strategies in all their subvariants.

FIRST STRATEGY: ELIMINATION

None of the four propositions creating the paradox carries its truth on its sleeve. Each one of them has been occasionally regarded by some as dubious and by others as outright false. Our task in this section is to see whether there are good reasons for rejecting any of these claims in particular. To anticipate a little, my conclusion will be that the examination largely bears out the truth of all four propositions, and that, consequently, the road to solution is better sought in the reconciliationist approach.

Claim (1)

Starting with claim (1)—"Altruism$_e$ is a selectively disadvantageous trait"—there are two phenomena that prima facie speak against it: (*a*) kin selection and (*b*) group selection.

a) The originator of the idea of kin selection was J. B. S. Haldane, although he did not use the term. As early as 1932 he wrote, "Insofar as it makes for the survival of one's descendants and near relations, altruistic behavior is a kind of Darwinian fitness, and may be expected to spread as a result of natural selection" (Haldane 1932, p. 131). So,

132   *Ethics   October 1995*

if we take account of the effects of behavior on the agent's relatives (or, more precisely, on the carriers of the same genes) it may happen, contrary to (1), that altruistic behavior becomes selectively advantageous, despite being harmful or even lethal for the individual in question.

Here all turns again on the definition of evolutionary altruism. Recall that it was defined as the behavior of an organism which decreases its own fitness while increasing the fitness of some other organisms. In evolutionary theory, at least since William D. Hamilton's (1964) important theoretical contribution, 'fitness' is often taken to mean inclusive fitness, and for the purposes of our discussion we shall stick to this usage. But adopting the inclusive fitness approach entails that our measurement of the fitness of a behavioral disposition is not limited solely to the consequences of that behavior on its emitter; "inclusive fitness" incorporates, *ex vi termini,* the effects of this behavior on close relatives. To take a concrete example, someone who sacrifices his own life and thereby, say, saves the lives of more than two of his full siblings is thereby actually increasing his (inclusive) fitness, and by so acting he is not behaving altruistically (according to this definition of altruism$_e$).[3] Therefore, when (1) is properly interpreted, it is in no way threatened by the existence of kin selection. (To evade the objection from kin selection the following concise formulation of [1] suggests itself: "The behavior that systematically decreases the inclusive fitness of its emitter is selectively disadvantageous.")

b) A more serious challenge to (1) is group selection. The groups consisting of altruists can fare better than the groups of selfish organisms, and consequently it seems that these groups can even be favored by selection despite the fact that such altruistic behavior, at the individual level, continues to decrease the inclusive fitness of any particular altruist organism. The range of group selection is a matter of great controversy in biology. In the earliest stage, group selection was being routinely invoked without much awareness of formidable problems concerning its way of operation. The central difficulty is that, although it is undoubtedly in the evolutionary interest of any individual to be a member of a group of altruists, he always gains a selective advantage by being an egoist himself. This fact that group selection is open to subversion from within (i.e., from the level of individual selection) is transparent in figure 1.[4]

---

3. "From a genetic perspective, you are helping part of your *self* (i.e., replicas of your genes) when you help your brothers and sisters. Faced with a decision between saving yourself or three full siblings, you save more of your (genetic) self by saving your siblings" (Krebs 1982, p. 453).

4. This kind of diagram is a standard way of presenting the basic structure of the group selection problem. See Sober 1984, p. 186; 1988, p. 80; 1993a, p. 206; 1993b, p. 98; Elster 1989, p. 127; Peressini 1993, p. 572).

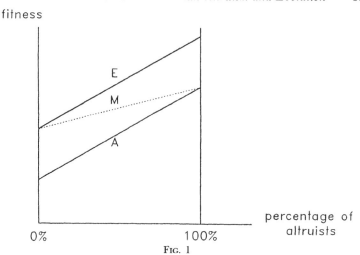

Fig. 1

Three lines, *E, M,* and *A,* represent, respectively, the fitness of an egoist, the mean fitness of the group, and the fitness of an altruist. Each of them is plotted as a function of the percentage of altruists in the group. Obviously, from the point of view of the group it is better to have as many altruists as possible (because line *M* rises if we move from 0 to 100 percent of altruists), but from the individual perspective it always pays to be selfish (because line *E* is always higher than *A*).

In order to explain altruism by group selection it is emphatically not enough to show that universal altruism is a collectively preferred state. There is an additional obligation (and not a trivial one at that) to specify a causal mechanism by which this state can be reached. (The property of being a collectively preferred state doesn't by itself cause anything.) The argument for group selection was elaborately worked out by Wynne-Edwards (1962), and this whole approach was then forcefully attacked by George C. Williams in his classic *Adaptation and Natural Selection* (Williams 1966). In overreaction to the once too facile use of group selection explanations there was a tendency later to dismiss them all as being pseudoexplanations. Today, however, it has become a matter of consensus that group selection is a possible evolutionary process, but still many biologists believe that it can work only under very special and relatively rarely satisfied conditions (e.g., if groups are small, if their extinction rate is high, if there is little intergroup migration, etc.). Assuming the fragility of group selection (i.e., its operating only under rather exceptional circumstances), our belief in (1) would be partly vindicated. That is, barring the unlikely concurrence of all the prerequisites for group selection it would remain true that altruism$_e$ is a selectively disadvantageous trait.

It should be noted, however, that there is a growing opinion in evolutionary circles that the skepticism toward group selection was

mainly motivated by bad arguments and elementary conceptual confu-
sions (involving the distinction between "replicators" and "vehicles"),
and that group selection should be recognized as an extremely important and strong causal force both in biology and in human behavioral sciences (see particularly Wilson and Sober 1994). Whatever the
outcome of this discussion about the general status of group selection,
it is by itself not likely to impinge so much on the issue of altruism
for the simple reason that, as Wilson and Sober themselves say, even
extreme individualists (the die-hard opponents of group selection)
"acknowledge group-level adaptations *when they are easily exploited
within groups*" (p. 599; emphasis added). In other words, the main
thrust of Wilson and Sober's criticism is that group-level adaptations
fail to be recognized in the absence of altruism (Wilson and Sober
1994), and it seems therefore that they would themselves agree that
the operation of group selection has been fairly transparent in those
situations (imagined or real) where it worked against the forces of
individual selection (i.e., in cases of altruism).

To recapitulate our discussion of (1), the objection from kin selection
was met by defining altruism$_e$ in terms of inclusive fitness, and by thus
showing that the alleged counterexample of selectively advantageous
altruism toward relatives does not count as altruism at all. The objection
from group selection was answered by pointing to its rare occurrence
and by concluding therefrom that it is improbable that a massive presence
of human altruism could be adequately accounted for by such a delicate
and extremely fine-tuned causal mechanism.[5]

Claim (2)

Claim (2)—that altruism$_p$ tends to lead to altruism$_e$—would be trivially
true if there were no difference in meaning between the two altruisms.
This would in fact amount to sliding back to the crude version of the
paradox of altruism, dismissed at the beginning but not wholly without
adherents. For example, even Edward O. Wilson did not hesitate to
state that "altruism is defined in biology, *as in everyday life*, as self-
destructive behavior for the benefit of others" (Wilson et al. 1973,
p. 953; emphasis added; repeated verbatim in Wilson et al. 1977,
pp. 458–59).

Most authors, however, are most of the time aware that a connection between evolutionary and psychological altruism is not so immediate and purely semantic. The misfit arises on two counts: evolutionary
altruism is defined in terms of actual effects with respect to fitness,

---

5. Later (see section titled Version [4.3]) I shall consider another approach which
I have classified as a reconciliationist view but which can with no less justification be
regarded as a way of denying (1).

whereas psychological altruism is defined in terms of intended effects with respect to personal interests. Obviously, then, a conjectured link between two kinds of altruism can break either because intended effects of human action do not in general correspond to actual effects or because the effects expressed in the currency of interests do not correspond to the effects as measured by fitness. Therefore, in order to make good claim (2)—that psychological altruism tends to lead to evolutionary altruism—we have to show (*a*) that human goals tend to be realized as intended, and (*b*) that there is some correlation between interests and fitness.

a) What matters evolutionarily is how intentions are actualized, not what they are inside the mind. If it were literally and massively true that the road to hell is paved with good intentions or, to put it less metaphorically, if human actions happened in general to have effects contrary to those envisaged by the subject, psychological altruism would cease to create any evolutionary puzzle. In that case, altruistic$_p$ acts would be favored by selection, for they would deviantly and perversely promote the self-interest of the agent against his own will. But as a matter of fact intentions are not so wholly impotent and disconnected from reality. More often than not they are realized according to the plans of the agent. Having an intention to ϕ leads to ϕing, other things being equal. And, frequently, other things are equal. So, although altruistic$_p$ acts are defined in terms of altruistic intentions rather than in terms of actual effects, altruistic intentions tend in reality to produce genuinely altruistic consequences.[6] For this reason, the gap between intentions and their realization offers a poor basis for attacking (2).[7]

b) An alternative and more promising route for disputing (2) is by driving deeper a wedge between interests and fitness. A thin end of the wedge is already there: the advancement of one's interests does not always coincide with the increase of one's fitness. For example, a couple's decision not to have children may serve their interests (say, because of the high probability of their life being destroyed by the birth of a seriously handicapped child), although this decision may significantly decrease their fitness (by their forgoing a nonzero chance of having perfectly healthy offspring). Conversely, too, one's interests may occasionally be harmed by acts which happen to increase one's fitness. All this, however, is not sufficient to undermine (2). It merely

---

6. I am here indebted to Kavka 1986.

7. There is a possibility, however, that despite intentions being typically realized according to the agent's plans, psychological altruism may still not lead to evolutionary altruism because, completely unbeknownst to the subject, these genuinely altruistic$_p$ acts just regularly happen to have fitness-increasing consequences as an altogether unintended by-product.

shows that there is no perfect match between interests and fitness, and this is fully compatible with (2) which was deliberately formulated as the statement of tendency. To vindicate (2), we do not even need to start by defining the concept of interest—which is very fortunate in the light of the notorious murkiness of this notion in the social science literature. Indeed, we can avoid the general issue of what interests are by simply confining our whole effort to showing that some important particular interests (counting as such on any definition of interest) have a systematic connection with fitness. Paradigmatic cases of this kind of interest are protection of one's health, avoidance of danger, keeping one's possessions, and so forth. So, again with a ceteris paribus clause, it is obviously against one's interest to act so as to destroy one's health, to bring oneself into a dangerous situation, to lose one's possessions, and so forth. In addition, it is easily recognized that such acts by themselves lead naturally to the loss of inclusive fitness (provided, of course, that they do not benefit close relatives of the agent). Therefore, we gain support for our belief that, at least with respect to some basic interests, the psychological propensity to sacrifice one's own interests tends to produce an evolutionarily self-defeating disposition to lower one's fitness.[8]

Claim (3)

Statement (3), the affirmation of the existence of psychological altruism, can be denied in two ways: by claiming (*a*) that it is necessarily false or (*b*) that it is contingently false.

a) The idea that there cannot be altruistic$_p$ acts is at the core of a philosophical thesis known as "psychological egoism." (The name derives from the need to distinguish it from another kind of egoism, "normative" or "ethical" egoism.) The main support for psychological egoism comes from a hedonistic interpretation of human motivation. Stripped to essentials, the argument is as follows: we are moved to action solely by the expectation of pleasure; attaining pleasure is a purely selfish aim; ergo, our behavior is never altruistic. In the opinion of many philosophers this view has been conclusively refuted already by Bishop Butler in his sermons (first published in 1726).[9] There are two basic difficulties for psychological egoism. On one hand, there are

---

8. It should be stressed here that the connection between interests and fitness is not only probabilistic but that it is also context dependent, and that it can easily break with changes in the environment. To borrow an example from Herbert Simon (1993, p. 158), although wealth was perhaps a major contributor to fitness in earlier centuries, in the contemporary Western societies there is a negative correlation between income level and reproduction rate; the statistics show that, as a matter of fact, the poor are fitter than the rich.

9. See Butler 1983, pp. 47–49.

many actions that we can only with great strain and implausibility reinterpret as being a search for pleasure and self-gratification. Who would not agree with Chesterton (quoted in Feinberg 1971, p. 498) that a philosopher is misusing the word 'self-indulgent' if he says that a man is self-indulgent when he wants to be burned at the stake? On the other hand, even with respect to some actions that do result in obtaining pleasure it seems demonstrably not true that they are undertaken in order to obtain pleasure. If I give $100 to charity this may give me a pleasant feeling of being a generous person. But I cannot be pleased with this act of generosity if I know that my contribution was exclusively stimulated by the anticipation of this pleasure. For in that case I am not being generous at all, and consequently I have nothing to be pleased with. In Jon Elster's terminology (Elster 1983, pp. 43–108), pleasure is often "a state that is essentially a by-product," and on that account hedonism cannot be the whole story about human motivation.

But taking this line is perhaps trying to prove too much. In order to put psychological egoism into doubt we are actually not obliged to make a positive step and produce a philosophical argument establishing the existence of at least some nonegoistically motivated desires. Rather, it is entirely sufficient to show, purely negatively, that the aprioristic argument for general egoism is unconvincing.[10] But this is obviously not such a formidable task anymore. Even those philosophers who are skeptical about the positive achievements of the Butlerian argument (see, e.g., Sober 1992) would certainly not be willing to argue that human altruism can be excluded on analytical grounds, by merely investigating the nature of practical reason.[11]

Psychological egoism is now too often serving just as the last resort to some ardent sociobiologists, when they find themselves confronted with an ostensibly altruistic act, a living counterexample to their simplistic theory of human behavior. Acutely challenged by this phenomenon, but unable to reduce it either to reciprocity ("soft-core altruism") or to helping one's kin ("hard-core altruism"), they simply fall back on the entirely aprioristic thesis of general egoism, and by appeal to this defunct philosophy they hope to explain away all these remaining

---

10. If the issue is to be decided empirically, purely conceptual or philosophical arguments (like psychological egoism) are *eo ipso* run out of court.

11. Bernard Williams suggests the following empirical test of altruism: "a man might be faced, by some manipulator, with the choice between the following: on the one hand, that *p* should be the case later but that he (the subject) should after a few minutes believe that not-*p;* on the other hand, that not-*p* should be the case later, but that he, after a few minutes, should believe that *p*. No conceptual manoeuvres could possibly persuade a man who wanted that *p* that he had to choose the latter alternative. If *p* involved someone else's welfare, this set-up could constitute something of a test for altruism" (1973, p. 262).

recalcitrant cases as well. For a typical move in this vein see Wilson (1978, p. 165), and for the criticism that is exactly on target see Kitcher (1985, pp. 402–3).

b) An alternative attack on (3), the attempt to show that it is contingently false, relies on empirical argument. In the case of any prima facie altruistic behavior the task is here to uncover the presence of some concealed motivation that changes completely the initial picture, and that, when taken into account, turns the behavior in question into a purely egoistic action. For instance, some acts of blood donation may on closer inspection prove to be motivated less by a wish to help others, and more by trying to improve one's social image. Surely, all that glitters is not altruism. Yet it is another thing, entirely, to claim that every appearance of altruism is deceptive and that some strong and dominating selfish motives are always there to be found to favor an egoistic interpretation. True, the route to a belief in universal egoism is facilitated by the fact that apparently altruistic acts regularly result in the reduction of negative arousal, the avoidance of external and internal punishments, or in a mood enhancement, and it is then surely legitimate to suspect that all the seemingly shining examples of human selflessness may in reality be motivated by a narrow-minded and purely egoistic anticipation of such likely consequences of our acts. But, surprisingly enough, the egoistic hypothesis is here meeting formidable empirical difficulties. In various ingeniously devised laboratory experiments (see Batson 1991, 1992; Batson and Oleson 1991; and in particular Batson and Shaw 1991, and the discussion of their target article in the same issue of *Psychological Inquiry*), Daniel C. Batson has created the situations where, exceptionally, each of the standard egoistical accounts gives opposite predictions from the altruism hypothesis, and he has also shown, most important, that it is the altruism hypothesis that consistently comes out as the winner in these decisive confrontations. Batson's research program is today the most serious challenge to psychological egoism, and quite possibly its burier too.

Claim (4)

If claim (4), that psychological altruism is a product of natural selection, is dropped psychological altruism ceases to have any puzzling evolutionary features. For like other traits that lower the fitness of their possessors, it presents no problem for evolution unless it is believed to have arisen by natural selection. Are there good reasons, though, to believe this?

Let us right away admit that at the present stage of the debate there is no orthodox or commonly accepted account of how psychological altruism was maintained by the forces of selection. (If there were such an account the paradox of altruism would immediately dissolve; for we would then be in the position to know how altruism$_p$ was

selected despite its connection with the selectively inferior altruism$_e$.)
What we do have at the moment are just various speculations about
the evolutionary origins of psychological altruism, the hypothesized
Darwinian histories with different degrees of plausibility. Most important, our belief in general thesis (4) does not draw its strength
from any of these particular selective accounts being very convincing
or imposing its truth on us. On the contrary, it seems that each of
these selective accounts was first and foremost prompted precisely by
the hunch that (4) must be true, that is, that there has to be some
evolutionary explanation for the genesis of psychological altruism.

To put the matter differently, although we lack direct evidence
in favor of (4)—because there is no generally accepted causal story
about how altruism$_p$ was produced by natural selection—there is
nevertheless weighty indirect evidence supporting it. The very nature
of the altruistic$_p$ predisposition in humans—the fact that it is so widespread, that it extends at least as far back in time as to the period of
hunter-gatherers, that it is sustained by powerful emotions, that it is
already present in very early childhood (Schwartz 1993, p. 322), and
that it occupies such a manifestly central place in human mental-
ity—all this taken together strongly suggests that altruism has its roots
in our evolutionary past. This is not a knockdown proof, but in the
opinion of many scholars such considerations carry enough weight to
regard (4) as much more than just a fruitful working hypothesis. By
way of illustration, John Rawls confidently states that "the theory of
evolution would suggest that [human nature] is the outcome of natural
selection," and that his postulated source of altruistic behavior, "the
capacity for a sense of justice and the moral feelings is an *adaptation*
of mankind to its place in nature" (Rawls 1972, pp. 502–3; emphasis
added). Alan Gibbard is just one among contemporary empiricist philosophers who view the fact that beings with a sense of justice seem
to fare worse than pure egoists as crying out for an evolutionary
explanation: "What kinds of psychological propensities are involved
[in a sense of justice], and how might evolutionary theory explain
humans' having those propensities?" (Gibbard 1982, p. 33). Alexander
Rosenberg expresses a widely shared view when he writes that "one
is tempted to . . . say that *the only likely* explanation of why Homo
sapiens cooperate, despite the temptations of costless free riding, must
be evolutionary" (Rosenberg 1988, p. 832; emphasis added). These
and similar quotations (see also Darwin 1874, pp. 149–50; Mackie
1977, pp. 113, 192) are not meant as an appeal to authority, but as
an illustration of a growing consensus that even if proposition (4) is
going eventually to be rejected the possibility of its truth should at
present be taken very seriously indeed.

In conclusion, it is worth stressing that our basic problem does
not go away even if we assume that human altruistic predispositions

have nothing to do with biology, and that they are merely a product of "culture." For in the process of cultural selection it is again the case that fitter cultural variants will tend to be preserved, so that it would still remain pretty mysterious how a selectively unfavorable trait like altruism was not stamped out a long time ago by the forces of cultural selection. Appealing to the claim that the cultural inheritance of altruistic traits fulfills a useful function at the social level (that it is "socially functional") is definitely insufficient, even when true; as already said, without pointing to possible mechanisms by which the allegedly functional state could be attained, such appeals easily degenerate into pseudoexplanations.[12] Expressed more concisely, the perseverance of an altruist "meme" in a population is no less puzzling than the perseverance of an "altruist" gene.

SECOND STRATEGY: RECONCILIATION

The consideration of the four propositions making the incongruous tetrad has shown that each of these claims is a serious candidate for truth. Moreover, many people are inclined to think that the evidence for them is collectively compelling, and that all these propositions ought to be jointly embraced. This leads to the increased pressure for finding a workable reconciliationist solution of the paradox. Perhaps the four propositions are in reality cotenable?

Pursuing this path, our attention is immediately directed to (4), the claim that altruism$_p$ is a product of natural selection. Referring to Sober's (1984) well-known and important distinction between "selection of" and "selection for," the truth of (4) obviously entails that there was selection *of* altruistic$_p$ organisms but not necessarily that altruism$_p$ was also selected *for*. To put it differently, a trait can be selected although it does not confer any selective advantage on its bearers, or, what comes to the same thing, without its having any adaptive value. This can happen in two crucially different ways: either a trait may have ceased to be adaptive, albeit it was itself one selected for in an earlier, different environment, or else it may never have been an adaptation but it was nevertheless selected through being tightly connected to some other adaptive trait that was the true target of selection. The first possibility is that a selected feature be only a relic of a past adaptation while the second possibility amounts to its being only a by-product of a still-adaptive trait. In both cases we see that a selectively not advantageous (or even disadvantageous) trait can be a product of selection.

---

12. For an interesting dual-inheritance scenario positing one such mechanism see section titled Version (4.2*b*), and for a purely cultural account of the emergence of altruistic norms of behavior see Allison (1992).

This is actually what makes room for a reconciliationist move. That is, even after granting (1) that altruism$_e$ is a fitness-decreasing trait, (3) that altruism$_p$ exists, and (2) that altruism$_p$ leads to altruism$_e$, it is still possible to claim without inconsistency (4) that altruism$_p$ is a product of selection. For, altruism$_p$ may have been selected in a different environment in the past when it did have fitness-increasing consequences, or it may have been selected, despite its selective disadvantage, by being inseparably tied to another, strongly adaptive trait (which more than compensated for its own harmful effects).

Here are these two versions of (4), concisely formulated:

> (4.1) *Vestige theory:* Altruism$_p$ is a product of natural selection, *but it was adaptive only long ago.*
>
> (4.2) *By-product theory:* Altruism$_p$ is a product of natural selection, *but it was never adaptive.*

There is also a third, strongest version of (4), according to which altruism$_p$ not only was adaptive but has until now preserved its adaptive qualities:

> (4.3) *Continuing adaptation theory:* Altruism$_p$ is a product of natural selection, *and it is adaptive.*

On the face of it, the set containing (1), (2), (3), and (4.3) may appear to be formally inconsistent. (How could altruism$_p$ possibly be selected for, if it leads to altruism$_e$ which is selectively disadvantageous?) Therefore, there is a strong temptation to see (4.3) as having no place in discussing the reconciliationist strategy. In a way this is exactly right. Yet in another sense (see n. 21 for clarification) it is just here that we encounter a most ingenious reconciliationist argument for resolving the paradox of altruism. The three different versions of (4) are given in figure 2. Let us consider them in turn.

(4.1) Vestige Theory

We can believe that a given feature was produced by natural selection and yet be completely in the dark about how it was produced. This point was made by Mark Ridley and Richard Dawkins: "Civilized human behavior has about as much connection with natural selection as does the behavior of a circus bear on a unicycle. . . . Similarly, there probably is a connection to be found between civilized human behavior and natural selection, but it is unlikely to be obvious on the surface" (Ridley and Dawkins 1981, p. 32). There is little doubt that the capacity which enables the bear to keep his balance on a unicycle was shaped by natural selection, but it can only be a joke to suggest that the very skill of riding a unicycle made bears better adapted to their environment.

Applying this to our case, one can argue that for the question whether altruism$_p$ is a product of selection it is irrelevant to consider

(4.1) **Vestige theory**

natural selection in ⎯⎯⎯⎯⎯→ altruism   (the once adaptive disposition,
past environments   misfiring in <u>contemporary</u>
                                         environments)

(4.2) **By-product theory**

(a)

natural selection ⎯⎯⎯→ reason ⎯⎯⎯→ altruism

(b)

natural selection ⎯⎯⎯→ conformity ⎯⎯⎯→ altruism

(4.3) **Continuing adaptation theory**

natural selection ⎯⎯⎯→ altruism (still adaptive)

FIG. 2

its consequences for fitness in the contemporary environment.[13] If this trait was selected at all, it must have been so tens of thousands years ago under greatly different circumstances. Hence the fact that altruism$_p$ is now indeed disadvantageous (via leading to altruism$_e$ which decreases fitness) does not tell decisively against the hypothesis of selection because it is possible that altruism$_p$ was selected by having been evolutionarily advantageous under very dissimilar conditions obtaining in the distant past.

The only story I know of that tries to flesh out this explanation sketch is the kin-selection account. The idea is simple: if humans once lived in small groups consisting mainly of close relatives, kin selection would favor indiscriminate altruism. For, under these circumstances the beneficiaries of altruistic acts would be almost exclusively the close relatives of the "altruist," and it is easily understood how such a trait could be brought to fixation by selection. Moreover, it can be shown that, against such a background, unrestricted altruism, might be *on informational grounds* (see Sober 1981, p. 104) fitter than discriminating between relatives and nonrelatives, and then helping only the former: in a group where any interaction is most likely to be with a close relative, a mechanism for distinguishing relatives from nonrelatives

13. "Our genes gave us the propensities we had at conception—propensities to have certain characteristics in various hunting-gathering environments. That tells us nothing directly about what we are like in fact, in our own environments" (Gibbard 1990, p. 27).

would be all but useless; besides, it could be acquired only at some cost (in evolution, no less than in economics, there's no such thing as a free lunch). With a sufficiently low probability of encountering a nonrelative the usefulness of being able to recognize relatives would become so small that the costs of acquiring such an ability would have to be greater than the gain. It would then always pay to be an indiscriminate altruist: better to make an occasional, very rare error (to aid a nonrelative) than to carry the costs of an expensive error-avoiding mechanism which practically does no useful work.

In the situation as described indiscriminate altruism is actually serving as a cheap ersatz for kin selection. Under altered circumstances, however, with humans living in large communities and with a lot of migration, altruism loses its evolutionary justification. It becomes a deleterious trait, a mere vestige of an adaptation. In this way, there is no more puzzle about how altruism was selected despite its presently being selectively disadvantageous. This explanation is gestured at by Tooby and Cosmides (1989).

It is a nice story, but it is doubtful whether it is anything more than that. The skepticism originates from two sources. First, and most important, the starting assumption that humans once lived in groups consisting mostly of relatives is, to use an understatement, very far from being generally accepted; various lines of evidence suggest that our Pleistocene ancestors lived in fairly large groups consisting of approximately 150 members (Dunbar 1994, p. 770). With the key assumption so empirically compromised the whole approach can hardly move off from the ground.

Second, it is not clear that indiscriminate altruism would be an evolutionarily best strategy, even if the postulated conditions obtained. True, under the conditions it would not be worth the effort to try to recognize relatives from nonrelatives (because ex hypothesi there would almost be no nonrelatives), but even then it would certainly make much evolutionary sense to distinguish between relatives with different degrees of relatedness (and, consequently, to help the closer kin more). However, once the behavior is guided by the variable degree of relatedness nonrelatives are automatically "perceived" and treated as having coefficient *r* of zero, and we are back to square one: altruism toward strangers remains inexplicable. The basic difficulty here is that in the situation where one interacts mainly with relatives one is well advised on evolutionary grounds to be an altruist, but not a nondiscriminating altruist. Unfortunately, the theory under consideration depends crucially on the presence of nondiscriminating altruism. For, it says that in the past it was fitness increasing to help anyone, whomever you met (without bothering to establish who the individual was), because, anthropomorphically speaking, you could have been practically sure in advance that he would turn out to be your relative.

Altruism is indeed explained by the fact that such a behavioral disposition misfires in a changed environment when your group expands and when it comes to consist predominantly of biological strangers. What is thereby not explained at all is why there was such a blanket altruism in the first place, that is, why the readiness to help others was not adjusted and apportioned according to the degree of genetic proximity. No doubt, this could be accounted for, too, by complicating the story and by introducing additional hypotheses. But this modified and more demanding explanation has yet to be satisfactorily elaborated in detail.[14]

(4.2) By-Product Theory

According to (4.1) the conflict between altruism$_p$ appearing to be counteradaptive and its also appearing to be a product of natural selection was resolved by the claim that, due to the change in the environment, altruism$_p$ ceased to be adaptive. In contrast, (4.2) claims that it was at no time an adaptation, but that it was nevertheless selected by having been inseparably connected with some other trait, which was adaptive. This amounts to the idea that altruism$_p$ is a by-product, or spin-off (or spandrel), of selection.

To defend this kind of approach one is under a threefold obligation: (i) to identify some other trait, (ii) to show that it was selected for, and (iii) to demonstrate that it is inextricably tied to altruism$_p$. I shall consider here two instances of such an approach.

a) According to an influential argument, the emergence of rationality is easy to incorporate into an evolutionary scenario. Those who were more rational (i.e., those who made fewer systematic errors when thinking or solving problems) were likely to cope better with their environment, and to leave more descendants than others. Hence, according to some philosophers (Daniel Dennett, for example) there is a strong case for regarding the faculty of rationality as an adaptation.[15] With rationality thus evolutionarily fortified one can then attempt to derive altruism as a consequence, by making use of a classical Kantian argument in practical philosophy. In a nutshell, the argument purports to show that being rational entails being concerned for the inter-

14. One possible move in that direction (suggested both by Elliott Sober and by the anonymous referee for *Ethics*) is to argue that the informational costs of distinguishing various degrees of relatedness would be much higher than making the "relative/ nonrelative" distinction, and that perhaps the costs would be too high to make it worth doing in an all-fairly-close-relative society. But then again, it seems that the theory would be shipwrecked on the hard empirical fact that, after all, primates actually happen to have the capacity to discriminate between relatives and nonrelatives (Dunbar 1994, p. 773).

15. For doubts about this line of reasoning see Stich (1990, pp. 55–74).

ests of all rational beings, or, by contraposition, that the lack of such
a concern (being completely self-interested) is a sign of irrationality.
On this view, a rational person cannot be a complete egoist.

The ablest contemporary defender of this Kantian line is Thomas
Nagel (1970). At one time he even went so far as to argue that one
who in full awareness did not care for the interests of others had to
be a solipsist. Put differently, he thought that by merely recognizing
the existence of other people a rational person is necessarily committed
to have at least some minimal concern for their interests. (Later, under
criticism, Nagel was forced to weaken his claim significantly; cf. Nagel
1986, p. 159.) Nagel was not concerned with an evolutionary account
of altruism; he only wanted to show that out-and-out egoism is incompatible with rationality. It was Colin McGinn who first openly combined this view that altruism is a consequence of rationality with the
thesis that human rationality is a biological adaptation, carrying
thereby an explicit suggestion that altruism is a by-product of natural
selection. There is also an obvious (and acknowledged) debt to
McGinn's ideas in Peter Singer's (1981) book *The Expanding Circle*,
where this hypothesis is worked out in more detail. The gist of
McGinn's argument stands out in the following sentences:

> If morality is founded upon naturally bestowed appetites in accor-
> dance with the principles of natural selection, and if these appe-
> tites simply cannot, consistently with the laws of evolutionary
> biology, extend as far as moralists have insisted, why then surely
> the idea of pure, disinterested altruism is a chimera which it is
> pointless to pursue. . . . If we want to secure morality against the
> forces of natural selection, *we need to associate it with possession
> of some characteristic whose evolutionary credentials are undisputed:* I
> suggest that *the cognitivist's associating it with reason meets this condi-
> tion*, while the noncognitivist's appetitive theory does not.
> (McGinn 1979, pp. 85, 93; emphasis added)

The argument has two steps: the first biological (that rationality
is an outcome of an evolutionary process), and the second philosophical (that rationality leads to altruism). It is the second step that does
main explanatory work, and that raises most questions. Since, for obvious reasons, this philosophical claim cannot here be given the full
consideration it deserves, the following brief comment must suffice.
Judged by the present state of metaethical discussions, the strong
Kantian version of cognitivism that ascribes concern for others to
any rational being qua rational being is highly controversial among
philosophers. The basic difficulty with it is well described by Peter
Railton: "Although rationalism in ethics has retained adherents long
after other rationalisms have been abandoned, the powerful philosophical currents that have worn away at the idea that unaided reason
might afford a standpoint from which to derive substantive conclusions

show no signs of slackening" (Railton 1986, p. 163). But if this kind of cognitivism is so problematic even in its own philosophical province, it is then surely all the more unsuitable as an instrument for reaching not strictly philosophical conclusions.[16]

In addition, its standing is not exactly improved by the fact that much social science literature (rational choice theory, decision theory, microeconomics, public choice theory) takes as its starting point the assumption that reason is motivationally inert. According to this widely shared view, rational considerations cannot move us to action by themselves, the main impulse always coming from some of our basic preferences that stand completely outside the jurisdiction of reason. In David Hume's memorable words (1888, p. 416): "'Tis not contrary to reason to prefer the destruction of the whole world to the scratching of my finger." Many contemporary cognitivists recognize the force of noncognitivist arguments, and they try to save cognitivism by weakening it significantly. Consequently, they are now often ready to concede that our moral attitudes cannot derive from rationality alone, and that what is needed besides is some kind of moral sensitivity, minimal concern for others, and the like (cf. Williams 1972, p. 26; Singer 1981; Lindley 1988, p. 528). After this concession, however, our fundamental question immediately reappears: How did these initial and rudimentary nonegoist dispositions (that are a prerequisite for full altruism) arise by the process of natural selection?

A possible answer might be that a budding altruistic disposition (in the psychological sense) could have evolved through the process of kin selection without being altruistic in the evolutionary sense.[17] According to our two definitions of altruism the tendency to help one's own relatives counts as altruism$_p$, but not as altruism$_e$. So, it seems after all that kin selection could provide a mechanism for injecting the first, minimal dose of altruism$_p$ and that by conferring a selective advantage on the carriers of this trait it could then set the stage for the circle of altruism expanding further afterward (see Singer 1981, p. 91).

The point that there is this cleavage between the two concepts of altruism (i.e., that altruism$_e$ is about genes whereas altruism$_p$ is about individuals) is both cogent and important. It shows that psychological

---

16. It is interesting to note here that a long time ago Charles Sanders Peirce tried to reach the same conclusion about the inherent irrationality of out-and-out egoism, by taking a completely different route. He developed a highly idiosyncratic argument from the nature of probability purporting to prove that "to be logical men should not be selfish" (see Peirce [1878] 1992, pp. 149–50). I am indebted to Michael Kremer for drawing my attention to this article.

17. This idea is defended in Kitcher (1993, pp. 508–9) and Sober (1994, pp. 18–19).

altruism toward relatives can be a purely Darwinian product. What remains highly dubious, however, is whether this kind of incipient nonegoism could serve as a foothold for our reason in its ascent to the completely generalized altruism. If we want to justify rationally the normative standpoint of universal altruism this surely cannot be achieved by relying solely on the purely factual premise that humans already display a kind of selective altruism. Those philosophers who want to claim that it is our reason that helps us to cross the border between the narrow-scope altruism and principled altruism are under obligation not only to show (*a*) that the border between the two altruisms is arbitrary, and (*b*) that we already happen to be narrow-scope altruists. They have also to establish (*c*) that this initial, minimal altruism is itself rationally justified. This is necessary simply because reason cannot generalize something on the ground that it exists, but only on the ground that it is reasonable in the first place. Therefore the cognitivist argument, labeled here (4.2*a*), cannot bootstrap itself by appeal to the factual premise that evolutionary forces have produced one kind of altruism$_p$, in the hope that it has then only to proceed further and broaden the scope of this other-benefiting behavioral tendency. No, reason has to take the uphill path and develop the rational defense of altruism all the way from the very beginning. A further difficulty for connecting altruism so closely with rationality is that, according to recent research in child psychology, the concern for others is in some forms already present at the developmental stage when children are definitely incapable of moral reasoning. (See Wilson 1993, p. 130, and references given therein.)

b) In a diametrically opposed approach, Robert Boyd and Peter J. Richerson (1985, pp. 204–40; 1990) suggested that it is actually a tendency not to use one's reason that links altruism with natural selection. They argued, first, that under certain conditions there is a selective advantage in uncritical conformity, that is, in simply copying the most common behavior in a subpopulation, without checking beforehand whether that behavior is also the most appropriate to the circumstances. It can be shown, particularly in heterogeneous environments, that due to selection the most fit behavioral variant tends to become the most common one in a subpopulation as well; therefore, by being a conformist (by merely imitating the behavior most frequently encountered in the environment) an individual would increase the chance of acquiring the best behavior without costly individual learning. In this way it is explained how a psychological disposition to conform (to behave according to the rule *A Roma alla romana*) is selected for.

The second step is to derive altruism as a consequence of conformity. The mechanism proposed for this purpose is cultural group selection. Assuming that egoism and altruism are behavioral traits that are culturally transmitted, Boyd and Richerson are in the position to demonstrate that the aforementioned genetic disposition to conform is a key factor which makes groups mainly consisting of altruists evolutionarily stable and resistant to change. Namely, if individuals have the built-in propensity to adopt whatever behavior is the most common in their environment, then groups with predominantly altruist members will in all likelihood, through the process of cultural conformist transmission, give rise to new generations consisting mainly of altruists. Conformity plays here a crucial role, making it possible for groups of altruists to persist through time long enough for the process of (cultural) group selection to take effect. As we saw earlier (pp. 132–33), one of the main obstacles to the genetic group selection of altruism was that altruist groups were open to subversion from within (they were easily invadable by the small number of genetic egoists). In the case of cultural group selection, as described by Boyd and Richerson, the obstacle disappears. True, a few occasional egoists would still fare consistently better than the more numerous altruists in the group, but they would not threaten the predominance of altruism: owing to the special character of cultural inheritance (conformity) even the descendants of egoists would tend to resemble the average type; that is, they would tend to be altruists. In short, egoism continues to be individually advantageous, but the main reason why it cannot spread in groups of altruists is that it tends not to be transmitted (because of the conformist bias).

A major virtue of the Boyd-Richerson approach is that the emergence of altruism is not seen as being the result of a single causal influence (biology or culture). Trying to do justice to the complexity of the issue, they proposed that the genesis of altruism should be explained by the combined operation of natural and culural selection. In their model, these two causal factors are combined in an original, elegant, and initially plausible way. It is an open question, however, whether all the empirical assumptions underpinning the model will be borne out by future scientific research.[18]

(4.3) Continuing Adaptation Theory

This theory says that altruism$_p$ is a product of natural selection, and that it is adaptive. As noted earlier, adding (4.3) to (1), (2), and (3) gets us on the brink of contradiction. Assuming (1) that altruism$_e$ is a selectively disadvantageous trait, (2) that altruism$_p$ leads to altruism$_e$, and (3) that altruism$_p$ exists, it is hard to see how it is possible (4.3) that altruism$_p$ is adaptive (i.e., selected for). Or, to put the question in general terms and in a more pointed way, can we coherently entertain the idea that it may be selectively advantageous to possess a selec-

18. A similar account is proposed by Herbert Simon (1990).

tively disadvantageous trait? With the caveat already mentioned (and spelled out in n. 21) this idea in fact proves to be explanatorily very fruitful and promising.

Consider first the simple Prisoner's Dilemma game between two players, A and B, exhibited in figure 3 in the so-called extensive form. Each player has a choice between cooperating (c) and defecting (d). There are four possible outcomes with different payoffs for A and B (with A's payoffs always being given first). Suppose that A acts first and that B makes his choice after gaining knowledge about whether A has cooperated or defected. This slightly modified version of the Prisoner's Dilemma doesn't make much difference. Obviously, defecting is still the dominant option for both, and hence the solution of the game is (0, 0). Because of the structure of this decision problem the outcome (1, 1), although preferred by both players, is simply unattainable.

Let us now change the situation in two respects. First, suppose that B can acquire a disposition to cooperate if A cooperates. And second, suppose that A has a relatively reliable (though not necessarily infallible) method for recognizing the presence of such a conditional behavioral disposition in B. That is, B cannot hope to deceive A that he is disposed to respond with cooperation to A's cooperation when he is not so disposed. Or, still differently but equivalently, if we call the disposition in question N, we are supposing that B's best way to persuade A that he has N is to really acquire N.

The conditions of choice are thereby essentially changed. Now A expects that B with a built-in N disposition will respond with cooperation to cooperation, and with defection to defection. So, confronted with such a "tit-for-tat" opponent, A actually faces a dilemma between cooperating, with the certain final outcome (1, 1), and defecting, with the certain final outcome (0, 0); clearly, if he is rational he chooses the former, and the collectively preferred state becomes attainable. But it is B's perspective that deserves our attention.

Assuming that payoffs are measuring fitness consequences of different outcomes for A and B, it is easily seen that manifestations of disposition N decrease B's fitness. Namely, after A has cooperated it is patently in B's evolutionary interest to defect (payoff of defection = 2) and not to cooperate (payoff of cooperation = 1). If, however, by possessing disposition N, B responds with cooperation this is an act of altruism$_e$ pure and simple, selectively disadvantageous and lacking any evolutionary justification. Nevertheless, despite the fact that disposition N is in a sense a systematically fitness-decreasing trait (i.e., every manifestation of it leads to the net loss of fitness) the possession of disposition N can still be selected for.

To see this, note that it is in B's interest to induce A to cooperate, because he is then sure of getting at least 1 whereas if A defects, B can get no more than 0. Indeed, let us assume that B can induce A to

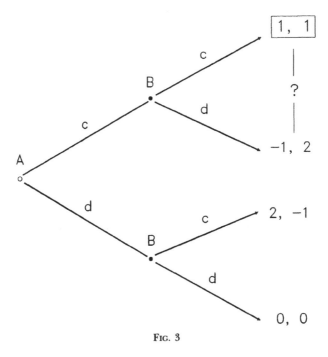

Fig. 3

cooperate only if he comes to persuade A that he (B) has an "altruistic" disposition N to respond in kind to A's cooperation. If, furthermore, the situation is as previously described, and B's best way of persuading A that he has disposition N happens to be the actual possession of N, B's fitness will in fact be enhanced by acquiring N. Namely, if B does not have N, the decision problem reduces to the standard Prisoner's Dilemma, and the equilibrium is (0, 0). If B does have N, however, he thereby makes A notice this and by assuring A that he is not a "cheater" B induces A to cooperate; finally, in accordance with N, B in turn also cooperates, and the outcome is (1, 1).

Disposition N is altruistic$_e$ (by our definition) because any manifestation of it (responding with cooperation to cooperation) results in the net loss of one point of fitness. By cooperating after A's cooperation B gets only 1 point whereas by defecting he could have acquired 2 points. So in a sense, N is evolutionarily self-defeating. But although any manifestation of N is fitness-decreasing, the possession of N can be fitness increasing. This can come about because good effects of having N can offset bad effects of manifesting N. In our situation, the optimal outcome for B is (−1, 2) but it is inaccessible to him. If he lacks N (if he is not disposed to cooperate after A's cooperation) this will induce A to defect, producing the result (0, 0). If he has N, however, this will motivate A to cooperate, but this will then also make B cooperate, leading to the suboptimal outcome from B's perspective $(1, 1)$, instead of the optimal one $(-1, 2)$.

B's predicament consists in the fact that by possessing N he is better off before A cooperates (because ex hypothesi it is only B's possession of N that can make A cooperate), but after A cooperates B would be better off if he lacked N (because, then, acting in accordance with disposition N he deprives himself of the optimal result). B's interest would be best served if he had disposition N only until the moment A cooperates, and if he lost it or at least did not manifest it afterward. But this hardly seems possible. For, as John Mackie remarked, "dispositions cannot be switched on and off in deference to the calculation of likely consequences on particular occasions" (1977, p. 192). By their very nature dispositions have some kind of persistence over time, and it may therefore well be that in our case, too, there is only a choice between taking or leaving the whole package (disposition N which has good effects first, but bad manifestations later). This reveals that even from a purely egoistic perspective it may sometimes be advisable to be altruistic. One's own self-interest may be best promoted by one's readiness to sacrifice it. The argument for this paradoxically sounding claim was first offered in David Gauthier's (1975) article "Reason and Maximization," and it was later more rigorously elaborated in his book *Morals by Agreement* (Gauthier 1986).[19] Finally, it was Robert Frank (1988) who in his *Passions within Reason* fleshed out this approach with rich empirical detail and showed that the whole idea of altruism being ultimately founded on egoism is not a mere abstract possibility but that it can have a very wide and surprisingly fruitful application in explaining human behavior.[20]

It is very important, however, not to confuse this approach with a completely different way of giving an egoistic rationale for altruism. That is, acting altruistically on a particular occasion can be egoistically justified by the fact that an agent who so acts gains thereby a reputation of an altruist, and this in turn may have good effects for him in making other agents more ready to interact cooperatively with him in the

---

19. "It would, after all, be paradoxical if the only way to justify a nonegoistic enterprise like morality were by the use of an egoistic argument" (Frankena 1980, p. 87).

20. Therefore, when Kenneth Binmore (1993, p. 138) says that "people cannot see inside each other's heads and [that] it is idle to examine models in which they can," he is simply wrong. Namely, beside many and various indications (usefully collected and described in Robert Frank's book) that the human mind systematically and unintentionally leaks information about its content to the outside it has recently been even experimentally demonstrated (Frank, Gilovich, and Regan 1993) that people can "see" inside each other's heads, i.e., that in playing the Prisoner's Dilemma game people somehow manage to recognize the presence of a cooperative disposition in others if allowed to interact with them even for as brief a period as half an hour!

152   *Ethics   October 1995*

future. Taking into account all the benefits that he could himself reap from these later cooperations just by first depriving himself of a much smaller immediate gain it is quite clear that only a very myopic egoist would refuse to cooperate under the circumstances. Seen from a wider perspective, such a conduct should not really be classified as altruism at all. Rather different but also purely egoistically inspired acts of cooperation in the iterated Prisoner's Dilemma (or in the so-called centipede game) would consist in cooperating with the sole purpose of appearing naive, stupid, or irrational "in the hope of tempting the opponent into an unwise attempt at exploitation" (Binmore 1988, p. 11).

In contrast, the behavior that interests us and that falls under description (4.3) is genuine altruism, for here we are assuming that the agent stands to gain nothing later by temporarily sacrificing his interests. For all that matters, there may simply be no interactions after the one we are considering. In that case, altruistic behavior is actually justified not by its subsequent effects, but instead by earlier beneficial effects of having the altruistic behavioral disposition. At the moment when the disposition manifests itself, however, the act is a genuinely altruistic one because by being "nice" the agent suffers a loss, never compensated afterward. To repeat, he would be best off if he could manage somehow both to acquire the altruistic disposition and to not let it ever be actualized. But this is not a feasible project. What is feasible and, indeed, best for the subject is to acquire the altruistic disposition although it is evident in advance that his interests will be harmed by this later. In this way a path is cleared for an evolutionary explanation of the genesis of altruism. The *nervus explanandi* is the claim that, under specified conditions, the possession of altruistic behavioral dispositions may maximize the fitness of its bearers.[21]

---

21. One might here object that this kind of "altruistic" disposition is no more selectively disadvantageous because organisms possessing such a disposition would definitely have a higher inclusive fitness than those lacking it. Consequently, it could be argued that the view (4.3) should actually be interpreted as denying the proposition (1) of the incongruous tetrad, and hence that it is more properly subsumed under the rubric of eliminativism than reconciliationism. This is a good point (made by an anonymous referee for *Ethics,* and by Gordon Belot in a discussion). Yet I have decided to retain my nomenclature for the following reason. Usually, the (dis)advantageousness of a behavioral disposition depends alone on whether its manifestations are advantageous or not. With the disposition in question, however, any organism possessing it would be better off never to manifest it (i.e., never to act in accordance with it): the good effects are in this case coming, not from the acts, but from the side effects of having the disposition. Here it is fitness increasing to have the tendency to produce behaviors that are all individually fitness decreasing. So, not in the least disputing the legitimacy of the proposed eliminativist interpretation of the view under consideration I want simply to point out there is also a secondary sense in which the crucial disposition

Someone could perhaps be tempted to argue here that there is a structurally similar decision-theoretical account of the emergence of altruism which is got by simply replacing 'behavioral disposition' with 'conditional intention', and 'fitness' with 'interest'. But what worked with dispositions does not work with conditional intentions.

Referring again to figure 3, let us assume now that A has a reliable way of recognizing not B's behavioral dispositions, but B's true intentions. Also, to cut verbiage, let us say that the condition C is fulfilled if A cooperates. Then, by the same argument as before, B is well advised as a rational agent to form a conditional intention to cooperate-if-C. Namely, if B forms that intention, A will recognize the presence of such an intention in B and this will motivate him to cooperate. B would thus gain at least 1 point, whereas otherwise (if B did not form the intention) A would defect, and B would be left with 0 points. But by forming the conditional intention to cooperate-if-C, B *eo ipso* insures that, after C is eventually fulfilled, he will then have the unconditional intention to cooperate. This is derived from the following intuitively plausible principle: (i) if at t_1 B forms a conditional intention to ϕ at t_3 in the event that condition C obtains at t_2, (ii) if C is realized at t_2, and (iii) if nothing intervenes, then B will have at t_3 an unconditional intention to ϕ at t_3.

It is precisely here that the basic difficulty comes. If we assume (as we did) that B is a completely self-interested and fully rational agent, then it is not clear how he can bring himself to intend to cooperate at t_3 when he definitely knows that at t_3 he can only lose by doing so. True, he knows that he can gain much by side effects of his forming the conditional intention to cooperate, but this is of no avail to him in the process of forming the intention: for, at t_3, good side effects (of A's cooperation) already belong to the past, and for B as a fully rational and self-interested chooser there is at that time simply no reason whatever to cooperate. But, of course, the fact that B knows all this in advance makes it impossible for him, even at t_1, to form the conditional intention to cooperate. For it is hard to see how B, who is driven only by his self-interest, could at t_1 form the intention to ϕ at t_3, when he is fully aware that when the time comes, at t_3, his interests would only be harmed by ϕing. (Illuminating discussions of this kind of decision-theoretic predicament are to be found in Kavka 1983; 1987, pp. 15–32; Bratman 1987, pp. 101–6.)

---

does not falsify (1), i.e., a sense in which it is selectively disadvantageous: namely, all its realizations are systematically and without exception selectively disadvantageous. I have let myself be guided by this secondary sense in classifying (4.3) because for expository purposes this solution to the paradox of altruism falls neatly into place at the end of the sequence of ever more astringent reconciliationist answers.

In the situation as described, B is, judging strictly by his interests, best off by forming the conditional intention in question, but the catch is that insofar as he is fully rational he cannot form that intention. Therefore, it is not only that reason doesn't pave the way for altruism; under the circumstances reason is a positive obstacle. To put it differently, although the indubitably best option for the agent, egoistically speaking, is to form a conditional intention to cooperate, a rational and self-interested person just cannot plan in full consciousness to form such an intention. If he remains both rational and self-interested when the time comes to act, his preferences, being as they are, will simply compel him to defect. But since he knows all this from the start there is something incoherent in the idea that he could (even conditionally) intend to cooperate.

In contrast to those who have hoped that only reason could bridge the gap between pure egoism and the moral point of view,[22] it is revealed here that, in some contexts at least, narrow selfishness can be transcended in no other way than by modifying the "nonrational" parts of the mind, and by natural selection working on the mental dispositions, habits, and emotions. This opens up an interesting possibility that, despite the notorious selfishness of its units of selection and its "blind" way of operation, biological evolution can still give rise to certain forms of altruism that are inherently unattainable even to infinitely intelligent selfish deliberators, as long as they remain fully rational. That is, a purely rational agent may happen to be stuck in the trough of myopic egoism, with his only chance of "tunnelling through" to the position of the enlightened self-interest (which here paradoxically coincides with genuine altruism) by Darwinian forces shaping his behavioral dispositions behind the back of his reasoning self.[23]

CONCLUSION

The main intention of this article was to propose a novel, "natural" classification of different approaches to the paradox of altruism, in the hope that by imposing the overarching structure on this continuing controversy the issue could be joined in a more fruitful way. I have argued that on the whole the so-called reconciliationist strategy holds more promise than the eliminativist one, and more specifically that among reconciliationist answers those designated as (4.2*b*) and (4.3) in my scheme are particularly well grounded and deserving further elaboration. I did not want, however, to exclude completely the possi-

---

22. Compare (4.2*a*).
23. These remarks about altruism and rationality are very sketchy, and they need a lot more spelling out. I hope to develop this line of thought in more detail on another occasion.

bility that some other of the discussed moves could account for the emergence of certain forms of human altruisms, although (very probably) only under very special or rarely satisfied conditions. We should also heed a warning (coming from Christopher Jencks) that "while it is analytically useful to label many different forms of behavior as ["altruistic"], the use of a single label encourages the illusion that there is a single underlying trait ["altruism"] that determines whether an individual engages in all these different forms of behavior" (Jencks 1990, p. 66). Although the opinion today prevails that at the core there is indeed one underlying, deep-seated behavioral disposition (perhaps shaped by evolutionary forces) that accounts for various manifestations of human altruism, at the present stage of our knowledge at least some room should be left for the possibility that the story will turn out to be more complicated and less orderly. For in the case (which cannot be ruled out a priori) that Jencks happens to be right in his hunch that different forms of altruism are only loosely connected to one another (and that there is simply no unifying trait or nucleus, below the surface), then searching for the explanation of the origins of human altruism would inevitably lead to Procrustean accounts of this multifaceted phenomenon.

REFERENCES

Allison, Paul D. 1992. The Cultural Evolution of Beneficient Norms. *Social Forces* 71:279–301.

Batson, Daniel C. 1991. *The Altruism Question: Toward a Social-Psychological Answer*. Hillsdale, N.J.: Lawrence Erlbaum.

Batson, Daniel C. 1992. Experimental Tests for the Existence of Altruism. *PSA* 2:69–78.

Batson, Daniel C., and Oleson, Kathryn C. 1991. Current Status of the Empathy-Altruism Hypothesis. Pages 62–85 in *Prosocial Behavior*, ed. M. S. Clark. Newbury Park, Calif.: Sage Publications.

Batson, Daniel C., and Shaw, Laura L. 1991. Evidence for Altruism. *Psychological Inquiry* 2:107–22.

Binmore, Kenneth. 1988. Modeling Rational Players: Part II. *Economics and Philosophy* 4:9–55.

Binmore, Kenneth. 1993. Bargaining and Morality. Pages 131–56 in *Rationality, Justice and the Social Contract*, ed. D. Gauthier and R. Sugden. Ann Arbor: University of Michigan Press.

Boyd, Robert, and Richerson, Peter J. 1985. *Culture and the Evolutionary Process*. Chicago: University of Chicago Press.

Boyd, Robert, and Richerson, Peter J. 1990. Culture and Cooperation. Pages 113–32 in *Beyond Self-Interest*, ed. Jane J. Mansbridge. Chicago: University of Chicago Press.

Bratman, Michael E. 1987. *Intentions, Plans, and Practical Reason*. Cambridge, Mass.: Harvard University Press.

Butler, Joseph. 1983. *Five Sermons*. Indianapolis: Hackett.

Darwin, Charles. 1874. *The Descent of Man and Selection in Relation to Sex*. Chicago: Rand McNally.

Dunbar, Robin I. M. 1994. Sociality among Humans and Non-Human Animals. Pages 756–82 in *Companion Encyclopedia of Anthropology*, ed. T. Ingold. London: Routledge.

Elster, Jon. 1983. *Sour Grapes*. Cambridge: Cambridge University Press.

156   *Ethics   October 1995*

Elster, Jon. 1989. *Nuts and Bolts for the Social Sciences.* Cambridge: Cambridge University Press.

Feinberg, Joel. 1971. Psychological Egoism. Pages 489–500 in *Reason and Responsibility,* ed. J. Feinberg. Encino, Calif.: Dickenson.

Frank, Robert H. 1988. *Passions within Reason.* New York: Norton.

Frank, Robert H., Gilovich, Thomas, and Regan, Dennis T. 1993. The Evolution of One-Shot Cooperation. *Ethology and Sociobiology* 14:247–56.

Frankena, William K. 1980. *Thinking about Morality.* Ann Arbor: University of Michigan Press.

Gauthier, David. 1975. Reason and Maximization. *Canadian Journal of Philosophy* 4:424–33.

Gauthier, David. 1986. *Morals by Agreement.* Oxford: Clarendon.

Gibbard, Allan. 1982. Human Evolution and the Sense of Justice. *Midwest Studies in Philosophy* 7:31–46.

Gibbard, Allan. 1990. *Wise Choices, Apt Feelings.* Oxford: Clarendon.

Haldane, John B. S. 1932. *The Causes of Evolution.* London: Longman.

Hamilton, William D. 1964. The Genetical Evolution of Social Behavior. *Journal of Theoretical Biology* 7:1–52.

Hume, David. 1888. *A Treatise of Human Nature,* Selby-Bigge ed. Oxford: Clarendon.

Jencks, Christopher. 1990. Varieties of Altruism. Pages 53–67 in *Beyond Self-Interest,* ed. Jane J. Mansbridge. Chicago: University of Chicago Press.

Kavka, Gregory S. 1983. The Toxin Puzzle. *Analysis* 43:33–36.

Kavka, Gregory S. 1986. *Hobbesian Moral and Political Theory.* Princeton, N.J.: Princeton University Press.

Kavka, Gregory S. 1987. *Moral Paradoxes of Nuclear Deterrence.* Cambridge: Cambridge University Press.

Kitcher, Philip. 1985. *Vaulting Ambition: Sociobiology and the Quest for Human Nature.* Cambridge, Mass.: MIT Press.

Kitcher, Philip. 1993. The Evolution of Human Altruism. *Journal of Philosophy* 90:497–516.

Krebs, Dennis. 1982. Psychological Approaches to Altruism: An Evaluation. *Ethics* 92:447–58.

Lindley, Richard. 1988. The Nature of Moral Philosophy. Pages 517–40 in *An Encyclopedia of Philosophy,* ed. G. H. R. Parkinson. London: Routledge.

Mackie, John L. 1977. *Ethics: Inventing Right and Wrong.* Harmondsworth: Penguin.

McGinn, Colin. 1979. Evolution, Animals, and the Basis of Morality. *Inquiry* 22:81–89.

Nagel, Thomas. 1970. *The Possibility of Altruism.* Princeton, N.J.: Princeton University Press.

Nagel, Thomas. 1986. *The View from Nowhere.* New York: Oxford University Press.

Peirce, Charles S. (1878) 1992. The Doctrine of Chances. Pages 142–54 in *The Essential Peirce,* ed. N. Houser and C. Kloesel. Bloomington: Indiana University Press.

Peressini, Anthony. 1993. Generalizing Evolutionary Altruism. *Philosophy of Science* 60:568–86.

Quine, Willard V. O. 1969. *Ontological Relativity and Other Essays.* New York: Columbia University Press.

Railton, Peter. 1986. Moral Realism. *Philosophical Review* 95:163–207.

Rawls, John. 1971. *A Theory of Justice.* Cambridge, Mass.: Harvard University Press.

Ridley, Mark, and Dawkins, Richard. 1981. The Natural Selection of Altruism. Pages 19–37 in *Altruism and Helping Behavior,* ed. J. P. Rushton and R. M. Sorrentino. Hillsdale, N.J.: Lawrence Erlbaum.

Rosenberg, Alexander. 1988. Grievous Faults in *Vaulting Ambition? Ethics* 98:827–37.

Schwartz, Barry. 1993. Why Altruism Is Impossible . . . and Ubiquitous. *Social Service Review* 67:314–43.

Simon, Herbert A. 1990. A Mechanism for Social Selection and Successful Altruism. *Science* 250:1665–68.

Simon, Herbert A. 1993. The Economics of Altruism. *American Economic Review* 83:156–61.

Singer, Peter. 1981. *The Expanding Circle: Ethics and Sociobiology.* Oxford: Oxford University Press.

Sober, Elliott. 1981. The Evolution of Rationality. *Synthese* 46:95–120.

Sober, Elliott. 1984. *The Nature of Selection.* Cambridge, Mass.: MIT Press.

Sober, Elliott. 1988. What Is Evolutionary Altruism? Pages 75–99 in *Philosophy and Biology,* ed. M. Mathen and B. Linsky. Calgary: University of Calgary Press.

Sober, Elliott. 1992. Hedonism and Butler's Stone. *Ethics* 103:97–103.

Sober, Elliott. 1993a. Evolutionary Altruism, Psychological Egoism, and Morality: Disentangling the Phenotypes. Pages 199–216 in *Evolutionary Ethics,* ed. M. H. Nitecki and D. V. Nitecki. Albany: SUNY Press.

Sober, Elliott. 1993b. *Philosophy of Biology.* Boulder, Colo.: Westview.

Sober, Elliott. 1994. Did Evolution Make Us Psychological Egoists. Pages 8–27 in his *From a Biological Point of View.* Cambridge: Cambridge University Press.

Stich, Stephen. 1990. *The Fragmentation of Reason.* Cambridge, Mass.: MIT Press.

Tooby, John, and Cosmides, Leda. 1989. Evolutionary Psychologists Need to Distinguish between the Evolutionary Process, Ancestral Selection Pressures, and Psychological Mechanisms. *Behavioral and Brain Sciences* 12:724–25.

Trivers, Robert. 1978. The Evolution of Reciprocal Altruism. Pages 213–26 in *The Sociobiology Debate,* ed. A. L. Caplan. New York: Harper & Row.

Williams, Bernard. 1972. *Morality: An Introduction to Ethics.* Cambridge: Cambridge University Press.

Williams, Bernard. 1973. *Problems of the Self.* Cambridge: Cambridge University Press.

Williams, George C. 1966. *Adaptation and Natural Selection.* Princeton, N.J.: Princeton University Press.

Wilson, David S. 1992. On the Relationship between Evolutionary and Psychological Definitions of Altruism and Selfishness. *Biology and Philosophy* 7:61–68.

Wilson, David S., and Sober, Elliott. 1994. Reintroducing Group Selection to the Human Behavioral Sciences. *Behavioral and Brain Sciences* 17:585–608.

Wilson, Edward O. 1978. *On Human Nature.* Cambridge, Mass.: Harvard University Press.

Wilson, Edward O., et al. 1973. *Life on Earth.* Stamford, Conn.: Sinauer.

Wilson, Edward O., et al. 1977. *Life: Cells, Organisms, Populations.* Sunderland, Mass.: Sinauer.

Wilson, James Q. 1993. *The Moral Sense.* New York: Free Press.

Wynne-Edwards, V. C. 1962. *Animal Dispersion in Relation to Social Behavior.* Edinburgh: Oliver & Boyd.

[7]

Evolution, altruism and cognitive architecture: a critique of Sober and Wilson's argument for psychological altruism

STEPHEN STICH

Department of Philosophy & Center for Cognitive Science, Rutgers University, 26 Nichol Avenue, New Brunswick, NJ 08901-2882, USA
(e-mail: stich@ruccs.rutgers.edu; phone: +732-932-9861; fax: +732-932-8617)

Received 3 August 2005; accepted in revised form 22 March 2006

Key words: Altruism, Cognitive architecture, Egoism, Evolution, Intrinsic and instrumental desire, Natural selection, Sub-doxastic states

Abstract. Sober and Wilson have propose a cluster of arguments for the conclusion that "natural selection is unlikely to have given us purely egoistic motives" and thus that psychological altruism is true. I maintain that none of these arguments is convincing. However, the most powerful of their arguments raises deep issues about what egoists and altruists are claiming and about the assumptions they make concerning the cognitive architecture underlying human motivation.

In their important book, *Unto Others: The Evolution and Psychology of Unselfish Behavior*, Elliott Sober and David Sloan Wilson offer a new and interesting evolutionary argument aimed at showing that in the venerable dispute between psychological altruism and psychological egoism, altruism is the likely winner. In this paper, I'll argue that Sober and Wilson's argument relies on an implicit assumption about the cognitive architecture subserving human action, that much recent work in cognitive science suggests the assumption may be mistaken, and that without the assumption, their argument is no longer persuasive. Before getting to any of that, however, we'll need to fill in a fair amount of background.

Preliminaries

Far too many discussions of evolution and altruism founder because they fail to draw a clear distinction between two very different notions of altruism which, following Sober and Wilson, I'll call *evolutionary altruism* and *psychological altruism*. One of the many virtues of Sober and Wilson's book is that they draw this distinction with exemplary clarity, and never lose sight of it.

A behavior is *evolutionarily altruistic* if and only if it decreases the inclusive fitness of the organism exhibiting the behavior and increases the inclusive fitness of some other organism. Roughly speaking, inclusive fitness is a measure of how many copies of an organism's genes will exist in subsequent generations.[1] Since an organism's close kin share many of its genes, an organism can increase its inclusive fitness either by reproducing or by helping close kin to reproduce. Thus many behaviors that help kin to reproduce are *not* evolutionarily altruistic, even if they are quite costly to the organism doing the helping.[2]

Evolutionary altruism poses a major puzzle for evolutionary theorists, since if an organism's evolutionarily altruistic behavior is heritable, we might expect that natural selection would replace the genes that influence the behavior with genes that did not foster altruistic behavior, and thus the altruistic behavior would disappear. In recent years, there has been a great deal of discussion of this problem. Some theorists, Sober and Wilson prominent among them, have offered sophisticated models purporting to show how, under appropriate circumstances, evolutionary altruism could indeed evolve, while others have maintained that the evolution of altruism is extremely unlikely, and that under closer examination all putative examples of altruistic behavior will turn out not to be altruistic at all. In the memorable words of biologist Michael Ghiselin (1974, 247) "Scratch an 'altruist' and watch a 'hypocrite' bleed." Since my focus, in this paper, is on psychological altruism, I'll take no stand in the controversy over the existence of evolutionary altruism.

A behavior is *psychologically altruistic* if and only if it is motivated by an ultimate desire for the well-being of some other organism, and as a first pass, we can say that a desire is *ultimate* if its object is desired for its own sake, rather than because the agent thinks that satisfying the desire will lead to the satisfaction of some other desire. Though I'll need to say more about ultimate desires and psychological altruism, what's already been said is enough to make the point that evolutionary altruism and psychological altruism are logically independent notions – neither one entails the other. It is logically possible for an organism to be evolutionarily altruistic even though it has no mind at all and thus can't have any ultimate desires. Indeed, since biologists interested in evolutionary altruism use the term 'behavior' very broadly, it is possible for paramecia, or even plants, to exhibit evolutionarily altruistic behavior. It is also logically possible for an organism to be a psychological altruist without being an evolutionary altruist. For example, an organism might have an ultimate desire for the welfare of its own offspring. Behaviors resulting from that desire will be psychologically altruistic though not evolutionarily altruistic, since typically such behaviors will increase the inclusive fitness of the parent.

---

[1] Giving a more precise account would raise some of the deepest issues in the philosophy of biology. (See, for example, Beatty 1992). Fortunately, for our purposes no more precise account will be needed.

[2] Some writers, including Sober & Wilson, define evolutionary altruism in terms of *individual* fitness rather than inclusive fitness. I prefer the inclusive fitness account since, as we'll soon see, it makes it easier to understand how Sober & Wilson's wise decision to focus on parental care sidesteps the debate over the existence of evolutionary altruism.

I've said that to be psychologically altruistic, a behavior must be motivated by an ultimate desire for the well-being of others. That formulation invites questions about what it is for a behavior to be *motivated* by an ultimate desire and about which desires are *for the well-being of others*. The second question, though it certainly needs to be considered in any full dress discussion of psychological altruism, can be put-off to the side here, since a rough and ready intuitive understanding of the notion is all I'll need to explain Sober and Wilson's argument and my concerns about it.[3] One interpretation of the traditional notion of *practical reasoning* provides a useful tool for explaining the relevant sense of a behavior being motivated by an ultimate desire. On this account, practical reasoning is a causal process via which a desire and a belief give rise to or sustain another desire. That second desire can then join forces with another belief to generate a third desire. And so on. Sometimes this process will lead to a desire to perform what Goldman calls a "basic" action, and that, in turn, will cause the agent to perform the basic action without the intervention of any further desires.[4] Desires produced by this process of practical reasoning are *instrumental* desires – the agent has them because she thinks that satisfying them will lead to something else that she desires. But not all desires can be instrumental desires. If we are to avoid circularity or an infinite regress there must be some desires that are *not* produced because the agent thinks that satisfying them will facilitate satisfying some other desire. These desires that are not produced or sustained by practical reasoning are the agent's ultimate desires. A behavior is *motivated* by a specific ultimate desire when that desire is part of the practical reasoning process that leads to the behavior. Figure 1 depicts some of these ideas in a format that will come in handy later on.

If a behavior is produced by a process of practical reasoning that includes an ultimate desire for the well-being of others, then that behavior is psychologically altruistic. Psychological egoism denies that there are any ultimate desires of this sort; it maintains that all ultimate desires are self-interested. According to one influential version of egoism, often called *psychological hedonism*, there are only two sorts of ultimate desires: the desire for pleasure and the desire to avoid pain. Another, less restrictive, version of egoism allows that people may have a much wider range of ultimate self-interested desires, including desires for their own survival, for wealth, for power and for prestige. Egoism acknowledges that people sometimes have desires for the well-being of others, but it insists that all these desires are instrumental. Psychological altruism, by contrast, concedes that many ultimate desires are self-interested but insists that there are also some ultimate desires for the well-being of others. Since psychological altruism maintains that people have both self-interested ultimate desires and ultimate desires for the well-being of others, Sober and Wilson sometimes refer to the view as *motivational pluralism*.

---

[3] For some substantive discussion of the question see Stich et al. (in preparation).

[4] For a classic statement of this account of practical reasoning, see Goldman (1970).

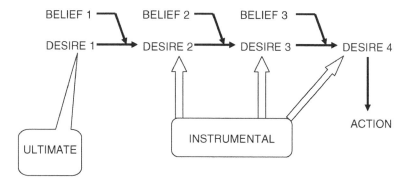

Figure 1. Practical reasoning, a causal process via which a desire and a belief give rise to or sustain another desire. An ultimate desire is one which is not produced by practical reasoning.

Sober and Wilson's evolutionary argument for psychological altruism

Sober and Wilson believe that there is an evolutionary argument for the existence of the sort of motivational structures required for psychological altruism. "Natural selection," they maintain, "is unlikely to have given us purely egoistic motives."[5] While granting that their case is "provisional" (8), they believe that their "analysis...provides evidence for the existence of psychological altruism" (12).

In setting out their argument, Sober and Wilson adopt the wise strategy of focusing on the case of parental care. Since the behaviors that organisms exhibit in taking care of their offspring are typically *not* altruistic in the evolutionary sense, we can simply put-off to the side whatever worries there may be about the existence of evolutionary altruism. Given the importance of parental care in many species, it is all but certain that natural selection played a significant role in shaping that behavior. And while different species no doubt utilize very different processes to generate and regulate parental care behavior, it is plausible to suppose that in humans *desires* play an important role in that process. Sober and Wilson believe that evolutionary considerations can help us determine the nature of these desires. Here is how they make the point:

> Although organisms take care of their young in many species, human parents provide a great deal of help, for a very long time, to their children. We expect that when parental care evolves in a lineage, natural selection is relevant to explaining why this transition occurs. Assuming that human parents take care of their children because of the desires they have, we also expect that evolutionary considerations will help illuminate what the desires are that play this motivational role." (301)

---

[5] Sober and Wilson (1998), p. 12. Hereafter, all quotes from Sober and Wilson (1998) will be identified by page numbers in parentheses.

Of course, as Sober and Wilson note, we hardly need evolutionary arguments to tell us about the content of some of the desires that motivate parental care. But it is much harder to determine whether these desires are instrumental or ultimate, and it is here, they think, that evolutionary considerations can be of help.

> We conjecture that human parents typically *want* their children to do well – to live rather than die, to be healthy rather than sick, and so on. The question we will address is whether this desire is merely an instrumental desire in the service of some egoistic ultimate goal, or part of a pluralistic motivational system in which there is an ultimate altruistic concern for the child's welfare. We will argue that there are evolutionary reasons to expect motivational pluralism to be the proximate mechanism for producing parental care in our species. (302)

Since parental care is essential in our species, and since providing it requires that parents have the appropriate set of desires, the processes driving evolution must have solved the problem of how to assure that parents would have the requisite desires. There are, Sober and Wilson maintain, three kinds of solutions to this evolutionary problem.

> A relatively direct solution to the design problem would be for parents to be psychological altruists – let them care about the well-being of their children as an end in itself. A more indirect solution would be for parents to be psychological hedonists[6] – let them care only about attaining pleasure and avoiding pain, but let them be so constituted that they feel good when their children do well and feel bad when their children do ill. And of course, there is a pluralistic solution to consider as well – let parents have altruistic *and* hedonistic motives, both of which motivate them to take care of their children. (305)

"Broadly speaking," they continue, "there are three considerations that bear on this question"(305). The first of these is *availability*; for natural selection to cause a trait to increase in frequency, the trait must have been available in an ancestral population. The second is *reliability*. Since parents who fail to provide care run a serious risk of never having grandchildren, we should expect that natural selection will prefer a more reliable solution to a less reliable one. The third consideration is *energetic efficiency*. Building and maintaining psychological mechanisms will inevitably require an investment of resources that might be used for some other purpose. So, other things being equal, we should expect natural selection to prefer the more efficient mechanism. There is, Sober and Wilson maintain, no reason to think that a psychologically altruistic mechanism

---

[6] Sober and Wilson cast their argument as contest between altruism and *hedonism* because "[b]y pitting altruism against hedonism, we are asking the altruism hypothesis to reply to the version of egoism that is most difficult to refute."(297)

would be less energetically efficient than a hedonist mechanism, nor is there any reason to think that an altruistic mechanism would have been less likely to be available. When it comes to reliability, on the other hand, they think there is a clear difference between a psychologically altruistic mechanism and various possible hedonistic mechanisms: an altruistic mechanism would be more reliable, and thus it is more likely that the altruistic mechanism would be the one that evolved.

To make their case, Sober and Wilson offer a brief sketch of how hedonistic and altruistic mechanisms might work, and then set out a variety of reasons for thinking that the altruistic mechanism would be more reliable. However, it has long been my conviction that in debates about psychological processes, the devil is often in the details. So rather than relying on Sober and Wilson's brief sketches, I will offer somewhat more detailed accounts of the psychological processes that might support psychologically altruistic and psychologically egoistic parental behavior. After setting out these accounts, I'll go on to evaluate Sober and Wilson's arguments about reliability.

Figure 2 is a depiction of the process underlying psychologically altruistic behavior. In Figure 2, the fact that the agent's child needs help (represented by the unboxed token of 'My child needs help' in the upper left) leads to the belief, *My child needs help*. Of course, formation of this belief requires complex perceptual and cognitive processing, but since this part of the story is irrelevant to the issue at hand, it has not been depicted. The belief, *My child needs help*, along with other beliefs the agent has leads to a belief that a certain action, A*, is the best way to help her child. Then, via practical reasoning, this belief and the *ultimate* desire, *I do what will be most helpful for*

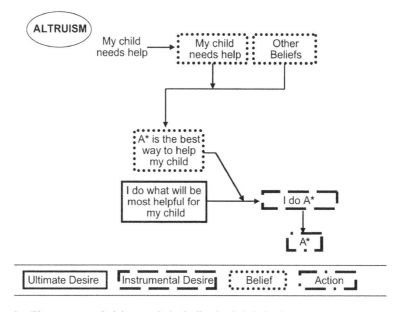

Figure 2.   The process underlying psychologically altruistic behavior.

my child, leads to the desire to do A*. Since in this altruistic account the desire, *I do what will be most helpful for my child*, is an ultimate desire, it is not itself the result of practical reasoning. The hedonistic alternatives I'll propose retain all the basic structure depicted in Figure 2, but they depict the desire that *I do what will be most helpful for my child* as an instrumental rather than an ultimate desire.

The simplest way to do this is via what I'll call *Future Pain Hedonism*, which maintains that the agent believes she will feel bad in the future if she does not help her child now. Figure 3 is my sketch of Future Pain Hedonism. In it, the content of the agent's ultimate desire is hedonistic: *I maximize my pleasure and minimize my pain*. The desire, *I do what is most helpful for my child*, is an instrumental desire, generated via practical reasoning from the ultimate hedonistic desire along with the belief that *If I don't do what is most helpful for my child I will feel bad*.

Figure 4 depicts another, more complicated, way in which the desire, *I do what is most helpful to my child*, might be the product of hedonistic practical reasoning, which I'll call *Current Pain Hedonism*. On this account, the child's need for help causes the parent to feel bad, and the parent believes that if she feels bad because her child needs help and she does what is most helpful, she will stop feeling bad. This version of hedonism is more complex than the previous version, since it includes an affective state – feeling bad – in addition to various beliefs and desires, and in order for that affective state to influence practical reasoning, the parent must not only experience it, but know (or at least believe) that she is experiencing it, and why.

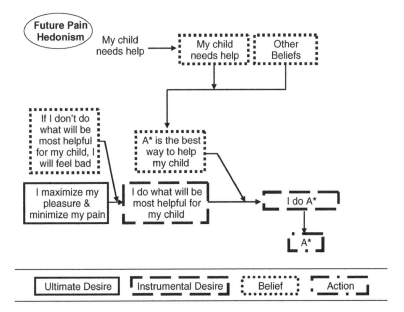

Figure 3. The process underlying future pain hedonism.

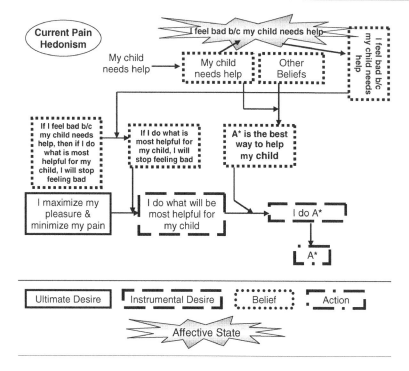

Figure 4.   The process underlying current pain hedonism.

In their attempt to show that natural selection would favor an altruistic process over the hedonistic alternatives, Sober and Wilson offer a number of arguments, all of them focused on the more complicated Current Pain Hedonism, though they think that "the argument would remain the same if we thought of the hedonist as acting to avoid future pain" (318). In discussing these arguments, I'll start with three that I don't find very plausible; I'll then take up one that I think poses a serious challenge to hedonism and leads to some important questions about how, exactly, psychological egoism and psychological altruism should be understood.

A first pair of arguments both focuses on the causal link between believing that one's child needs help and feeling an appropriate level of distress or pain. The worry raised by the first argument is that the link could occasionally fail.

> If the fitness of hedonism depends on how well correlated the organism's pleasure and pain are with its beliefs about the well-being of its children, how strong is this correlation apt to be? (315)...[W]e think it is quite improbable that the psychological pain that hedonism postulates will be *perfectly* correlated with believing that one's children are doing badly. One virtue of ALT [altruism] is that its reliability does not depend on the strength of such correlations." (316, emphasis in the original)

The second argument focuses on the fact that, to do its job appropriately, the mechanism underlying the belief-to-affect link must not only produce pain or distress, it must produce *lots* of it.

> Hedonism assumes that evolution produced organisms – ourselves included – in which psychological pain is strongly correlated with having beliefs of various kinds. In the context of our example of parental care, the hedonist asserts that whenever the organism believes that its children are well off, it tends to experience pleasure; whenever the organism believes that its children are doing badly, it tends to feel pain. What is needed is not just that *some* pleasure and *some* pain accompany these two beliefs. The amount of pleasure that comes from seeing one's children do well must exceed the amount that comes from eating chocolate ice cream and from having one's temples massaged to the sound of murmuring voices. This may require some tricky engineering... To achieve simplicity at the level of ultimate desires, complexity is required at the level of instrumental desires. This complexity must be taken into account in assessing the fitness of hedonism.[7] (315)

Sober and Wilson are certainly right that current pain hedonism requires the affect generated by the belief that one's child is doing well or badly be of an appropriate magnitude, and that this will require some psychological engineering that is not required by the altruist process. They are also right that the mechanism responsible for this belief-to-affect link will not establish a perfect correlation between belief and affect; like just about any psychological mechanism it is bound to fail now and then.

However, I don't think that either of these facts offers much reason to think that natural selection would favor the altruistic process. To see why, let's first consider the fact that the belief-to-affect link will be less than perfectly reliable. It seems that natural selection has built lots of adaptively important processes by using links between categories of belief and various sorts of affective states. Emotions like anger, fear and disgust, which play a crucial role in regulating behavior, are examples of states that are often triggered by different sorts of beliefs. And in all of these cases, it seems (logically) possible to eliminate the pathway that runs via affect, and replace it with an ultimate desire to behave appropriately when one acquires a triggering belief. Fear, for example, might be replaced by an ultimate desire to take protective action when you believe

---

[7] It is perhaps worth noting that, *pace* Sober and Wilson, neither of these arguments applies to Future Pain Hedonism, since that version of hedonism does not posit the sort of belief-to-affect link that Sober and Wilson are worried about. I should also note that, for simplicity, in discussing these arguments I'll ignore the pleasure engendered by the belief that one's child is well off and focus on the pain or distress engendered by the belief that one's child is doing badly.

that you are in danger. Since natural selection has clearly opted for an emotion mediation system in these cases rather than relying on an ultimate desire that avoids the need for a belief-to-affect link, we need some further argument to show that natural selection would not do the same in the case of parental care, and Sober and Wilson do not offer one.

The second argument faces a very similar challenge. It will indeed require some "tricky engineering" to be sure that beliefs about one's children produce the right amount of affect. But much the same is true in the case of other systems involving affect. For the fear system to work properly, seeing a tiger on the path in front of you must generate quite intense fear – a lot more than would be generated by your belief that if you run away quickly you might stub your toe. While it no doubt takes some tricky engineering to make this all work properly, natural selection was up to the challenge. Sober and Wilson give us no reason to think natural selection was not up to the challenge in the case of parental care as well.[8]

A third argument offered by Sober and Wilson is aimed at showing that natural selection would likely have preferred a system for producing parental care, which they call 'PLUR', in which *both* hedonistic motivation and altruistic motivation plays a role, over a "monistic" system that relies on hedonism alone. The central idea is that, under many circumstances, two control mechanisms are better than one.

> PLUR postulates two pathways from the belief that one's children need help to the act of providing help. If these operate at least somewhat independently of each other, and each on its own raises the probability of helping, then the two together will raise the probability of helping even more. Unless the two pathways postulated by PLUR hopelessly confound each other, PLUR will be more reliable than HED [hedonism]. PLUR is superior because it is a *multiply connected control device*. (320, italics in the original)

Sober and Wilson go on to observe that "multiply connected control devices have often evolved." They sketch a few examples, then note that "further examples could be supplied from biology, and also from engineering, where intelligent designers supply machines (like the space shuttle) with backup systems. Error is inevitable, but the chance of disastrous error can be minimized by well-crafted redundancy" (320).

Sober and Wilson are surely right that well-crafted redundancy will typically improve reliability and reduce the chance of disastrous error. They are also right that both natural selection and intelligent human designers have produced lots of systems with this sort of redundancy. But, as the disaster which

---

[8] Edouard Machery has pointed out another problem with the "tricky engineering" argument. On Sober and Wilson's account, altruists will have many ultimate desires in addition to the desire to do what will be most helpful for their children. So to insure that the desire leading to parental care usually prevails will *also* require some tricky engineering.

befell the Columbia space shuttle vividly illustrates, human engineers also often design crucial systems *without* backups. So too does natural selection, as people with damaged hearts or livers, or with small but disabling strokes, are all too well aware. One reason for lack of redundancy is that redundancy almost never comes without costs, and those costs have to be weighted against the incremental benefits that a backup system provides. Since Sober and Wilson offer us no reason to believe that, in the case of parental care, the added reliability of PLUR would justify the additional costs, their redundancy argument lends no support to the claim that natural selection would prefer PLUR to a monistic hedonism, or, for that matter, to a monistic altruism.

Sober and Wilson's fourth argument raises what I think is a much more troublesome issue for the hedonistic hypothesis.

> Suppose a hedonistic organism believes on a given occasion that providing parental care is the way for it to attain its ultimate goal of maximizing pleasure and minimizing pain. What would happen if the organism provides parental care, but then discovers that this action fails to deliver maximal pleasure and minimal pain? If the organism is able to learn from experience, it will probably be less inclined to take care of its children on subsequent occasions. Instrumental desires tend to diminish and disappear in the face of negative evidence of this sort. This can make hedonistic motivation a rather poor control device." (314) ...[The] instrumental desire will remain in place only if the organism ... is trapped by an unalterable illusion. (315)

Sober and Wilson are not as careful as they should be here. When it turns out that parental care does not produce the expected hedonic benefits, the hedonistic organism needs to have some beliefs about *why* this happened before it can effectively adjust its beliefs and instrumental desires. If, for example, the hedonist portrayed in Figures 3 or 4 comes to believe (perhaps correctly) that it was mistaken in inferring that A* was the best way to help, then it will need to adjust some of the beliefs that led to that inference, but the beliefs linking helping to the reduction of negative affect will require no modification. But despite this slip, I think that Sober and Wilson are onto something important here. Both versions of hedonism that I've sketched rely quite crucially on beliefs about the relation between helping behavior and affect. In the case of Future Pain Hedonism, as elaborated in Figure 3, the crucial belief is: *If I don't do what will be most helpful for my child, I will feel bad.* In the version of Current Pain Hedonism sketched in Figure 4, it's: *If I feel bad because my child needs help, then if I do what is most helpful for my child, I will stop feeling bad.* These beliefs make empirical claims, and like other empirical beliefs they might be undermined by evidence (including misleading evidence) or by more theoretical beliefs (rational or irrational) that a person could acquire by a variety of routes. This makes the process underlying parental care look quite vulnerable to disruption and suggests that natural selection would likely opt for some more reliable way to get this crucial job done.[9] The version of altruism depicted in Figure 2 fits the bill nicely. By making the desire, *I do what will be most helpful for my child*, an ultimate desire, it sidesteps the need for empirical beliefs that might all too easily be undermined.

I think this is both a powerful argument for psychological altruism and an original one, though ultimately I am not persuaded. To explain why, we'll have to clarify what the altruist and the egoist are claiming. Altruists, recall, maintain that people have ultimate desires for the well-being of others, while egoists believe that all desires for the well-being of others are instrumental, and that all of our ultimate desires are self-interested. An instrumental desire is a desire that is produced or sustained by a process of practical reasoning like the one depicted in Figure 1 in which a desire and a belief give rise to or sustain another desire. In the discussion of practical reasoning, in the 'Preliminaries' section, nothing was said about the notion of *belief*; it was simply taken for granted. Like other writers in this area, including Sober and Wilson, I tacitly adopted the standard view that beliefs are inferentially integrated representational states that play a characteristic role in an agent's cognitive economy. To say that a belief is *inferentially integrated* is to say (roughly) that it can be both generated and removed by inferential processes that can take any (or just about any) other beliefs as premises.

While inferentially integrated representational states play a central role in many discussions of psychological processes and cognitive architecture, the literature in both cognitive science and philosophy also often discusses belief-like states that are "stickier" than this. Once they are in place, these "stickier" belief-like states are harder to modify by acquiring or changing other beliefs. They are also typically unavailable to introspective access. In Stich (1978), they were dubbed *sub-doxastic states*. Perhaps the most familiar example of sub-doxastic states are the grammatical rules that, according to Chomsky and his followers, underlie speech production, comprehension and the production of linguistic intuitions. These representational states are clearly not inferentially integrated, since a speaker's explicit beliefs about them typically has no effect on them. A speaker can, for example, have thoroughly mistaken beliefs about the rules that govern his linguistic processing without those beliefs having any effect on the rules or on the linguistic processing that they subserve. Another important example are the *core beliefs* posited by the psychologists Susan Carey and Elizabeth Spelke (Carey and Spelke 1996; Spelke 2000, 2003). These

---

[9] Note that the vulnerability to disruption we're considering now is likely to be a much more serious problem than the vulnerability that was center stage in Sober and Wilson's first argument. In that argument, the danger posed for the hedonistic parental care system was that "the psychological pain that hedonism postulates" might not be "*perfectly* correlated with believing that one's children are doing badly" (316, emphasis in the original). But, absent other problems, a hedonistic system in which belief and affect were highly – though imperfectly – correlated would still do quite a good job of parental care. Our current concern is with the stability of the crucial belief linking helping behavior and affect. If that belief is removed the hedonistic parental care system simply crashes, and the organism will not engage in parental care at all, except by accident.

are innate representational states that underlie young children's inferences about the physical and mathematical properties of objects. In the course of development, many people acquire more sophisticated theories about these matters, some of which are incompatible with the innate core beliefs. But, if Carey and Spelke are correct, the core beliefs remain unaltered by these new beliefs and continue to affect people's performance in a variety of experimental tasks. Although sub-doxastic states are sticky and hard to remove, they do play a role in *inference-like* interactions with other representational states, though their access to other representational premises and other premises' access to them is limited. In *The Modularity of Mind*, Fodor (1983) notes that representational states stored in the sorts of mental modules he posits are typically sub-doxastic, since modules are "informationally encapsulated". But not all sub-doxastic states need reside in Fodorian modules.

Since sub-doxastic states can play a role in inference-like interactions, and since practical reasoning is an inference-like interaction, it is possible that sub-doxastic states play the belief-role in some instances of practical reasoning. So, for example, rather than the practical reasoning structure illustrated in Figure 1, some examples of practical reasoning might have the structure shown in Figure 5. What makes practical reasoning structures like this important for our purposes is that, since SUB-DOXASTIC STATE 1 is difficult or impossible to remove using evidence or inference, DESIRE 2 will be reliably correlated with DESIRE 1.

Let's now ask whether, in Figure 5, DESIRE 2 is instrumental or ultimate? As we noted earlier, the objects of ultimate desires are typically characterized as "desired for their own sakes" while instrumental desires are those that agents have only because they think that satisfying the desire will lead to the satisfaction of some other desire. In Figure 5, the agent has DESIRE 2 only because he thinks that satisfying the desire will lead to the satisfaction of DESIRE 1. So it looks like the natural answer to our question is that DESIRE 2 is instrumental; the only ultimate desire depicted in Figure 5 is DESIRE 1.

If this is right, if desires like DESIRE 2 are instrumental rather than ultimate, then Sober and Wilson's evolutionary argument for psychological altruism is in trouble. The central insight of that argument was that both

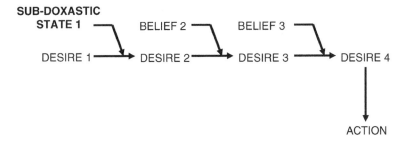

Figure 5. An episode of practical reasoning in which a sub-doxastic state plays a role.

versions of hedonism rely on empirical beliefs which might all too easily be undermined by other beliefs the agent might acquire. Suppose, however, that in Figures 3 and 4, the representations

> *If I don't do what will be most helpful for my child, I will feel bad*

and

> *If I feel bad because my child needs help, then if I do what is most help-
> ful for my child, I will stop feeling bad*

are not beliefs but sticky sub-doxastic states. If we grant that desires produced or sustained by a desire and a sub-doxastic state count as instrumental desires, not ultimate desires, then the crucial desire whose presence Sober and Wilson sought to guarantee by making it an ultimate desire, viz.

> *I do what will be most helpful for my child*

is no longer at risk of being undermined by other beliefs. Since the crucial desire is reliably present in both the Altruistic model and in both versions of the Hedonist model, natural selection can't prefer Altruism because of its greater reliability in getting a crucial job done.

In a passage from Sober and Wilson quoted earlier, they noted that when an instrumental desire does not lead to the expected hedonic pay-off, the "desire will remain in place only if the organism ... is trapped by an unalterable illusion" (315). But as a number of authors have noted, some illusions – or as I would prefer to put it, some belief-like representational states that are not strictly true – are conducive to fitness (Stich 1990; Plantinga 1993; Sober 1994; Godfrey-Smith 1996). In a variety of domains, it appears that natural selection has used sub-doxastic states and processes that have some of the features of mental modules to insure that those representations stay put and are not undermined by the systems that revise beliefs. Since natural selection often exploits the same trick over and over again, it is entirely possible that, when faced with the problem of assuring that parents were motivated to care for their children, this was the strategy it selected. My conclusion, of course, is *not* that parental care is subserved by an egoistic psychological process, but rather that Sober and Wilson's argument leaves this option quite open. Their analysis does not "provide... evidence for the existence of psychological altruism" (12).

Acknowledgements

I'm grateful to Elliott Sober, Edouard Machery, Kim Sterelny and an anonymous referee for helpful comments on earlier drafts of this paper.

References

Beatty J. 1992. Fitness: theoretical contexts. In: Keller E. and Lloyd E. (eds), Keywords in Evolutionary Biology, Harvard University Press, Cambridge, MA.

Carey S. and Spelke E. 1996. Science and core knowledge. Philos Sci 63(4): 515–533.

Fodor J. 1983. The Modularity of Mind. MIT Press. Cambridge, MA.

Ghiselin M. 1974. The Economy of Nature and the Evolution of Sex. University of California Press, Berkeley.

Godfrey-Smith P. 1996. Complexity and the Function of Mind in Nature. Cambridge University Press, Cambridge.

Goldman A. 1970. A Theory of Human Action. Prentice-Hall, Englewood-Cliffs, NJ.

Plantinga A. 1993. Warrant and Proper Function. Oxford University Press, Oxford.

Sober E. 1994. The adaptive advantage of learning and a priori prejudice. In: From a Biological Point of View. Cambridge University Press, Cambridge.

Sober E. and Wilson D.S. 1998. Unto Others: The Evolution and Psychology of Unselfish Behavior. Harvard University Press, Cambridge, MA.

Spelke E. 2000. Core knowledge. Am Psychol 55: 1233–1243.

Spelke E. 2003. Core knowledge. In: Kanwisher N. and Duncan J. (eds), Attention and Performance, Vol. 20: Functional Neuroimaging of Visual Cognition. Oxford University Press.

Stich S. 1978. Beliefs and sub-doxastic states. Philos Sci 45: 499–518.

Stich S. 1990. The Fragmentation of Reason. MIT Press, Cambridge, MA.

Stich S., Doris J. and Roedder E. (in preparation). Egoism *vs.* altruism. In: Doris J. et al. (eds), The Handbook of Moral Psychology. Oxford University Press (to be published).

Part IV
Meta-ethics

[8]

Darwin's nihilistic idea: evolution and the meaninglessness of life

TAMLER SOMMERS and ALEX ROSENBERG*
*Department of Philosophy, Center for Philosophy of Biology, Duke University, Durham, NC 27708, USA; *Author for correspondence*

Received 6 January 2002; revised 15 February 2002; accepted in revised form 2 July 2002

Introduction

No one has expressed the destructive power of Darwinian theory more effectively than Daniel Dennett. Others have recognized that the theory of evolution offers us a universal acid, but Dennett, bless his heart, coined the term. Many have appreciated that the mechanism of random variation and natural selection is a substrate-neutral algorithm that operates at every level of organization from the macromolecular to the mental, at every time scale from the geological epoch to the nanosecond. But it took Dennett to express the idea in a polysyllable or two. These two features of Darwinism undermine more wishful thinking about the way the world is than any other brace of notions since mechanism was vindicated in physics.

The solvent algorithm deprives nature of purpose, on the global and the local scale. Both evolutionary phylogeny and organismic ontology can be explained as the operation of passive environmental filtration on variations produced by real (not just epistemic) randomness. Types and tokens are built by the iteration of this same process at multiple levels. Even when you get to the *locus classicus* of purposeful phenomena in human cognition and its consequences in action, natural selection explains both capacities and performance in a way that dispenses with purpose even here. Darwinism thus puts the capstone on a process which since Newton's time has driven teleology to the explanatory sidelines. In short it has made Darwinians into metaphysical Nihilists denying that there is any meaning or purpose to the universe, its contents and its cosmic history. But in making Darwinians into metaphysical nihilists, the solvent algorithm should have made them into ethical nihilists too. For intrinsic values and obligations make sense only against the background of purposes, goals, and ends which are not merely instrumental. But the leading Darwinian philosophers have shied away from this implication and instead have embraced ethical naturalism. And this despite the ever-increasing power of Darwinism to explain away normative ethics as a local adaptation.

One might well expect tenderhearted scientists and others faced by a forced choice between naturalism and nihilism to choose the former, if only because acknowledging Darwinism's commitment to nihilism makes it an even less attractive theory to the unconverted. But to see ethical naturalism exuberantly defended by no less a steely eyed Darwinian than Dennett is something of a surprise. In the

654

conclusion to *Darwin's Dangerous Idea: Evolution and the Meanings of Life*, Dennett waxes thus:

> There is no denying, at this point, that Darwin's idea is a universal solvent, capable of cutting right to the heart of everything in sight. The question is: what does it leave behind? I have tried to show that once it passes through everything, we are left with stronger, sounder versions of our most important ideas. Some of the traditional details perish, and some of these are losses to be regretted, but good riddance to the rest of them. What remains is more than enough to build on.Does Darwin's idea turn out to be, in the end, just what we need in our attempt to preserve . . . the values we cherish? I have completed my case for the defense (Dennett (1995), p. 521. Page references below are all to this work)

We cannot think we are alone in finding this the most surprising and unsettling conclusion of Dennett's defense of Darwinism against all comers.. Most of those who fear Darwin's dangerous idea reject it owing to their recognition that it is universal acid, eating through every available argument for the values people cherish. We differ from those who fear Darwinism because we believe it is true. But we do not think that we can or need hide our countenances from the nihilism it underwrites. In this paper we seek to show that the first fifteen chapters of Dennett's work is a case for the nihilist's prosecution, not the naturalist's defense. We will argue that chapters 16 and 17 of *Darwin's Dangerous Idea* do not fit within the framework of the 453 pages that precede them; nor does the thesis they defend pass Dennett's own intellectual muster. If we are right the only alternative to the naturalism Dennett hopes to embrace is either an intuitionism that deprives Darwin's idea of its danger by denying its universal writ, or a sort of "nice" nihilism. It is this latter, which in the end Dennett and the rest of us Darwinians must embrace.

Nihilism, naturalism, intuitionism

For present purposes, we need a characterization of nihilism, naturalism, and intuitionism. Ethical naturalism combines three theses: a) normative terms that name intrinsic (as opposed to instrumental) values, and categorical obligations denote real properties of things or acts, b) there are true or at least well justified synthetic propositions about which items or actions instantiate these properties,[1] and c) There exists a set of scientifically accessible natural facts about us or the world that can

---

[1] Naturalism is a wider doctrine than "cognitivism"—the metaethical thesis that ethical claims have a truth-value. For if it turned out that "noncognitivism" is correct, that the fundamental ethical claims are imperative statements, or norms that lack truth value, naturalism would be still be committed to grounding them as obligatory in accordance with condition c). It is for this reason that condition b) and c) are expressed in terms of the disjunctive "true or well justified or well grounded". Hereafter, when we write of ethical truths, we mean to be understood in terms that do not beg the question between cognitivism and noncognitivism.

justify a) and explain the truth or well groundedness of b). To qualify as a naturalist one must treat the theory of natural selection as well established, but it may be supposed that one need not ground the truth of ethical claims on this theory. Darwinian naturalism is a restriction on the more general version of naturalism only in one respect: it holds that the natural facts about the world that explain and underwrite ethical truths are reported by the theory of natural selection and its application to biological data. In section 3, we return to the important question of whether one can be a naturalist without also being a Darwinian naturalist.

Nihilism consists in the following claims: a) normative terms-good, bad, right, duty, etc-do not name real properties of events or things, either natural nor non-natural ones; b) all claims about what is good in itself, or about categorical moral rights or duties, are either false or meaningless; c) the almost universal beliefs that there are such properties and that such claims are true can be "explained away" by appropriate scientific theory. Nihilism takes the form of what Mackie (1977) calls an "error theory." It does not deny that beliefs about norms and values can motivate people's actions. It does not deny the felt "internalism" of moral claims, nor does it deny that normative beliefs confer benefits on the people who hold them. Indeed nihilism is consistent with the claim that such beliefs are necessary for human survival, welfare and flourishing. Nihilism only claims that these beliefs, where they exist, are false. It treats morality as instrumentally useful—instrumentally useful for our nonmoral ends or perhaps the nonmoral ends of some other biological systems, such as our genes for example. As such, it must undermine the values we cherish. If Darwinism underwrites nihilism, Dennett cannot be right about Darwinism's salubrious effects for "the meanings of life". However nihilism can be, as one might say, "nice", provided that in its explaining away of ethics, it also shows that we are in fact disposed to behave nicely—to cooperate, be altruistic, show guilt and shame, anger and resentment in just the way we would if some morality were true, right, or real. *Darwinian* nihilism is the thesis that the theory of natural selection and its application to biological data explains why morality is at most an instrumentally useful illusion. According to the Darwinian nihilist, the theory of natural selection can both show that we are in error about the status of moral claims, and, perhaps more importantly, can explain this why the error is so widespread.

Intuitionism is the tendentious label we shall use for a common view about ethics. It shares with naturalism a commitment to two theses: a) normative terms that name intrinsic (as opposed to instrumental) values, and categorical obligations denote real properties of things or acts; b) there are true or at least well justified synthetic propositions about which items or actions instantiate these properties. But unlike naturalism, intuitionism's epistemology makes it possible for us to have knowledge of a) and b) without the aid of Darwinian, biological, or any other sort of scientifically warranted means, but rather by direct inspection or perception, or "intuition," whence the label.

Intuitionism of course rejects naturalism's thesis that the facts which make ethical claims true or objectively binding are scientifically accessible. Accordingly, it constitutes a serious threat to Dennett's claim that Darwinism is a universal acid. For suppose ethical values or truths may be identified by direct inspection, perception, pure reason or other means unsanctioned by the epistemology employed by natural science. Then it is open to claim that intuition can provide us with further non-ethical information about factual matters, including the falsity of Darwinism. On the other hand, if Darwinism is truly a universal acid, then it works on epistemologies as well as metaphysical theories. As such it dissolves the epistemology of intuitionism, by excluding the emergence of an epistemic modality that can directly secure access to concrete matters not realized by the natural arrangement of things. We return to this point in section 3 below. So when it comes to ethical value, Darwinians must choose between naturalism and nihilism. There is no *tertium quid*. As noted above, however, almost every Darwinian has preferred naturalism (or a dignified silence) to nihilism and Dennett is no exception. But Dennett more than other Darwinians should be sensitive to the difficulties of naturalism, difficulties that have made naturalism problematic in philosophy since the time of David Hume.

Darwinian nihilism shares with Darwinian naturalism common commitments in metaethics: indeed, they can embrace the same program of research in the explanation of how cooperation arose and morality emerged. The salient features of this program, which Dennett cogently reviews, include Hamilton (1967), Trivers (1971) treatments of the selective advantage of kin and reciprocal altruism in iterated prisoner's dilemmas, Axelrod (1984) accounts of how tit-for tat triumphs in certain simulations, Ruse and Wilson (1989)'s demonstration of how even division in cut-the-cake games constitutes the largest basin of attraction in a selective environment, Wilson and Sober (1998)'s elaboration of how between group selection for cooperation may swamp within-group selection for selfishness, Frank (1989) account of the role of moral emotions as commitment-strategies in prisoner's dilemmas-iterated or not, and Gibbard (1992)'s theory of the content and structure of moral norms. Darwinian naturalism departs from Darwinian nihilism when it goes on to suggest that the natural selection of cooperation, justice, and other normative institutions underwrites some moral claims as true or correct. The nihilist will deny that adaptational explanations can preserve "the values we cherish", and still less that they enable us to construct "sounder versions of our most important ideas" [, p. 521]

We may illustrate the differences between naturalism and nihilism by examining Dennett's critique of Thomas Hobbes. Dennett views Hobbes as a greedy sociobiological reductionist; that is, one who seeks to naturalize ethical values and obligations "without the use of cranes", apparently without "whole layers of theory" [p. 81] that will ground ethical value in biological fact while doing justice to the complexities of human behavior and its causes. It is this greediness that leads Dennett to deem Hobbes naturalizing project a failure.

On our view, however, Hobbes had no naturalizing project. He was in fact a moral nihilist. The relevant Hobbesian passages are infamous: "these words of good and evil . . . are ever used with relation to the person that uses them, there being nothing simply and absolutely so, nor any common rule of good and evil to be taken from the nature of the objects themselves [*Leviathan*, Bk 1, chapter 6, *Good, Evil*]." Rights and obligations simply reflect men's "conclusions or theorems concerning what conduces to the conservation and defense of themselves," "*Good* and *Evil* are names that signify our appetites and aversion, which in different tempers, customs, and doctrines of men are different" [Leviathan, chapter 15]. As an analysis of ordinary moral language, Hobbes' account may be found wanting. But his intentions are clear enough: he is giving an instrumental justification of these norms, or as he called them, "laws of nature," revealing them to be hypothetical imperatives conducive to human survival and satisfaction. On this view Hobbes was not hoping simultaneously to explain "how . . . right and wrong came into existence in the first place", and to "*justify* a set of ethical norms." Only the first project was un-controversially a part of Hobbes' program. If we agree that *justifying* morality involves more than just showing that morality is a conditionally advantageous policy, then we may say that this justification-project was not among Hobbes' aims. For this reason he is better treated as a nihilist than a naturalist. And as a mere nihilist, aiming to explain (away) morality, Hobbes would also stand acquitted of the charge of seeking "greedily" to reduce ethical statements and predicates wholly to natural facts and properties.

Ambivalence about the naturalistic fallacy

Scholars rightly hold that we should distinguish Moore (1903) "naturalistic fal-lacy", and the "open question" argument on which it is based from Hume's point about the fallaciousness of inferring ought from is. However, it is this latter Humean claim that Dennett seems to treat as the "naturalistic fallacy" and we will follow this practice. Dennett does not so much reject the "naturalistic fallacy" as regret it. In other words, he does not deny the cogency of the point that you cannot infer "ought' from "is", rather he prunes the claims of naturalism back just because the inference is fallacious:

> According to the standard doctrine, if we stay firmly planted in the realm of facts about the world as it *is*, we will never find any collection of them, taken as axioms, from which any particular ethical conclusion *can be conclusively proven.* You can't get there from here, any more than you can get from any consistent set of axioms about arithmetic to all the true statements of arithmetic. [p. 467]

This passage expresses acceptance of Hume's dictum, at least the claim that there are no valid deductive inference from premises about what is the case to conclusions about what ought to be the case. Adding Moore's open-question argument to Hume's dictum is sometimes supposed to produce a stronger version according to which factual premises cannot even inductively support normative conclusions. It is in part for this reason that they are often though to work in tandem. Naturalism must reject either the deductive or the inductive version of Hume's (and Moore's) dictum. The naturalist must show that normative conclusions are actually justified either by deduction from factual premises, or inductive inference from them. Dennett accepts the obligation to do this very thing. But what he actually does is something much weaker, as we shall see.

658

Having conceded that you cannot infer "ought" from "is" with deductive validly, Dennett writes,

> Well, so what? We may bring out the force of this rhetorical question with another one, rather more pointed: If "ought" cannot be derived from "is" just what can "ought" be derived from? Does it float, untethered to facts from any other discipline or tradition? Do our moral intuitions arise from some inexplicable ethics module implanted in our brains? That would be a dubious skyhook on which to hand our deepest convictions about what is right and wrong. [p. 467]

Here intuitionism, as we have called it, is rightly rejected. But without pausing to even consider the alternative of nihilism—Darwinian or otherwise–an apparently inductive inference to naturalism is embraced:

> From what can "ought" be derived? The most compelling answer is this: ethics must be *somehow* based on an appreciation of human nature-on a sense of what a human being is or might be like, and on what a human being might want to have or want to be. If *that* is naturalism, then naturalism is no fallacy. No one could seriously deny that ethics is responsive to such facts about human nature. [p. 468]

A lot depends here on what Dennett means by "based on."Does he merely express a demand that ethics be responsive to human nature? Then it is the obvious and uninteresting point that since "ought to do x" implies "can do x" by contraposition "can't do x" implied "is not (morally) required to do x". Thus, for example, there is no obligation to fly unaided because human beings cannot fly unaided. If that is all that is at stake, Dennett's claim is innocent of the move from 'is' to 'ought'.

But naturalism requires more than "responsiveness to human nature." It is the thesis that "ought" is derivable (deductively or inductively) from "is". Naturalism must insist that the naturalistic fallacy is no fallacy, and Darwinian naturalism must show that the theory of natural selection provides the (inductive) support for the truth of ethical claims. Dennett wants to show this, and insists that the only fallacy here is "is not naturalism but rather, any simple minded attempt to rush from facts to values. In other words, the fallacy is *greedy* reductionism of values to facts, rather than reductionism considered more circumspectly, as the attempt to unify our world-view so that our ethical principles don't clash irrationally with the way the world is." [p. 468] But can naturalism require no more than that our ethical beliefs not clash irrationally with the way the world is? This is too low a standard for something to count as naturalism. For naturalism to be vindicated, Darwinism must do more than merely reconcile morality and natural selection. Darwinism must underwrite morality and work to justify its claims.

The mere compatibility of ethical truths with the theory of natural selection would of course be acceptable for the intuitionist, but as noted, intuitionism is not a permissible outcome for the Darwinian naturalist. For if there are ethical truths which escape explanation by naturalism, then at a minimum there will be a range of emergent irreducible facts, forces, causes, that make a difference to human life beyond those that Darwinian science deals with. If such facts and forces exist and can be known to us, there is more in heaven and earth than is dreamed of in Darwinism. If there are ethical truths which escape naturalism, then there are facts about biological systems (us and perhaps other creatures) that are not the result of the operation of a mindless substrate-neutral algorithm Dennett rightly identifies as Darwinism's core. The admission of this one class of exceptions which won't succumb to the universal acid must open the flood-gates to other exceptions–the immaterial mind, vital forces, hidden purposes, omega points, the deity of the theists–that trimmers of and qualifiers on Darwinism (along with its deniers) have sought from A.R. Wallace's day to John Paul II's. It is in fact this implicit and sometimes explicit recognition that mere compatibility of ethical claims and Darwinian theory is not sufficient, coupled with the horror of nihilism, that have driven Darwinians, who should have known better, to commit the naturalistic fallacy.

Of course, the mere compatibility of Darwinism with the claims of ethics we endorse would suffice, if these claims were like the claims, say, of arithmetic. Mathematical truths must be unified into the same worldview with Darwinism, but need not all be naturalized by Darwinism alone or even in part. If we cannot effect this unification as yet, owing to problems about the metaphysics and epistemology of mathematical abstracta, it will be sufficient for the nonce if mathematical claims are shown, in Dennett's words, "not to clash irrationally" with the way Darwinism says "the world *is*." Why then, it may be argued, should we impose a higher standard than mere compatibility with Darwinism on the claims of ethics?

The answer to this question lies in the fundamental difference between ethical truths—specific or general–and mathematical ones. It is a difference that demands that the Darwinian naturalist establish more than mere consistency between Darwinism and the ethical judgments we believe to be true. The difference also shows that in the end Darwinian naturalism is the only game in town for naturalists. There is no scope for a non-Darwinian naturalism, no scope for a theory that treats ethical facts as natural, but accounts for them on the basis of some other natural, factual, scientific but non-Darwinian considerations. This difference between moral truths, if there are any, and other truths, such as for example mathematical ones, shows that all naturalists must in the end be Darwinian naturalists. If Darwinian naturalism fails, there is no alternative non-Darwinian naturalism available. The only alternatives left will be nihilism and intuitionism.

The reason that there is no scope for non-Darwinian naturalism in ethics is that ethical beliefs, whether general or specific, guide conduct that makes a difference for survival and reproduction. In this respect ethical beliefs are quite unlike all but a few truths of arithmetic. Naturalists are committed to the thesis that if there are ethical truths, it must often have been adaptive to believe them. Our capacity to make *true* moral judgments must therefore be wholly or largely the result of the

660

operation of natural selection. To see why this is the case, consider the two alternatives: (1) our capacity to form true ethical judgments has resulted in spite of the strong deflecting force of natural selection. To believe this alternative is tantamount to denying that capacity to form ethical beliefs has consequences for survival and reproduction or denying the truth of Darwinism in general. Neither move is open to a naturalist of any sort. Or (2) The adaptiveness of our capacity to make moral judgments and their truthfulness are not causally related. On this view the fact that any of our moral judgments are both adaptive and true is a cosmic coincidence. For if natural selection did not connect the truth-makers of our ethical beliefs with our capacity to form such beliefs, it must have shaped us to adopt ethical beliefs because they were adaptive but *not* because they are true. We should accordingly expect that only a vanishingly small fraction of our moral judgments are correct.[2] And such a conclusion no naturalist—Darwinian or otherwise–who seeks to vindicate our most important and widely shared ethical views can endorse

The upshot of all this is that neither a Darwinian naturalist nor any other kind of naturalist can be satisfied merely to show that ethical truths are consistent with the theory of natural selection. They must show how the adaptationist scenarios of the theory of natural selection constitute at least a significant portion of the truth-makers for the ethical truths they allege we know. For example, suppose one adopted an "ideal-observer" version of naturalism, according to which the natural truth-makers for true ethical judgments are facts about how each and every normal Homo sapiens would respond psychologically to some event under so-called "ideal conditions". Without reference to the Darwinian connection between the "natural" fact about ideal observers and our actual moral beliefs, such a theory cannot claim to be naturalistic. Why? Because there is an undeniable causal connection between on the one hand, psychological response to stimuli, especially of the sort which result in expressions of the (moral) emotions of anger, shame, guilt and disdain (under ideal conditions or otherwise)—and, on the other hand, survival and reproduction. Our dispositions to react affectively to certain situations have been formed, at least in large part, by natural selection. Even the ideal observer naturalists cannot be satisfied with mere consistency between their theory and Darwinism.

This is why, short of simply denying the truth of the theory of natural selection in general, Darwinian naturalism is the only game in town for any naturalist. If Darwinian naturalism fails, the Darwinian is left with a choice between nihilism and intuitionism. And no Darwinian can be satisfied with intuitionism.

The critique of greedy reductionism

Dennett criticizes greedy reductionists like BF. Skinner and E.O. Wilson because, like emergentist sky-hookers, they too seek to dispense with cranes. "What is wrong with Skinner," Dennett writes, "is not that he tried to base ethics on scientific facts about human nature, but that his attempt was too simplistic! The same defect can be

---

[2] We owe this argument to Sharon Street. No agreement with any of the further conclusions we draw from it should however be attributed to her.

seen in another attempt at ethics by another Harvard professor, E.O. Wilson . . . " [p. 469] The criticism leads one to suppose that Dennett will provide the required crane, once he has shown that Skinner's and Wilson's ethics are unsupported by their science. Alas, the implicit promissory note remains unredeemed.

Together with Michael Ruse, Wilson has argued that "Morality, or more strictly our belief in morality, is merely an adaptation put in place to further our reproductive ends" Dennett quotes them:

> In an important sense, ethics as we understand it is an illusion fobbed off on us by our genes to get us to cooperate . . . Furthermore the way our biology enforces its ends is by making us think that there is an objective higher code to which we are all subject. (Ruse and Wilson (1989), p 51)

Notice that this is an argument for nihilism, not greedy naturalistic reductionism. It gives us an explanation for why we would believe ethics to be real or objective when it is not. As such, it sits uncomfortably with the patently naturalistic view of ethics Wilson adopts elsewhere. But what is interesting in Dennett's attack on this claim is that he does not argue directly against the claim that ethics is an illusion, or that we have been selected to think it objective when it isn't (as the Darwinian nihilist believes). Instead he argues against the claim that ethical beliefs obtain *for the benefit of the genes* (as principal beneficiaries) on the grounds that once persons and their memes are on the scene, these too are potential beneficiaries of the emergence of ethical dispositions. This is true enough, and not something any nihilist or greedy reductionist need deny (not even the most monomaniacally selfish gene theorist). However, from this observation Dennett infers that therefore "the truth of an evolutionary explanation [of the emergence of ethics] would not show that our allegiance to ethical principles or a "higher code" was an "illusion". Also correct as far as it goes. But when we combine an evolutionary account of ethical beliefs with the conception of Darwinian theory as a "universal acid" (which Dennett has argued for in the first 15 chapters of his book) the result is moral nihilism. If all apparently purposive processes, states, events, and conditions are in reality the operation of a purely mechanical substrate neutral algorithm, then as far explanatory tasks go, the only values we need attribute to biological systems are instrumental ones. An evolutionary account of moral belief will not only explain ethics but it will explain it away.

Of course, the Darwinian nihilist accepts that like other phenotypes, once the ethical dispositions emerge, as a result of natural selection, they may variously confer benefits (and costs) on the creatures that evince them independent of their impact on reproductive fitness. All the nihilist requires is that these dispositions initially emerged and persist over evolutionary time scales as a result of an etiology of selection, and that they will disappear if they reduce reproductive fitness for long enough, regardless of the non-reproductive fitness benefits they confer. Persons, as Dennett says, "are not at all bound to answer to the interests of their genes alone-or their memes alone". True, but all the Darwinian nihilist needs is the admission that under the aspect of evolution, persons must sooner or later always answer to their

662

genes. As Wilson wrote, "The genes hold culture on a leash. The leash is very long, but inevitably values will be constrained in accordance with their effects on the human gene pool. [On Human Nature, p.167]" He could have said this by way of a retrospective explanation: values have been constrained by their effects on the gene pool, and the constraints are narrow enough to explain and explain away much of their common features across time and tide.

Dennett however believes that culture-memes-have the power to snap the genetic leash Wilson identifies. But the reason he gives for this conclusion is far too weak: It begins strongly enough, accusing Wilson, and others, like Richard Alexander, who advance a similar argument, of committing not the naturalistic fallacy but the *genetic fallacy*, no pun intended. However, it isn't the genetic fallacy those who explain away ethics biologically are accused of committing.

The traditional genetic fallacy criticism points out that the historical origins of a belief are by themselves no reason to either to substantiate or undercut it. Thus, it is fallacious to infer directly from the fact that selection for reproductive fitness is the cause of our dispositions to advance and honor moral claims, to the claim that ethical propositions are either false or unjustified. Of course Darwinian nihilism makes no such *direct* inference. It makes it indirectly by adding to the historical claims that explain why we embrace our values the further claims that a) one cannot derive moral values from natural facts (the naturalistic fallacy), b) Darwinism is well supported; c) the existence of un-naturalized facts about values is incompatible with Darwinism's universal acidity. From its historical explanation of the emergence of ethical beliefs, together with these auxiliary premises, nihilism infers the conclusion that ethical propositions are false or otherwise unjustified. No genetic fallacy here.

But in Dennett's hands the genetic fallacy turns out to be something different anyway. It is the mistaken claim that since morality arose for the benefit of the genes, it can only benefit the genes: "the massively misleading idea that the *summum bonum* at the source of every chain of practical reasoning is the imperative of our genes. [p. 473]" But this claim is no part of the sociobiologist's claim when he argues that "ethics is an illusion fobbed off on us by our genes to get us to cooperate". Explaining away ethics is quite different from the suggestion sometimes made by E.O. Wilson, when he is not in nihilist mode, that the survival of the human genome is the "cardinal value", the *summum bonum*, and that the history of genic selection can ground this valuation. This conclusion is indeed the callow naturalistic fallacy which, in their ambivalence some sociobiologists- like Wilson-commit when they seek to avoid the conclusions of nihilism. Of course the idea that the diversity of our gene pool-whether 'our' means the genes only of Homo sapiens, or mammals, vertebrates, . . . or animals, or the biosphere as a whole-is the *summum bonnum* can never be grounded in any matter of fact. However, the important point here is that the naturalistic fallacy isn't the genetic fallacy. You can commit either one without the other. And the mistake which Dennett is really accusing Wilson, Alexander, Skinner, and other greedy reductionists of, is not the genetic fallacy but the plain old Naturalistic Fallacy, the one his own commitment to naturalism is going to require Dennett to commit.

Dennett's naturalistic non-fallacy

Dennett might make good use of the logical point that just because morality emerged owing to the advantages it conferred on the replicating lineage of genes of the organisms who hearken to its teachings, it does not follow that morality fails to advantage these people as well as their genes. Indeed he might go on to identify the benefits to such organisms, and to show how the benefit to people, instead of just their genes, provides the factual basis from which to infer the justification of morality. But note, it cannot just be the instrumental justification for morality; it cannot be the claim that being moral serves the subjective, selfish, personal interests, tastes, preferences, or other adventitious goals of humanity. For that is what no nihilist-Darwinian or otherwise—will deny. Dennett's project has to show that morality secures some intrinsically, not just instrumentally valuable goals, ends, purposes identified as such by Darwinism, or biology, or science. This is Dennett's task.

What Dennett is obliged to produce for us is a crane or cranes that can erect what those who hanker for skyhooks demand: ethical truth, intrinsic value, categorical obligation. How does he attempt to do this? He asks the question explicitly, as the title of the first section of the penultimate chapter of *Darwin's Dangerous Idea*: "Can ethics be naturalized?" He begins by pointing out that there are

> no discoverable and confirmable ethical truths, no forced moves or Good tricks. Great edifices of ethical theory have been constructed, criticized and defended, revised and extended by the best methods of rational inquiry . . . but they do not yet command the untroubled assent of all those who have studied them carefully." [p. 495]

He suggests:

> Perhaps we can get some clues about the status and prospects of ethical theory by reflecting on what we have seen to be the limitations of the great design process that has ethicists among its products to date. What follows, we may ask, from the fact that ethical decision making, like all actual processes of exploration in design space, must be to some degree myopic and time-pressured." [p. 495]

The answer is nothing. Nothing of relevance to the present question of naturalizing ethics follows unless an inference can move from is—from actual processes in design space—to ought, to normative claims of one or more ethical theories. What the reflection proposed can teach us is that, as with other biological phenomena, the environmental irregularities, the reflexive character of selection and counter-selection, the ever increasing volume of design space occupied and attainable, all undercut any useful generalizations about universally optimal or even generally satisfactory instrumental stratagems in our strategic games against nature or against

664

other strategizers. "No remotely compelling system of ethics has ever been made *computationally tractable*, even indirectly, for real world moral problems." This is not only true, it is what evolutionary biology should lead us to expect. Given the variegated character of the environment over even short time spans, no single behavioral strategy for attaining a given outcome will always work, or even usually work to attain it. And even if something works to attain the aim of some organism often enough over long enough a period of time, some other organism will evolve a strategy that takes advantage of the first bit of behavior to the disadvantage of the first organism. If anything, these considerations suggest that there is nothing to naturalize by way of moral theory. If there are ethical truths to be naturalized they will not be the systematic claims of Mill or Kant, but the singular, particular claims we make about the rightness or wrongness of individual actions and outcomes. But that fact does not free Dennett from the obligation to show how biological facts underwrite the truth of even these singular judgments.

But rather than discharging this obligation, Dennett instead turns to the task of showing us, by a well-wrought analogy, how we *actually make* these particular confident local moral judgments. The example, of choosing the best qualified applicant from 250,000 candidates, "is meant to illustrate, enlarged and in slow motion, the ubiquitous features of real-time decision making." [p. 502] Decision processes, Dennett tells us, are matters of satisficing, in Simon's now famous term. "Time-pressured decision making is like that all *the way down*. Satisficing extends even back behind the fixed biological design of the decision-making agent, to the design "decisions" that Mother Nature settled for when designing us and other organisms." [p. 503]. This appreciation of the nature of actual decision making leads Dennett to the question "how, then, can we hope to regulate, or at least improve, our ethical decision making, if it is irremediably heuristic, time pressured, and myopic." Dennett will go on to provide an answer to this question. But in thinking about it, we need to bear firmly in mind that this is still not the project he has set himself. Showing how we can improve our moral decision making pre-supposes some standard-whether truth, correctness, justifiability–to which the improvement will enable our actual ethical decision making more nearly to approach. But that standard needs to be naturalized to begin with.

What we need in moral decision-making, Dennett tells us, are "conversation-stoppers"-devices that put an end to internal debate, reflection, calculation. Because of natural variation in the environment and our own cognitive and emotional equipment, "we cannot expect there to be a single stable solution to such a design problem [of creating the internal conversation stopper], but rather a variety of uncertain and temporary equilibria, with conversation-stoppers tending to accrete pearly layers of supporting dogma which themselves cannot withstand extended scrutiny but do actually serve on occasion, blessedly, to deflect and terminate consideration". For example, "But that would violate a person's *right*." is such an internal conversation/deliberation stopper. And here Dennett recalls the infamous and accurate remark of Bentham that rights are nonsense and natural rights are nonsense on stilts. It's nonsense, Dennett agrees but, "*good* nonsense-and good only because it is on stilts, only because it happens to have the "political" power to keep rising about the meta-reflections-not indefinitely, but usually "high enough"-to reassert itself as a compelling-that is, conversation-stopping-first principle." [p. 507]

It will not have escaped the reader that what Dennett is arguing is that "the moral claim of rights, natural or otherwise" is *instrumentally* good nonsense, good for putting an end to deliberation about which among prospective actions to choose. But this isn't the required task. It is true that Buridan's ass lacked a conversation stopper which might have enhanced its fitness, reproductive or otherwise. But lacking such a stopper was not immoral, only ineffective. This is not the sort of endorsement of rights which is wanted in the present case. "Having rules works-somewhat-and not having rules doesn't work at all," [p. 507] concludes Dennett. And having rules must be a matter, not of rational acceptance on the basis of well-supported evidence, but "some unquestioned dogmatism that will render agents impervious to the subtle invasion of hyper-rationality." [p. 508] Explaining rules as the reflection of a convenient but unquestioned dogmatism does not advance any naturalization project.

Still owing us a naturalization of value or obligation, Dennett comes to the final chapter of *Darwin's Dangerous Idea*, where he again invokes his apt account of natural selection using cranes to lift lineages of biological systems and their traits through the subregions of design space to ones where good, better and (locally) best adaptations lie. The trajectory of this path through design space produced, among other things J.S. Bach, who is "precious not because he had within his brain a magic pearl of genius-stuff, a skyhook, but because he was, or contained, an utterly idiosyncratic structure of cranes, made of cranes, made of cranes, made of cranes." [p. 512]. But is it impertinent to ask why the fact that his genius is the result of cranes within cranes within cranes makes it precious? Surely the assertion without argument that its origins in the algorithm of natural selection make Bach's genius precious to us or precious *simpliciter is* another instance of the genetic fallacy. To see the point quite palpably, note that Stalin or Osama bin Laden, or Michael Behe, or your favorite villain, is also "an utterly idiosyncratic structure of cranes, made of cranes, made of cranes, made of cranes."

Dennett asks the question of "how much of what we value is explicable in terms of its designedness". There is an ambiguity here. Are we seeking a Darwinian explanation of our valuations or a Darwinian explanation of the emergence of the objects that our values lead us *correctly* to prize? In the discussion of Bach it sounds like the latter. What Dennett seeks is the path of random variation and natural selection through design space that resulted in those things we correctly value-correctly because they have value above and beyond our subjective appraisals of them. He says that asking this question runs the "omnipresent risk of greedy reductionism". This is the risk of trying to show how objective value can be underwritten by natural facts and without cranes. Presumably, therefore, Dennett can be expected finally to introduce or identify the cranes that will do this job ungreedily. For this is the job that needs doing. But instead he re-identifies or mis-identifies the project as one of explaining our subjective appraisals instead of the objective values of the objects we in fact prize. He tells us, "the Design Space perspective certainly doesn't explain everything about value, but it lets us see what happens when we try to unify our sense of value in a single perspective. On the one

666

hand, it helps to explain our intuition that uniqueness or individuality is "intrinsical-ly" valuable. On the other hand it lets us confirm all the incommensurable values that people talk about." [p. 513]

But explaining our intuition that something is intrinsically valuable is not what is required here. To vindicate naturalism Dennett owes us a set of considerations which underwrite that intuition as *correct*. In fact Dennett recognizes that Darwin-ism cannot provide answers to questions about absolute or comparative intrinsic value. But he doesn't recognize that its silence on these questions is due to the non-existence of the requisite cranes, not our failure hitherto to detect them: "I do not suggest that Darwinian thinking gives us answers to such questions [as the preferability of euthanasia over heroic measures for deformed infants]; I do suggest that Darwinian thinking helps us see why the traditional hope of solving these problems (finding a moral algorithm) is forlorn." [p. 514] This hardly sounds like naturalism.

In the last few pages of *Darwin's Dangerous Idea* Dennett advocates the value of many things: "Among the precious artifacts worth saving are whole cultures themselves . . . several thousand languages", species, religions traditions (including those with a previous or current track record of intolerance and anti-Darwinism), even a certain song he learned as a child. He writes:" . . . we can take steps to conserve what is valuable in every culture without keeping alive (or virulent) all its weaknesses." But what purely naturalistic (and non-instrumental) considerations can provide grounds—inductive or deductive—for the conclusion that these things ought to be preserved for reasons independent of their local adaptiveness?

The nihilist holds that there is no answer to this question and that is why Dennett cannot answer it, despite the need he feels to do so. The nihilist goes on to observe that Dennett doesn't really have to answer it, for in the universal acid he has a doctrine that renders superfluous the need to infer from "is" to "ought". Indeed, the way Darwinism cuts through all the cant about intrinsic value-simple, complex, evident or hidden, widely instanced or uniquely realized, is evident on the very last pages of *Darwin's Dangerous Idea*. There Dennett confronts for the last time the wishful thinking of ambivalent theologian and physicist Paul Davies. Davies simply cannot accept that a cosmically purposeless universe could have, let alone did, find its way through design space to something so important as the human mind. Davies conclusion is that the human mind is too important to be the result of the long-term operation of a substrate neutral algorithm. Here is Dennett's last chance to grasp the nettle of nihilism. Again, he declines. Instead he first agrees with Davies on the overwhelming importance, excellence, presumably intrinsic value of the human mind. Then he notes, with undoubted logic, that Davies assumes without argument that what has been produced through the operation of blind variation and natural selection cannot have the required importance, excellence, or value. True enough, but again the mere consistency of natural selection and intrinsic value is not enough to ground Dennett's conclusion. He needs to show that Darwinian processes do actively confer upon the mind this value, not merely that they are compatible with its having this value. But Dennett seems blind to this lacuna in his argument, for he writes, "I have argued that Darwin has shown us how, in fact, *everything* of importance is just such a product."

Having described so effectively the systematic deprivation of design in nature by the dint of uncovering the substrate-algorithm which eats through all purposes, Dennett concludes:

> Darwin offers us an explanation of how God is distributed in the whole of nature: it is in the distribution of Design throughout nature, creating in the tree of life, an utterly unique and irreplaceable creation, an actual pattern in the immeasurable reaches of design space that could never be exactly duplicated in its many details. Is this Tree of Life a God one could worship? . . . Is it something sacred? Yes, say I with Nietzsche. I could not pray to it. But I can stand in affirmation of its magnificence. This world is sacred." [p. 520]

The prose is moving and sincere. But in the cold light of philosophical scrutiny, it won't wash. It doesn't do the work Dennett requires. Sacred to us, for example is not the same as sacred *tout court* or sacred *simpliciter*. The indubitable fact that chance and necessity together have produced one or a vast number of unique things is by itself no reason that these things have value or impose obligation just in virtue of their origin, their existence, or their uniqueness. All this is obvious enough. But what may be harder to focus on in the moving prose is that Dennett has really provided no argument for thinking to the contrary. What he has produced are strong arguments for thinking that everything of importance to us, including (indeed especially) our ethical beliefs, is just a product of mindless purposeless forces. If, as Dennett must agree, these are the only forces that there are in the universe, importance is explainable without residue in terms of mindless purposelessness. And if there is no residue left to explain, we can answer the challenge of the anti-Darwinian, "You'll *never* explain *this*!" [p. 521] at least to our own Darwinian satisfaction. There is nothing left of morality to explain.

Conclusion: Nice nihilism

Dennett's failure to substantiate naturalism does not by itself establish nihilism. But in light of every naturalist's obligation to be a Darwinian about how ethical beliefs emerged in the first place, Dennett's failure makes Darwinian nihilism a good bet. Darwinian nihilism departs from naturalism only in declining to endorse our morality or any other as true or correct. It must decline to do so because it holds that the explanation of how our moral beliefs arose also explains away as mistaken the widespread belief that moral claims are true. The Darwinian explanation becomes the Darwinian nihilist's "explaining away" when it becomes apparent that the best explanation—blind variation and natural selection– for the emergence of our ethical belief does not require that these beliefs have truth-makers. To turn the Darwinian explanation into an "explaining away" the nihilist need only add the uncontroversial scientific principle that if our best theory of why people believe P does not require that P is true, then there are no grounds to believe P is true.

It is worth noting by way of conclusion, that nihilism need not be a particularly disquieting doctrine. Embracing nihilism is not, as is commonly believed, a prescription for a-morality or immorality. Nihilism is not a prescription or proscrip-

668

tion of any conduct. The nihilist may well admit that accepting categorical and hypothetical imperatives may often serve the parochial interests of oneself and others. To be an ethical nihilist commits one to nothing more than the denial of objective or intrinsic moral values and categorical imperatives.

Darwinian nihilism explains away ethics by showing that our ethical beliefs reflect dispositions very strongly selected for over long periods, which began well before the emergence of hominids, or indeed perhaps primates (*vide* the vampire bat). These dispositions are so "deep" that for most people most of the time, it is impossible to override them, even when it is in our individual self-interest to do so, still less when there is no self-interested reason to do so. Hence, the Darwinian nihilist expects that most people are conventionally moral, and that even the widespread acceptance of the truth of Darwinian nihilism would have little or no effect on this expectation. Most of us just couldn't persistently be mean, even if we tried. And we have no reason to try.

But nice nihilism is hardly "a stronger, sounder version of our most important ideas." If it is the right conclusion then we must respond to Dennett's final question "Does Darwin's idea turn out to be, in the end, just what we need in our attempt to preserve . . . the values we cherish? [p. 521]" with a simple "no."

Acknowledgements

Thanks to a referee for *Biology and Philosophy*, David Kaplan, Frederic Bouchard, Geoff Sayre-McCord, and his graduate seminar at the University of North Carolina, Chapel Hill, for helpful comments. Rosenberg gratefully acknowledges support of research on this subject by the Social Policy and Philosophy Center at Bowling Green State University. Since nihilism is true, no one is morally responsible for the remaining errors.

References

Axelrod R. 1984. *The Evolution of Cooperation*. University of Michigan Press, Ann Arbor.
Dennett D.C. 1995. Darwin's Dangerous Idea: Evolution and the Meanings of Life. Simon and Schuster, New York.
Frank R. 1989. *Passions Within* Reason. W W Norton, New York.
Gibbard A. 1992. *Wise Choices, Apt Feelings*. Harvard University Press, Cambridge.
Hamilton W.D. 1964. "The genetic evolution of social behavior". *Journal of Theoretical Biology* 7: 1–52.
Hamilton W.D. 1967. "Extraordinary sex ratios". *Science* 156: 477–488.
Hobbes T. 1651. *Leviathan*.
Mackie J.L. 1977. *Ethics: Inventing Right and Wrong*. Harmondsworth, Penguin.
Moore G.E. 1903. *Principia Ethica*. Routledge, London.
Ruse and Wilson 1989. "Evolution of ethics". *New Scientist* 17: 51.
Skyrms B. 1996. *Evolution of the Social Contract*. Cambridge University Press, Cambridge.
Street S. 2002. "Normativity and Life: The Metaethical Implications of Darwinism.".
Trivers R.L. 1971. "The evolution of reciprocal altruism". *Quarterly Review of Biology* 46: 35–57.
Wilson D.S. and Sober E. 1998. *Unto Others*. Harvard University Press, Cambridge.

[9]

EVOLUTIONARY ETHICS: A PHOENIX ARISEN

by Michael Ruse

Abstract. Evolutionary ethics has a (deservedly) bad reputation. But we must not remain prisoners of our past. Recent advances in Darwinian evolutionary biology pave the way for a linking of science and morality, at once more modest yet more profound than earlier excursions in this direction. There is no need to repudiate the insights of the great philosophers of the past, particularly David Hume. So humans' simian origins really matter. The question is not whether evolution is to be linked to ethics, but how.

We humans are modified monkeys, not the favored creation of a benevolent God, on the sixth day. The time has therefore come to face squarely our animal nature, particularly as we interact with others. Admittedly, so-called evolutionary ethics has a bad reputation. However, the question is not whether evolution is connected with ethics, but how. Fortunately, thanks to recent developments in biological science, the way is now becoming clear.

I begin this discussion with a brief historical introduction to the topic. Then I move to the core of my scientific and philosophical case. I conclude by taking up some central objections.

SOCIAL DARWINISM

In 1859 Charles Darwin published his *On the Origin of Species by Means of Natural Selection*. In that work he argues that all organisms (including ourselves) came through a slow, natural process of evolution. Also, Darwin suggested a mechanism: more organisms are born than can survive and reproduce; this leads to competition; the winners are thus "naturally selected," and hence change ensues in the direction of increased "adaptiveness." It is hardly true that Darwin, or even science generally, brought about the death of Christianity; but after the *Origin*

Michael Ruse, professor of history and philosophy, University of Guelph, Ontario, Canada N1G 2W1, presented this paper at the Thirty-first Annual Conference ("Recent Discoveries in Neurobiology—Do They Matter for Religion, the Social Sciences, and the Humanities?") of the Institute on Religion in an Age of Science, Star Island, New Hampshire, 28 July-4 August 1984.

increasing numbers turned from the Bible towards evolution, in some form, for moral insight and guidance (Ruse 1979a; Russett 1976). The product was generally known as social Darwinism, the traditional form of evolutionary ethics—although, as many have noted, despite its name, it owed its genesis more to that general man of Victorian science, Herbert Spencer, than to Darwin himself (Russett 1976).

A full moral system needs two parts. On the one hand, you must have the "substantival" or "normative" ethical component. Here, you offer actual guidance as in, "Thou shalt not kill." On the other hand, you must have (what is known formally as) the "metaethical" dimension. Here, you are offering foundations or justification as in, "That which you should do is that which God wills." Without these two parts, your system is incomplete (Taylor 1978).

To the social Darwinians, the metaethical foundations they sought lay readily at hand. They exist in the perceived nature of the evolutionary process. Supposedly, we have a progression from simple to complex, from amoeba to man, from (as Spencer happily pointed out) savage to Englishman (Spencer 1852; 1857). This progress is a good thing and conveys immediate worth. We need no further justification of what ought to be. And now, at once, we have the substantival directives of our system. Morally, we should aid and promote—and not hinder—the evolutionary process. Furthermore, if, as was supposedly claimed by Darwin and certainly echoed by Spencer, the evolutionary process begins with a bloody struggle for existence and concludes with the triumph of the fittest, then so be it. Our obligation is to prize the strong and successful and to let the weakest go to the wall (Ruse 1985).

Of course, as many pointed out—most splendidly Darwin's great supporter and ardent co-evolutionist, Thomas Henry Huxley (1901)—none of this will do. Metaethically speaking, evolution simply is not progressive (Williams 1966). Apart from anything else, it branches all over the place, making it quite impossible to offer true assessments of top and bottom, higher and lower, better and worse. Among today's organisms, venereal disease thrives, whereas the great apes stand near extinction. Is gonorrhea really superior to the chimpanzee? And, following up the metaethical inadequacies, at the substantival level, if anything is false, social Darwinism is false. Morality does not consist in walking over the weak and the sick, the very young and the very old. Someone who tells you otherwise is an ethical cretin.

Social Darwinism (and, so many concluded, any kind of evolutionary ethics) is wrong—not just mistaken but fundamentally misguided. Why? The answer was pinpointed by such philosophers as David Hume (in the eighteenth century) and G. E. Moore (in the twentieth century). Hume (1978) noted that you simply cannot go straight from talk of facts (like evolution) to talk of morals and obligations, from "is" language to "ought" language.

In every system of morality, which I have hitherto met with, I have always remark'd, that the author proceeds for some time in the ordinary way of reasoning, and establishes the being of a God, or makes observations concerning human affairs, when of a sudden I am surpriz'd to find, that instead of the usual copulations of propositions, is, and is not, I meet with no proposition that is not connected with an ought, or an ought not. This change is imperceptible; but is, however, of the last consequence. For as this ought, or ought not, expresses some new relation or affirmation, 'tis necessary that it shou'd be observ'd and explain'd; and at the same time that a reason should be given, for what seems altogether inconceivable, how this new relation can be a deduction from others, which are entirely different from it (Hume 1978, 469).

Then, in 1903, Moore backed up this point, in his *Principia Ethica*, arguing that all who would derive morality from the physical world stand convicted of the "naturalistic fallacy." Explicitly Moore noted that the evolutionary ethicizer is a major offender, as he goes from talk of the facts and process of evolution to talk of what one ought (or ought not) do.

At all levels, therefore, traditional evolutionary ethics ground to a complete stop. It promoted a grotesque distortion of true morality and could do so only because its foundations were rotten (Flew 1967). So matters have rested for three-quarters of a century. Now, however, the time has come for the case to be reopened. Let us see why.

THE EVOLUTION OF MORALITY

We must begin with the science, most particularly with the evolution of the human moral sense or capacity. In fact, as Darwin pointed out, contrary to the Spencerian interpretations of the evolutionary process, although the process may start with competition for limited resources—a struggle for existence (more strictly, struggle for reproduction)—this certainly does not imply that there will always be fierce and ongoing hand-to-hand combat. Between members of the same species most particularly, much more personal benefit can frequently be achieved through a process of cooperation—a kind of enlightened self-interest, as it were (Darwin 1859; 1871). Thus, for instance, if my conspecific and I battle until one is totally vanquished, no one really gains, for even the winner will probably be so beaten and exhausted that future tasks will overwhelm. Whereas, if we cooperate, although we must share the booty, there will be no losers and both will benefit (Trivers 1971; Wilson 1975; Dawkins 1976; Ruse 1979b).

All such cooperation for personal evolutionary gain is known technically as "altruism." I emphasize that this term is rooted in metaphor, even though now it has the just-given formal biological meaning.

There is no implication that evolutionary "altruism" (working together for biological payoff) is inevitably associated with moral altruism (where this is the original literal sense, implying a conscious being helping others because it is right and proper to do so). The connection is no more than that between the physicist's notion of "work" and what you and I do in the yard on Saturday afternoons when we mow the grass.

However, just as mowing the lawn does involve work in the physicist's sense, so also today's students of the evolution of social behavior ("sociobiologists") argue that moral (literal) altruism might be one way in which biological (metaphorical) "altruism" could be achieved (Wilson 1978; Ruse & Wilson 1986). Furthermore, they argue that in humans, and perhaps also in the great apes, such a possibility is a reality. Literal, moral altruism is a major way in which advantageous biological cooperation is achieved. Humans are the kinds of animals which benefit biologically from cooperation within their groups, and literal, moral altruism is the way in which we achieve that end (Lovejoy 1981).

There was no inevitability in altruistic inclinations having developed as one of the human adaptations. Judging from what we know of ourselves and other animals, there were a number of other ways in which biological "altruism" might have been effected (Lumsden & Wilson 1983). Most obviously, humans could have gone the route of the ants. They are highly social, having taken "altruism" to its highest pitch through what one might call "genetic hardwiring." Ants are machine-like, working in their nests according to innate dispositions, triggered by chemicals (pheromones) and the like (Wilson 1971).

There are great biological advantages to this kind of functioning: it eliminates the need for learning, it cuts down on the mistakes, and much more. Unfortunately, however, this is all bought at the expense of any kind of flexibility. If circumstances change, individual ants cannot respond. This does not matter so much in the case of ants, since (biologically speaking) they are cheap to produce. Regretfully, humans require significant biological investment, and so apparently the production of "altruism" through innate, unalterable forces, poses too much of a risk.

Since the ant option is closed, we humans might theoretically have achieved "altruism" by going right to the other extreme. We might have evolved superbrains, rationally calculating at each point if a certain course of action is in our best interests. "Should I help you prepare for a difficult test? What's in it for me? Will you pay me? Do I need help in return? Or what?" Here, there is simply a disinterested calculation of personal benefits. However, we have clearly not evolved this way. Apart from anything else, such a superbrain would itself have high biological cost and might not be that efficient. By the time I have decided whether or not to save the child from the speeding bus, the dreadful event has occurred (Lumsden & Wilson 1981; Ruse & Wilson 1986).

It would seem, therefore, that human evolution has been driven towards a middle-of-the-road position. In order to achieve "altruism," we are altruistic! To make us cooperate for our biological ends, evolution has filled us full of thoughts about right and wrong, the need to help our fellows, and so forth. We are obviously not totally selfless. Indeed, thanks to the struggle for reproduction, our normal disposition is to look after ourselves. However, it is in our biological interests to cooperate. Thus we have evolved innate mental dispositions (what the sociobiologists Charles Lumsden and Edward O. Wilson call "epigenetic rules") inclining us to cooperate, in the name of this thing which we call morality (Lumsden & Wilson 1981). We have no choice about the morality of which we are aware. But, unlike the ants, we can certainly choose whether or not to obey the dictates of our conscience. We are not blindly locked into our courses of action like robots. We are inclined to behave morally but not predestined to such a policy.

This, then, is the modern (Darwinian) biologist's case for the evolution of morality. Our moral sense, our altruistic nature, is an adaptation—a feature helping us in the struggle for existence and reproduction—no less than hands and eyes, teeth and feet. It is a cost-effective way of getting us to cooperate, which avoids both the pitfalls of blind action and the expense of a superbrain of pure rationality.

SUBSTANTIVE ETHICS

But what has any of this to do with the questions that philosophers find pressing and interesting? Let us grant the scientific case sketched in the last section. What now of substantival ethics, and most particularly what of metaethics? If we think that what has just been said has any relevance to foundations, then surely we violate Hume's law and smash into the naturalistic fallacy, no less than does the Spencerian.

Turning first to the moral norms endorsed by the modern evolutionist, there is little to haunt us from the past. As we have just seen, the whole point of today's approach is that we transcend a rugged struggle for existence—in thought and deed. Of course, humans are selfish and violent at times. This has been admitted. But, no less than the moralist, the evolutionist denies that this darker side to human beings has anything to do with moral urges. What excites the evolutionist is the fact that we have feelings of moral obligation laid over our brute biological nature, inclining us to be decent for altruistic reasons.

What is the actual content (speaking substantively) of a modern evolutionary ethic? At this point we turn to philosophers for guidance! After all, these are the people whose intent it is to uncover the basic rules which govern our ethical lives. The evolutionist may modify or even reject the philosophers' claims; but, given the central (empirical) hypothesis that normal, regular morality is that which our biology uses to promote "altruism," the presumption must be that the findings of the philosophers will tell much.

In fact, there is little need for apprehension. Claims of some of today's leading thinkers sound almost as if they were prepared expressly to fill the evolutionist's bills—a point which these thinkers themselves have acknowledged. In particular, let me draw your attention to the ideas of John Rawls, whose *A Theory of Justice* deservedly holds its place as the major work in moral philosophy of the last decade. Rawls writes:

The guiding idea is that the principles of justice for the basic structure of society are . . . the principles that free and rational persons concerned to further their own interests would accept in an initial position of equality as defining the fundamental terms of their association. These principles are to regulate all further agreements; they specify the kinds of social cooperation that can be entered into and the forms of government that can be established. This way of regarding the principles of justice I shall call justice as fairness (Rawls 1971, 11).

How exactly does one spell out these principles that would be adopted by "free and rational persons concerned to further their own interests"? Here, Rawls invites us to put ourselves behind a "veil of ignorance," as it were. If we knew that we were going to be born into a society and that we would be healthy, handsome, wise, and rich, we would opt for a system which favors the fortunate. But we might be sick, ugly, stupid, and poor. Thus, in our ignorance, we will opt for a just society, governed by rules that would best benefit us no matter what state or post we might have in that society.

Rawls argues that, under these conditions, a just society is seen to be one which, first, maximizes liberty and freedom, and, second, distributes society's rewards so that everyone benefits as much as possible. Rawls is not arguing for some kind of communistic, totally equal distribution of goods. Rather, the distribution must help the unfortunate as well as the fortunate. If you could show that the only way to get statewide, good quality medical care is by paying doctors twice as much as anyone else, then so be it.

I need hardly say how readily all of this meshes with the evolutionary approach. For both the biologist and the Rawlsian, the question is that of how one might obtain right action from groups of people whose natural inclination is (or rather, of whom one would expect the natural inclination to be) that of looking after themselves. In both cases the answer is found in a form of enlightened self-interest. We behave morally because, ultimately, there is more in it for us than if we do not.

·Where the evolutionist picks up and goes beyond the Rawlsian is in linking the principles of justice to our biological past, via the epigenetic rules. This is a great bonus, for Rawls himself admits that his own analysis is restricted to the conceptual level. He leaves unanswered major questions about origins. "In justice as fairness the original position of equality corresponds to the state of nature in the traditional theory of the social contract. This original position is not, of course, thought of as an actual historical state of affairs, much less as a primitive condition of culture. It is understood as a purely hypothetical situation characterized so as to lead to a certain conception of justice" (Rawls 1971, 12).

This is all very well. But, "purely hypothetical situations" are hardly satisfying. Interestingly, as hinted above, Rawls himself suggests that biology might be important. "In arguing for the greater stability of the principles of justice I have assumed that certain psychological laws are true, or approximately so. I shall not pursue the question of stability beyond this point. We may note however that one might ask how it is that human beings have acquired a nature described by these psychological principles. The theory of evolution would suggest that it is the outcome of natural selection; the capacity for a sense of justice and the moral feelings is an adaptation of mankind to its place in nature" (Rawls 1971, 502-3). This is precisely the evolutionist's approach. There is no need to suppose hypothetical contracts. Natural selection made us as we are.

FOUNDATIONS—METAETHICS

I expect that many traditional philosophers will feel able to go this far with the evolutionist. But now the barriers will come up. The argument will run like this: The evolution of ethics has nothing to do with the status of ethics. I may be kind to others because my biology tells me to be kind to others and because those protohumans who were not kind to others failed to survive and reproduce. But is it right that I be kind to others? Do I really, objectively, truly have moral obligations? To suppose that the story of origins tells of truth or falsity is to confuse causes with reasons. In a Spencerian fashion, it is to jumble the way things came about with the way things really are.[1] Since Rawls has been quoted as an authority, let us recall what he says at the end of his speculations on the evolution of morality: "These remarks are not intended as justifying reasons for the contract view" (Rawls 1971, 504).

This is a powerful response, but today's evolutionary ethicist argues that it misses entirely the full force of what biology tells us. It is indeed true that you cannot *deduce* moral claims from factual claims (about origins). However, using factual claims about origins, you can give moral claims the only foundational *explanation* that they might possibly have. In particular, the evolutionist argues that, thanks to our science, we see that claims like "You ought to maximize personal liberty" are no more than subjective expressions, impressed upon our thinking because of their adaptive value. In other words, we see that morality has no philosophically objective foundation. It is just an illusion, fobbed off on us to promote biological "altruism."

This is a strong claim, so let us understand it fully. The evolutionist is no longer attempting to derive morality from factual foundations. His/her claim now is that there are no foundations of any sort from which to derive morality—be these foundations evolution, God's will, or whatever. Since, clearly, ethics is not nonexistent, the evolutionist locates our moral feelings simply in the subjective nature of human psychology. At this level, morality has no more (and no less) status than that of the terror we feel at the unknown—another emotion which undoubtedly has good biological adaptive value.

Consider an analogy. During the First World War, many bereaved parents turned to spiritualism for solace. Down the Ouija board would come the messages: "It's alright Mum. I've gone to a far better place. I'm just waiting for you and Dad." I take it that these were not in fact the words of the late Private Higgins, speaking from beyond. Rather they were illusory—a function of people's psychology as they projected their wishes. (We can, I think, discount universal fraud.)

The moral to be drawn from this little story is that we do not need any further justificatory foundation for "It's alright Mum" than that just given. At this point, we do not need a reasoned underpinning to the words of reassurance. ("Why is it alright?" "Because I'm sitting on a cloud, dressed in a bedsheet, playing a harp.") What we need is a causal explanation of why the bereaved "heard" what they did. The evolutionist's case is that something similar is very true of ethics. Ultimately, there is no reasoned justification for ethics in the sense of foundations to which one can appeal in reasoned argument. All one can offer is a causal argument to show why we hold ethical beliefs. But once such an argument is offered, we can see that this is all that is needed.

In a sense, therefore, the evolutionist's case is that ethics is a collective illusion of the human race, fashioned and maintained by natural selection in order to promote individual reproduction. Yet, more must be said than this. Obviously, "Stamping on small children is wrong," is not really illusory like "It's alright Mum, I'm okay!" However, we can easily show why the analogy breaks down at this point. Morality is a shared belief (or set of beliefs) of the human race, unlike the messages down the Ouija board. Thus, we can distinguish between "Love little children," which is certainly not what we would normally call illusory, and "Be kind to cabbages on Fridays," which certainly is what we would normally call illusory. We all (or nearly all) believe the former but not the latter.

Perhaps we can more accurately express the evolutionist's thesis by drawing back from a flat assertion that ethics is illusory. What is really important to the evolutionist's case is the claim that ethics is illusory inasmuch as it persuades us that it has an objective reference. This is the crux of the biological position. Once it is grasped everything falls into place.

This concession about the illusory status of ethics in no way weakens the evolutionist's case. Far from it! If you think about it, you will see that the very essence of an ethical claim, like "Love little children," is that, whatever its true status may be, we think it binding upon us *because we think it has an objective status*. "Love little children" is not like "My favorite vegetable is spinach." The latter is just a matter of subjective preference. If you do not like spinach, then nothing ensues. But we do not take the former (moral) claim to be just a matter of preference. It is regarded as objectively binding upon us—whether we take the ultimate source of this objectivity to be God's will, or (if we are Platonists) intuited relations between the forms, or (like G. E. Moore) apprehension of nonnatural properties, or whatever.

The evolutionist's claim, consequently, is that morality is subjective—it is all a question of human feelings or sentiments—but he/she admits that we "objectify" morality, to use an ugly but descriptive term. We think morality has objective reference even though it does not. Because of this, a causal analysis of the type offered by the evolutionist is appropriate and adequate, whereas a justification of moral claims in terms of reasoned foundations is neither needed nor appropriate.

Furthermore, completing the case, the evolutionist points out that there are good (biological) reasons why it is part of our nature to objectify morality. If we did not regard it as binding, we would ignore it. It is precisely because we think that morality is more than mere subjective desires, that we are led to obey it.[2]

RECIPROCATION

This completes the modern-day case for evolutionary ethics. A host of questions will be raised. I will concentrate on two of the more important. First, let us turn to a substantival question.

Many of the queries at this level will be based on misunderstandings of the evolutionist's position. For instance, although the evolutionist is subjectivist about ethics, this does not in any sense imply that he/she is a relativist—especially not a cultural relativist. The whole point about the evolutionary approach to ethics is that morality does not work unless we are all in the game (with perhaps one or two cheaters—so-called criminals or sociopaths). Moreover, we have to believe in morality; otherwise it will not work. Hence, the evolutionist looks for shared moral insights, and cultural variations are dismissed as mere fluctuations due to contingent impinging factors.

Analogously, there is no question of simply breaking from morality if we so wish. Even though we have insight into our biological nature, it is still *our* biological nature. We can certainly do immoral things. We do them all the time. But, a policy of persistently and consistently breaking the rules can only lead to internal tensions. Plato had a good point in the *Republic* when he argued that only the truly good person is the truly happy person, and the truly happy person is the one whose parts of the personality ("soul") function harmoniously together.

A much more significant question, on which I will focus, concerns the question of reciprocation. No one should be misled into thinking that the evolutionist proclaims the virtues (moral or otherwise) of selfishness or that the evolutionist's position imples that, as a matter of contingent fact, we are totally selfish. It has been admitted that human beings have a tendency towards selfishness; but, you did not need an evolutionist to tell you that. What is surprising is that we are not totally selfish. Humans have genuinely altruistic feelings towards their fellows. The fact that, according to the evolutionist, we are brought to literal, moral altruism by our genes acting in our biological self-interests says nothing against the genuineness of our feelings. Would you doubt the goodness of Mother Theresa's heart, were you told that she was strictly disciplined as a child?

Nevertheless, while this is indeed all true, a nagging doubt remains. Let us look for a moment at the actual causal models proposed by sociobiologists in order to explain the evolution of altruism. First, it is suggested that *kin selection* is important. Relatives share copies of the same genes. Hence, inasmuch as a relative reproduces, you yourself reproduce vicariously, as it were. Therefore, help given to relatives leading to survival and reproduction rebounds to your own benefit. Second, there is *reciprocal altruism*. Simply, if I help you (even though you be no relative), then you are more likely to help me—and conversely. We both gain together, whereas apart we both lose.[3]

Now, surely, with both of these mechanisms, the possibility of genuine altruism seems precluded. With kin selection, the rewards come through your relatives' reproduction, so there is no need for crude overt returns. But, would not mere nonmoral love do all that is needed? I love my children, and I help them not because it is right but because I love them. As Immanuel Kant (1959) rightly points out, unless you are actually heeding the call of duty, there is no moral credit. A mother happily suckling her baby is not performing a moral act.

In the case of reciprocal altruism, the problems for the evolutionary ethicist are even more obvious. You do something in hope of return. This is not genuine altruism but a straight bargain. There is nothing immoral in such a transaction. If I pay cash for a kilo of potatoes, there is no wrongdoing. But there is nothing moral in such a transaction, either. Morality means going out on a limb, because it is right to do so. Morality vanishes if you hope for payment.

The evolutionist has answers to these lines of criticism—answers which strengthen the overall position. First, it is indeed true that much we do for our family stems from love, without thought of duty. But, only the childless would think moral obligations never enter into intrafamilial relations. Time and again we have to drive ourselves on, and we do it because it is right. Without the concepts of right and wrong, we would be much less successful parents (uncles, aunts, etc.) than we are. Humans require so much child care that they make the case for a biological backing to morality particularly compelling. If parental duties were left to feelings of kindliness, the system would break down. (I am sure there has been a feedback causal process at work here. Because we have a moral capacity, child care could be extended; and extensive child care needs set up selective pressure towards increased moral awareness.)

Second, it is agreed that reciprocal altruism would fail if there were no returns—or ways of enforcing returns. However, it is not necessary to suppose that such reciprocation requires a crude demand of returns for favors granted. Apart from anything else, morality is clearly more like a group insurance policy than a person-to-person transaction. I help you, but do not necessarily expect you personally to help me. Rather, my help is thrown into the general pool, as it were, and then I am free to draw on help as needed.

Furthermore, enforcement of the system comes about through morality itself! I help you, and I can demand help in return, not because I have helped you or even because I want help, but because it is *right* that you help me. Reciprocation is kept in place by moral obligations. If you cease to play fair, then before long I and others will chastise you or take you out of the moral sphere. We do not do this because we do not like you but because you are a bad person or too "sick" to recognize the right way of doing things. Morality demands that we give freely, but it does not expect us to make suckers of ourselves. (What about Jesus' demand that we forgive seven times seventy times? The moral person responds that forgiveness is one thing, but that complacently letting a bad act occur four-hundred and ninety times borders on the criminally irresponsible. We *ought* to put a stop to such an appalling state of affairs.[4])

Thus far there should be little in the evolutionist's approach to normative ethics, properly understood, which would spur controversy. But, let me conclude this section by pointing to one implication which will certainly cause debate. Many moralists argue that we have an equal obligation to all human beings, indifferently as to relationship acquaintance, nationality, or whatever (Singer 1972). In principle, my obligations to some unknown child in (say) Ethiopia are no less than to my own son. Nevertheless, although many (most?) would pay lip service to some such view as this, my suspicion is that, sincerely meant, this doctrine makes the evolutionist decidedly queasy. Biologically, our major concern has to be towards our own kin, then to those in at least some sort of relationship to us (not necessarily a blood relationship), and only finally to complete strangers. And, feelings of moral obligation have to mirror biology.

I speak tentatively now. You could argue that biology gives us an equal sense of obligation towards all and that this sense is then filtered across strong (nonmoral) feelings of warmth towards our own children, followed by diminishing sentiments towards nonrelatives, ending with a natural air of suspicion and indifference towards strangers. But my hunch is that the care we must bestow on our children is too vital to be left to chance, and therefore we expect to find, what we do in fact find, namely that our very senses of obligation vary. Therefore, whatever we may sometimes say, truly we have a stronger feeling of moral obligation towards some people than towards others.

It is perhaps a little odd to speak thus hesitantly about our own feelings, including moral feelings. You might think that one should be able to introspect and speak definitively. However, matters are not always quite this simple, particularly when (as now) we are faced with a case where our technology has outstripped our biology and our consequent morality. A hundred years ago it would have made little sense to talk of moral obligations to Ethiopians. Now we know about Ethiopians and, at least at some level, we can do something for them. But what should we do for them? Within the limits of our abilities, as much for each one as for each one of our own children? I suspect that most people would say not. I hasten to add that no evolutionist says we have no obligations to the world's starving poor. The question is whether we have a moral obligation to beggar our families and to send all to Oxfam.

In closing this section, let me at least note that, over this matter of varying obligations, the evolutionist takes no more stringent a line than does Rawls. Explicitly, Rawls treats close kin as a case meriting special attention, and as he himself admits it is far from obvious that his theory readily embraces relations with the Third World (Rawls 1980). It is not intuitively true that, even hypothetically, we were in an original position with the people of Africa—or India, or China. Hence, although the evolutionist certainly does not want to hide behind the cloth of the more conventional moral philosopher, he can take comfort from the fact that he is in good company.

OBJECTIVITY

We turn now to metaethical worries. The central claim of the evolutionist is that ethics is subjective, a matter of feelings or sentiment, without genuine objective referent. What distinguishes ethics from other feelings is our belief that ethics is objectively based, and it is because we think this that ethics works.

The most obvious and important objection to all of this is that the evolutionist has hardly yet really eliminated the putative objective foundation of morality. Of course, ethics is in some way subjective. How could it not be? It is a system of beliefs held by humans. But this does not in itself deny that there is something more. Consider, analogously, the case of perception. I see the apple. My sensations are subjective, and my organs of vision (eyes) came through the evolutionary process, for excellent biological reasons. Yet, no one would deny that the apple is independently, objectively real. Could not the same be true of ethics? Ultimately, ethics resides objectively in God's will, or some such thing. (Nozick [1981] pursues a line of argument akin to this.)

Let us grant the perception case although, parenthetically, I suspect the evolutionist might well have some questions about the existence of a real world beyond the knowing subject. The analogy with ethics still breaks down. Imagine two worlds, identical except that one has an objective ethics (whatever that might mean) and one does not. Perhaps, in one world God wants us to look after the sick, and in the other He could not care less what we do. The evolutionist argues that, in both situations, we would have evolved in such a way as to think that, morally, we ought to care for the sick. To suppose otherwise, to suppose that only the world of objective ethics has us caring about the sick, is to suppose that there are extrascientific forces at work, directing and guiding the course of evolution. And this is a supposition which is an anathema to the modern biologist (Ruse 1982).

In other words, in the light of what we know of evolutionary processes, the objective foundation has to be judged redundant. But, if anything is a contradiction in terms, it is a redundant objective morality: "The only reason for loving your neighbor is that God wants this, but you will think you ought to love your neighbor whether or not God wants it." In fact, if you take seriously the notion that humans are the product of natural selection, the situation is even worse than this. We are what we are because of contingent circumstances, not because we necessarily had to be as we are. Suppose, instead of evolving from savannah-living primates (which we did), we had come from cave dwellers. Our nature and our morality might have been very different. Or, take the termites (to go to an extreme example from a human perspective). They have to eat each other's feces, because they lose certain parasites, vital for digestion, when they molt. Had humans come along a similar trail, our highest ethical imperatives would have been very strange indeed.

What this all means is that, whatever objective morality may truly dictate, we might have evolved in such a way as to miss completely its real essence. We might have developed so that we think we should hate our neighbors, when really we should love them. Worse than this even, perhaps we really should be hating our neighbors, even though we think we should love them! Clearly, this possibility reduces objectivity in ethics to a mass of paradox.

But does it? Let us grant that the evolutionist has a good case against the person who would argue that the foundations of morality lie in sources external to us humans, be these sources God's will, the relations of Platonic forms, nonnatural properties, or whatever. However, there is at least one well-known attempt to achieve objectivity (of a kind) without the assumption of externality. I refer, of course, to the metaethical theorizing of Immanuel Kant (1949; 1959). He argued that the supreme principle of morality, the so-called categorical imperative, has a necessity quite transcending the contingency of human desires. It is synthetic *a priori*, where by this Kant meant that morality is a condition which comes into play, necessarily, when rational beings interact. He argued that a disregard of morality leads to "contradictions," that is to a breakdown in social functioning. Thus, we see that morality is not just subjective whim but has its being in the very essence of rational interaction. To counter an example offered above, we could not have evolved as pure haters, because such beings simply could not interact socially.

Since, more than once in this paper, the evolutionist has invoked the ideas of Rawls in his own support, a critic might reasonably point out that (having left matters dangling in *A Theory of Justice*), more recently

Rawls has tried explicitly to put morality on a Kantian foundation. At a general level he writes as follows: "What justifies a conception of justice is not its being true to an order antecedent to and given to us, but its congruence with our deeper understanding of ourselves and our aspirations, and our realization that, given our history and the traditions embedded in our public life, it is the most reasonable doctrine for us" (Rawls 1980, 519). Then, spelling matters out a little more, Rawls claims that: "[A] Kantian doctrine interprets the notion of objectivity in terms of a suitably constructed social point of view that is authoritative with respect to all individual and associational points of view. This rendering of objectivity implies that, rather than think of the principles of justice as true, it is better to say that they are the principles most reasonable for us, given our conception of persons as free and equal, and fully cooperating members of a democratic society" (Rawls 1980, 554). Thus, in some way we try to show both that morality is reasonable and that it is more than a matter of mere desire or taste, like a preference for vegetables.

Responding to the Kantian/Rawlsian, so-called constructivist position, the evolutionist will want to make two points. First, there is much in the position with which he/she heartily sympathizes! Both constructivist and evolutionist agree that morality must not be sought outside human beings, and yet both agree that there is more to morality than mere feelings. Additionally, both try to make their case by pointing out that morality is the most sensible strategy for an individual to pursue. Being nice pays dividends—although, as both constructivist and evolutionist point out, one behaves morally for good reasons, not because one is consciously aware of the benefits.

Second, for all of the sympathy, the evolutionist will feel compelled to pull back from the full conclusions of the constructivist position. The evolutionist argues that morality (as we know it) is the most sensible policy, as we humans are today. However, he/she draws back from the constructivist claim that (human-type) morality must be the optimal strategy for *any* rational being. What about our termite-humans, for instance? They might be perfectly rational. Possibly, the response will be that the termite-humans' sense of obligation to eat rather strange foodstuffs is covered by a prohibition against suicide, which Kant certainly thinks follows from the categorical imperative. Hence, the constructivist admits that one's distinctive (in our case, human) nature gives one's actual morality a correspondingly distinctive appearance; but he/she argues that underlying the differences is a shared morality. The principle is the same as when everyone (including the evolutionist) explains differences in cultural norms as due to special circumstances, not to diverse ultimate moral commitments (Taylor 1958).

Yet, the evolutionist continues the challenge. If the constructivist argues that the only thing which counts is rational beings working together and that their contingent nature is irrelevant, then it is difficult to see why morality necessarily emerges at all. Suppose that we had evolved into totally rational beings, like the above-mentioned superbrains, and that we calculated chances, risks, and benefits at all times. We would be neither moral not immoral, feeling no urges of obligation at all.

Obviously, we are not like this. Apparently, therefore, we must take account of a being's contingent nature—no matter how rational it may be—in order to get some kind of morality. But this is the thin end of the wedge for moralities other than human morality. Think, for instance, how we might patch up the society of pure haters so that a kind of morality could emerge—and this is a kind quite different from ours. Suppose that it is part of our nature to hate others, and that we think we have an obligation to hate others. A Kantian "contradiction," that is, breakdown in sociality, might still be avoided and cooperation achieved, because we know that others hate us and so we feel we had better work warily together to avoid their wrath. If this sounds far-fetched, consider how today's supposed superpowers function. Everything would be perfectly rational and could work (after a fashion). Yet, there would be little that we humans would recognize as "moral" in any of this.

Of course, you might still point out that such a society of pure haters would end up with rules much akin to those that the constructivist endorses, about liberty and so forth. But these rules would not be *moral* in any sense. They would be, explicitly, rules of expediency, of self-interest. I give you liberty not because I care for you, or respect you, or think I ought to treat you as a worthwhile individual. I hate your guts! And, I think I *ought* to hate you. I give you liberty simply because it is in my consciously thought-out interests to do so. This may be a sensible prudent policy. It is not a moral policy.

The evolutionist concludes, against the constructivist, that our morality is a function of our actual human nature and that it cannot be divorced from the contingencies of our evolution. Morality, as we know it, cannot have the necessity or objectivity sought by the Kantian and Rawlsian.

CONCLUSION

Our biology is working hard to make the evolutionist's position seem implausible. We are convinced that morality really is objective, in some way. However, if we take modern biology seriously, we come to see how we are children of our past. We learn what the true situation really is.

Michael Ruse 111

Evolution and ethics are at last united in a profitable symbiosis, and this is done without committing all of the fallacies of the last century.

NOTES

1. Versions of this argument occur in Raphael (1958), Quinton (1966), Singer (1972), and—I blush to say it—Ruse (1979b).

2. See Murphy (1982) for more on the argument that a causal explanation might be all that can be offered for ethics, and Mackie (1977) for discussion of "objectification" in ethics.

3. These two mechanisms are discussed in detail in Ruse (1979b). They are related to human behavior, in some detail, in Wilson (1978).

4. This criticism assumes that the Christian is obligated to forgive endlessly, without response. Modern scholarship suggests that this is far from Jesus' true message. See Betz (1985) for more on this point, and Mackie (1978) for more on the sociobiologically inspired criticism that Christianity makes unreasonable demands on us. This latter line of argument obviously parallels that of Sigmund Freud in *Civilization and its Discontents* (1961).

REFERENCES

Betz, D. 1985. *Essays on the Sermon on the Mount*. Philadelphia: Fortress Press.
Darwin, C. 1859. *On the Origin of Species by Means of Natural Selection*. London: John Murray.
_____. 1871. *The Descent of Man*. London: John Murray.
Dawkins, R. 1976. *The Selfish Gene*. Oxford: Oxford Univ. Press.
Flew, A. G. N. 1967. *Evolutionary Ethics*. London: Macmillan.
Freud, S. 1961. *Civilization and its Discontents*. In vol. 21 of *Complete Psychological Works of Sigmund Freud*, ed. J. Strachey, 64-145. London: Hogarth Press. First published 1929-30.
Hume, D. 1978. *A Treatise of Human Nature*. Oxford: Clarendon Press.
Huxley, T. H. 1901. *Evolution and Ethics, and Other Essays*. London: Macmillan.
Kant, I. 1949. *Critique of Practical Reason*. Trans. L. W. Beck. Chicago: Univ. of Chicago Press.
_____. 1959. *Foundations of the Metaphysics of Morals*. Trans. L. W. Beck. Indianapolis: Bobbs-Merrill.
Lovejoy, O. 1981. "The Origin of Man." *Science* 211:341-50.
Lumsden, C. J. and E. O. Wilson. 1981. *Genes, Mind and Culture: The Coevolutionary Process*. Cambridge, Mass.: Harvard Univ. Press.
_____. 1983. *Promethean Fire*. Cambridge, Mass.: Harvard Univ. Press.
Mackie, J. L. 1977. *Ethics: Inventing Right and Wrong*. Harmondsworth, England: Penguin.
_____. 1978. "The Law of the Jungle." *Philosophy* 53:553-73.
Moore, G. E. 1903. *Principia Ethica*. Cambridge: Cambridge Univ. Press.
Murphy, J. G. 1982. *Evolution, Morality, and the Meaning of Life*. Totowa, N.J.: Rowman & Littlefield.
Nozick, R. 1981. *Philosophical Explanations*. Cambridge, Mass.: Harvard Univ. Press.
Quinton, A. 1966. "Ethics and the Theory of Evolution." In *Biology and Personality*, ed. I. T. Ramsey. Oxford: Blackwell.
Raphael, D. D. 1958. "Darwinism and Ethics." In *A Century of Darwin*, ed. S. A. Barnett, 355-78. London: Heinemann.
Rawls, J. 1971. *A Theory of Justice*. Cambridge, Mass.: Harvard Univ. Press.
_____. 1980. "Kantian Constructivism in Moral Theory." *Journal of Philosophy* 77:515-72.
Ruse, M. 1979a. *The Darwinian Revolution: Science Red in Tooth and Claw*. Chicago: Univ. of Chicago Press.
_____. 1979b. *Sociobiology: Sense or Nonsense?* Dordrecht, Holland: Reidel.

————. 1982.   *Darwinism Defended: A Guide to the Evolution Controversies*. Reading, Mass.: Addison-Wesley.

————. 1985.   *Taking Darwin Seriously: A Naturalistic Approach to Philosophy*. Oxford: Blackwell.

Ruse, M. and E. O. Wilson. 1986.   "Darwinism as Applied Science." *Philosophy*.

Russett, C. E. 1976.   *Darwin in America*. San Francisco: W. H. Freeman.

Singer, P. 1972.   "Famine, Affluence, and Morality." *Philosophy and Public Affairs* 1:229-43.

————. 1981.   *The Expanding Circle: Ethics and Sociobiology*. New York: Farrar, Straus, & Giroux.

Spencer, H. 1852.   "A Theory of Population, Deduced from the General Law of Animal Fertility." *Westminster Review* 1:468-501.

————. 1857.   "Progress: Its Law and Cause." *Westminster Review*. Reprinted in *Essays: Scientific, Political, and Speculative*. 1868. 1:1-60. London: Williams & Norgate.

Taylor, P. W. 1958.   "Social Science and Ethical Relativism." *Journal of Philosophy* 55:32-44.

————. 1978.   *Problems of Moral Philosophy*. Belmont, Calif.: Wadsworth.

Trivers, R. L. 1971.   "The Evolution of Reciprocal Altruism." *Quarterly Review of Biology* 46:35-57.

Williams, G. C. 1966.   *Adaptation and Natural Selection*. Princeton, N.J.: Princeton Univ. Press.

Wilson, E. O. 1971.   *The Insect Societies*. Cambridge, Mass.: Belknap.

————. 1975.   *Sociobiology: The New Synthesis*. Cambridge, Mass.: Harvard Univ. Press.

————. 1978.   *On Human Nature*. Cambridge, Mass.: Harvard Univ. Press.

[10]

Darwinian Ethics and Error

R. JOYCE
Department of Philosophy
University of Sheffield
Sheffield
UK
E-mail: r.j.joyce@sheffield.ac.uk

> *We give to necessity the praise of virtue.*
>
> Quintilian, *Institutiono Oratoria*, I, 8, 14

> *Poor virtue! A mere name thou art, I find,*
> *But I did practise thee as real!*
>
> Unknown; cited by Plutarch, *Moralia*, 'De superstitione'

Abstract. Suppose that the human tendency to think of certain actions and omissions as morally required – a notion that surely lies at the heart of moral discourse – is a trait that has been naturally selected for. Many have thought that from this premise we can justify or vindicate moral concepts. I argue that this is mistaken, and defend Michael Ruse's view that the more plausible implication is an error theory – the idea that morality is an illusion foisted upon us by evolution. The naturalistic fallacy is a red herring in this debate, since there is really nothing that counts as a 'fallacy' at all. If morality is an illusion, it appears to follow that we should, upon discovering this, abolish moral discourse on pain of irrationality. I argue that this conclusion is too hasty, and that we may be able usefully to employ a moral discourse, warts and all, without believing in it.

Key words: Darwin, error theory, ethics, evolution, evolutionary ethics, Mackie, naturalistic fallacy, Ruse

Introduction

Michael Ruse argues that morality is fundamentally a product of natural selection, and that the correct metaethical conclusion to draw from this is a moral error theory (Ruse 1986a, 1986b). I am strongly inclined to agree on both counts, and here wish to address some recent opposition. I will not argue for the premise – it will be discussed at the outset only in so far as we need to understand it – rather, it is the movement from the premise to an error theory that interests me here.

714

Ruse argues that the content of morality is objective – we treat our moral claims as claims about the world. I think that the correct manner of expressing this is to focus on the fact that we consider morality as something that 'binds' us, that we cannot opt out of; in other words, the content of morality is that of *categorical* (as opposed to hypothetical) *imperatives*. A hypothetical imperative is the familiar, everyday 'You ought to catch the 2.30 train' – the utterer and addressee understand that there is a tacit suffix: '. . . if you want to get to so-and-so in good time'. If it turns out that the addressee lacks that end, then the imperative is withdrawn. A categorical imperative, by contrast, 'declares an action to be objectively necessary in itself without any reference to any purpose' (Kant 1993: 78). Categorical imperatives can be seen as 'about the world' inasmuch as they apparently appeal to rules of conduct 'which are simply there, in the nature of things, without being the requirements of any person or body of persons' (Mackie 1977: 59).[1] It seems quite correct that moral discourse is objective in this sense: when we condemn a moral villain, we do not first check that he has the appropriate interests or desires. If he is guilty of something repugnant, like stealing from innocent people on a whim, we would not dream of retracting our judgment 'He ought not do it' upon discovering that he has a conflict-free desire to steal (and desires all likely consequences of stealing too); there is nothing he can assert (however truly) concerning his ends and interests that will get him off the hook.

From an evolutionary point of view, there is a good explanation for our treating morality as consisting of categorical imperatives. The actions that morality prescribes with categorical force are those that constitute or promote, roughly speaking, *cooperation*. To cooperate with those who may return the favour (reciprocal altruism), and those who share a substantial portion of one's genetic material (kin altruism), enhances reproductive fitness. Therefore evolutionary forces have favoured cooperation.[2] Evolution might have simply 'made us' cooperate (and refrain from defecting), or might have granted us powerful epistemic abilities whereby we can calculate the reproductive advantages of cooperation on a case by case basis. But neither option is optimally efficient. Of the former, Ruse writes: 'we would have wasted the virtues of our brain power, and the flexibility which it gives us'; of the latter: 'this would have required massive brain power to calculate probabilities and the like' (Ruse 1986a: 221). A 'middle road' is selected for: we have evolved an innate disposition in favour of certain types of action, against certain others. This disposition is not merely the development of appropriate *emotions* or *desires*: it's not merely that I *want* to look after my children – but I feel that I *ought* to. I feel, if you will, that there is a *requirement* upon me to look after my children; that I *must*. Desires, after all, are unreliable things: after a long day, a parent might not particularly *want* to care for the children, and this is where a sense of *requirement* kicks in. Since the desire is absent – since the long-term satisfactions of child-rearing are being under-appreciated due to distraction, weakness of will, or simple exhaustion – it is important that the requirement is not conceived of in hypothetical terms. Morality as a system of categorical imperatives compensates for the limitations of desire.

Though the above raises a great many questions and, no doubt, objections, here I wish neither to defend it nor elaborate it, but rather ask, from a metaethical point of view, what follows from it. My contention is that Ruse is correct in holding that the most plausible consequence is a moral error theory. There do not really exist *any* categorical requirements binding our actions, enjoining cooperation and proscribing defection. It's all an illusion which, in evolutionary terms, has served us very effectively. Thus all our judgments of the form 'ϕ is morally obligatory' are untrue: *moral obligatoriness* is a property that no actual action instantiates.[3] In practical terms, cooperation is fostered most effectively if we have a disposition to see it as categorically required: 'morality simply does not work ... unless we believe that it is objective' (Ruse 1986a: 253). But, in metaphysical terms, there is no need to think that there *are* such requirements: everything that needs explaining is explained by the thesis of evolutionary error. The further hypothesis, that these judgments are *true* – that there is a realm of moral facts – is redundant. (And if Ockham's Razor doesn't do the trick, then categorical imperatives can be tackled head-on: Mackie argues that they are 'queer'; Philippa Foot argues that they depend for their legitimacy on 'a magical force', etc. (Foot 1972))

As a final preliminary, let me stipulate a distinction between a moral error theory and what I shall dub 'moral abolitionism'. A moral error theory states that our moral judgments are fundamentally flawed – that our moral discourse contains few, if any, true judgements. But an error theory does not entail anything concerning what we ought to *do* with our discourse once we've uncovered its flaws. (Note that the previous 'ought' did not pose as a *moral* 'ought', so there is no circularity lurking in the question 'Given that there is nothing that we morally ought to do, what ought we to do?') Moral abolitionism is one way of answering that question: it is the view that we ought to *do away* with it. Thus the error theoretic stance is a philosophical position, whereas what I am calling 'abolitionism' is the result of a practical decision. That one leads to the other is a natural enough thought. Elizabeth Anscombe – believing that our moral deontological concepts (concerning what we *ought* to do, what we *must not* do, etc.) are 'survivals, or derivatives from survivals, from an earlier conception of ethics which no longer generally survives', and are unintelligible outside that framework – concludes that they must 'be

716

jettisoned if this is psychologically possible' (Anscombe 1958). But the move from error to abolition is by no means mandatory, and Ruse stoutly resists it:

> I hasten to add that I am not now suggesting that morality is in any way a sign of immaturity. Nor would I have those of us who see the illusory nature of morality's objectivity throw over moral thought. ... Morality is a part of human nature, and ... an effective adaptation. Why should we forego morality any more than we should put out our eyes? I would not say that we could not escape morality – presumably we could get into wholesale, anti-morality, genetic engineering – but I strongly suspect that a simple attempt to ignore it will fail. This is surely the (true) message of Dostoevsky in *Crime and Punishment*. Raskolnikov tries to go beyond conventional right and wrong, but finds ultimately that this is impossible (1986a: 253).

I will discuss the move from error to abolition in the final section, but first I shall address the question of the passage from morality being an evolved trait to morality being in error.

Evolutionary ethics and success

Earlier enthusiasts of 'evolutionary ethics' sought in natural selection a *vindication* of a kind of morality: *moral goodness* might be identified with (something like) *is* [or *has been*] *naturally selected for*. Since there is a plausible case to be made that certain types of action and psychological trait have been naturally selected for, there is a plausible case to be made that certain actions and traits are morally good. The error theory disappears, to be replaced with an evolutionary *success* theory! Ruse will have none of this, and is particularly sensitive to the concern that any such theory will fall foul of the *naturalistic fallacy* (of which, more later). Nevertheless, several commentators, accepting that some of the attitudes we have towards cooperative actions are born of natural selection, still think that a kind of evolutionary success theory is on the cards. (Just to be clear, by 'a success theory', I mean one that holds that our moral discourse is not fundamentally in error, that many of our utterances – such as 'ϕ is morally wrong', 'You must ψ', etc. – are true. An *evolutionary* success theory shall hold that the kind of fact in virtue of which such judgments are true is, in some manner, a fact about human evolution.)

Ruse, in Humean spirit, sees morality as a matter of our 'objectifying' our moral sentiments (Ruse 1986a: 253). Says Hume: 'Vice and virtue may be compared to sounds, colours, heat and cold, which, according to modern philosophy, are not qualities in objects but perceptions in the mind' (Hume

1978: 469) – moral judgments are a matter of the 'gilding and staining [of] natural objects with the colours borrowed from internal sentiment.' (Hume 1983, Appendix I). Regarding the ontology of colour, it has been a popular strategy in recent years to accept Hume's basic projectivist premise, yet to place colours in the world, as a dispositional property of the surfaces of objects. Redness, for example, is said to be the dispositional property of producing the phenomenological response *redness* in normal human viewers (as they are actually constituted) under good viewing conditions (i.e., in broad daylight) (McDowell 1985; Johnston 1992; Campbell 1993). There is a kind of objectivity here, since had a tomato ripened fifty million years ago it would still be red, in so far as *were* a normal human to observe it in good viewing conditions (never mind that there weren't any humans in existence) that human *would* have a certain response. We might say that this analysis makes colours existentially independent of, though conceptually dependent on, human minds.

William Rottschaefer and David Martinsen attempt the same move for morality: we can accept that positive attitudes towards cooperative actions have been naturally selected for, yet identify *moral rightness* (for example) with a relational property instantiated by these actions: (something like) *such that humans have evolved to respond with favour* (or even: *such that humans have evolved to have a response of 'moral objectification'*) (Rottschaefer and Martinsen 1990; Rottschaefer 1998). Now if we're accepting the premise that the attitude favouring cooperative activity is an evolved trait, then it cannot be denied that such activity *does* instantiate the kind of relational property gestured at, but a crucial question remains: 'Is that property the referent of the term *rightness*?' Regarding colour, the point is put succinctly by Michael Smith (1993: 239): 'Someone who denies that colours are properties of objects need not deny that objects *have* these dispositions, all he has to deny is that colours *are* such dispositions.' The mere availability of a dispositional account of a concept does not force that analysis upon us. After all, for *any* predicate we can find a dispositional property had by all and only the items in the predicate's extension. All and only the objects satisfying '. . . is a manatee' are (trivially) such that they *would prompt the response 'There's a manatee!' in an infallible manatee spotter.*

Let's allow that cooperative actions of a certain kind have a 'Darwinian' dispositional property – they are such that humans have, through the pressures of natural selection, come to favour them. (That may be vague, but it's adequate for our general purposes.) Would there be reason to *resist* thinking that this property is the referent of a familiar moral term of positive appraisal? Yes. For such a property cannot (at least as far as I can see) underwrite the notion of moral *requirement* – and what is moral rightness, if not something

718

we are *required* to pursue? Consider again the unrepentant moral villain
earlier mentioned. We can allow that the action he performed had the following relational property: being such that humans have evolved to respond
with disfavour. According to Rottschaefer and Martinsen, then, the action
was *wrong* – really, objectively wrong. Unfortunately, moral naturalism does
not come that easily. For at the heart of our moral discourse is the idea that
the criminal *ought not* to have performed the action, that he was somehow
required to refrain. And it would be very odd if we thought that he ought to
ϕ while admitting that he has no reason to ϕ; therefore canons of ordinary
moral thinking will also suppose the criminal to have had a *reason* to refrain
(regardless of his desires, and regardless of whether he is aware of the fact).
But why do the things favoured by natural selection *bind* him, or provide him
with *reasons*? Moreover, many have thought that if a person makes a moral
judgment (that some action is wrong), it follows of necessity that she has
some prima facie *motivation* against that action.[4] But the criminal may note
with utter indifference that an action is such that humans have been naturally
selected to disfavour it – what's that fact to *him*?

It would be tempting, but futile, to appeal to the fact that our criminal *is*
a human, with all the natural human dispositions, and therefore has reason
to act in accordance with natural selection. This is, in effect, how Robert
Richards argues in presenting his evolutionary success theory (Richards
1986). Since, according to Richards, all humans have evolved to act for the
community good, we may say to any human: 'Since you are a moral being,
constituted so by evolution, you ought to act for the community good'. He
likens this derivation of an 'ought' to that occurring in 'Since lightning has
struck, thunder ought to follow'. This is surprising, since the 'ought' of the
latter is an *epistemic* or *predictive* one. Such 'ought's still, arguably, entail
reasons: 'That lightning has struck gives one reason for believing that thunder
will follow' (Harman 1975; Mackie 1977: 74). But the *moral* 'ought' that
Richards hopes to derive surely is not an epistemic one: when we say that the
villain ought not steal, we are not saying that we are able to predict, on the
basis of some antecedent concerning evolution, that he will not steal; and, by
the same token, the *reason* entailed by the 'ought' pertains to *his* reasons for
not stealing, not *our* reasons for believing that he won't steal!

Presumably what Richards hopes to do is to make moral imperatives *hypothetical*, depending for their legitimacy on an end which all humans, as a
matter of fact, have been assigned by natural selection: the good of the community. If our moral villain has this end, then he ought to do (*ceteris paribus*)
whatever will satisfy it; he has a (*prima facie*) reason to do whatever will
satisfy it. Now evolutionary forces have certainly not bestowed upon us all an
active *desire* to promote community good – at most, we are endowed with a

disposition, or *capacity*, in favour of its promotion (as Richards recognizes). But why does a mere disposition provide an 'end', or ground an 'ought' statement? In general, 'oughts' may be grounded by desires – if Sally wants coffee, then, *ceteris paribus*, she ought to head to the café – and *interests* (if they are distinct from desires) may also ground 'oughts' – if it is in Sally's interests to stop smoking, despite her having conflict-free desires to carry on, then, *ceteris paribus*, she ought to stop. But I cannot see that the same goes for dispositions. Allow that evolution has endowed Jack with a disposition to favour the promotion of the community's good, but imagine that his upbringing was such that the disposition went quite undeveloped, and now has been effectively quashed. Why *ought* he still act for the community's good? Why does he still have a *reason* to?[5]

Richards toys with the idea of simply branding Jack a 'sociopath', therefore not fully human, and therefore not a proper subject of moral injunctions. Perhaps this would stick if Jack lacked the disposition altogether, as the result of a genetic aberration, but we are not claiming any genetic anomaly – Jack still *has* the disposition, it has just gone utterly undeveloped, and now, let's imagine, it is too late for Jack to develop it, in much the same way as it is now too late for him to become a concert pianist.[6] It's important to note that our 'villain', despite earlier characterizations, need not be the serial killer stalking back streets, need not be the suicidal teenager heading to school with an automatic rifle in his bag. The kind of self-centred person we encounter every day – one who regulates his or her actions consciously and solely in terms of perceived self-gain – will suffice perfectly well as an example of someone whose altruistic dispositions have been quelled. Bearing this in mind, talk of 'sociopaths' who fall short of satisfying the criteria for *being human* seems wildly overstated.

Consider such a character: pleasant enough to interact with, has a successful career, a family, etc. But if she has made a promise that will be inconvenient to keep, and she sees that she can break it without incurring penalty (perhaps she can make a decent excuse), then, despite her knowledge that doing so will seriously penalize others, and, say, harm the community in general, she will not hesitate to break the promise. Let us point out to her that the action of promise-breaking has a certain 'Darwinian' dispositional property: it is such that humans have evolved to disfavour it. She accepts this, but notes it with unconcern (along with facts about the evolution of manatees). Let us inform her that she herself has this disposition, in the sense that had she received a certain kind of upbringing she would have favoured the good of the community (and may pass this disposition on to her offspring). But, given that she *didn't* receive that upbringing, but one that left the disposition dormant, why does she now have a reason to refrain from promise-breaking? To say

720

that the disposition *must* have some manifestation, such that in some sense she, in acting against the community's good, *must* be subtly undermining her own projects and interests, is just desperate.

These observations disclose my doubt concerning Ruse's view that morality is unavoidably with us, and that only genetic engineering could eliminate it (see his quote about Raskolnikov above). If the employment of moral concepts is a genetically present *disposition*, then it is perfectly possible that certain socialization processes (or perhaps merely a course of metaethics) could dampen or completely nullify the moral sentiment. In my experience, there are more people around who do not properly participate in moral thinking (but who are hardly thereby 'psychopathic') than philosophers like to admit. If this is correct, then it would be possible to eliminate moral discourse without resorting to genetic tampering if we wanted to (as we can, arguably, disable the manifestation of aggressive or xenophobic dispositions). Whether we *ought* to do so is the subject of Section IV.

I have re-iterated the question of why facts about evolution provide persons with *reasons*, why they ground moral 'ought' statements. And it should be clear that my answer is: 'As far as I can see, they don't.' Of course, if evolution has endowed me with a disposition to favour cooperation, and my upbringing was such that this disposition *has* developed fully, then indeed I have a (*prima facie*) reason to cooperate. But now all the work is being done by the fact that my upbringing provided me with certain attitudes and traits that are now actively operative – and these attitudes would ground 'ought' statements even they had nothing to do with evolution. It will not do to maintain that any agent in whom such dispositions lie untapped (and now 'untappable') is simply a *sociopath*, who lies beyond the pale of moral injunctions. We have already seen that such agents are possibly quite common, and they certainly remain subject to the dictates of moral discourse. We think that a person – regardless of an upbringing that left her intractably selfish – morally *ought not* break promises for the sake of convenience. Pointing to a relational property pertaining to natural 'fitness', indicating that natural selection provides humans with certain dispositions against promise-breaking, does not help. And if an ethical theory cannot account for so central and familiar a moral judgment – that a selfish person ought not break an inconvenient promise – it has not gotten off the ground.

Robust Darwinian naturalism and the naturalistic fallacy

Rottschaefer and Martinsen anticipate an accusation from Ruse that their theory blunders into the dreaded naturalistic fallacy, and go to some effort (as does Richards) to show that it does not. But it is not the naturalistic fallacy that I accuse such theorists of, for, I hereby admit, I have little idea what that fallacy is, nor why ethicists – especially those interested in evolution – seem so fearfully mesmerized by it.[7] It has become commonplace to assume that G.E. Moore's notorious fallacy does for 'good' what Hume did for 'ought', but no part of *Principia Ethica* that I am familiar with bears resemblance to Hume's claim that one cannot derive an 'ought' from an 'is' (Moore 1903). It is true that nothing like the following is formally valid (if by this we mean 'is an instance of a theorem of the predicate calculus'):

(1)   Things of type ϕ are such that humans, by the process of natural selection, are disposed to have attitude A towards ϕ.

Therefore: Things of type ϕ are morally good.

But no naturalist would claim such a thing. Rather, she will treat the above as an enthymeme, inserting a major premise if required:

(2)   If things of type ϕ are such that humans, by the process of natural selection, are disposed to have attitude A towards ϕ, then things of type ϕ are morally good.

It is not good complaining that (2) reproduces, in conditional form, a formally invalid argument, for the naturalist does not claim that (2) is 'valid', merely that it is *true*. Nor can it be simply insisted that (2) commits 'the naturalistic fallacy' in virtue of relating a fact to a value, and therefore must be false. That's just begging the question. It is also important to remember that the 'fallacy', according to Moore, is committed no less by statements of the following kind:

(Yellow) Having the natural properties P, Q, R, *is* what it is to be yellow.

So he evidently did not think that it is the 'evaluativeness' of goodness that powers the fallacy, but its *indefinability*. But again, we cannot simply *assume* that goodness is indefinable (or unanalyzable), for that is precisely a point at issue. When we look at the heart of Moore's description of the fallacy (in §12), what we actually find seems to be advice that we ought not confuse the 'is' of identity with the 'is' of predication. Moore thinks that the hedonic naturalist, when he claims 'Pleasure is good' may be saying something true so long as it's an 'is' of predication; but to mistake it for an 'is' of identity (a *definition*, by Moore's lights) leads to absurdity. In the same way, if I say 'The book is red' and 'The book is square' – but these are taken as identity claims – I'm left with the crazy conclusion that redness is squareness.

722

Keeping track of one's 'is's is surely good advice – perhaps to confuse them may even be called a kind of 'fallacy' – but Moore is quite mistaken if he thinks that the naturalist *must* be confused over 'is'. (2) can be seen as entailed by a naturalistic thesis:

(Naturalism) For any ϕ, ϕ is a type of thing towards which humans, by the process of natural selection, are disposed to have attitude A if and only if things of type ϕ are morally good.

There is one 'is' of predication there. With rewording, the biconditional might be strengthened into an 'is' of identity flanked by property names. Thus naturalism might be an *a posteriori* claim, comparable to 'Water is H_2O', or an *a priori* (but covert) thesis, like 'Knowledge is justified true belief'. But in neither case need the naturalist fall foul of the problem that *Moore* called 'the naturalistic fallacy'. Moore does allow that *some* things may be defined without trouble: his stock example is a definition of horse. (That's Moore's syntax; I'd much rather speak of a definition of 'horse' or a definition of *horseness*. Since he's adamant he does not intend the former, I assume he means the latter.) So what is it about *yellowness* and *goodness* that makes them different from *horseness*? Moore's answer is that they are 'simple', 'non-natural', and 'indefinable' – but this cannot be treated as a self-evident datum, for it is exactly what the naturalist, in offering something like (Naturalism), denies. Oddly, of this all-important premise, Moore writes: 'As for the reasons why good [sic] is not to be considered a natural object, they may be reserved for discussion in another place.' It appears that this 'further discussion' is the very next section of *Principia Ethica* – where the Open Question Argument is deployed. But the woes of that argument are well-documented, and won't be rehearsed here. For effective criticism, see, e.g. Harman (1977: 19), Frankena (1973: 99ff.) and Putnam (1981: 205ff.). (I'll merely note that it doesn't even work for Moore's favourite example of the definition of *horseness* – for the analysis he offers is *a posteriori* in nature – making mention of a horse's *heart* and *liver*, etc. – such that a perfectly competent speaker might be certain that X is a horse, but uncertain that X has property N [where 'N' stands for the 'definition' Moore offers, involving hearts and livers].)

Consider something like (Naturalism) – what Rottschaefer and Martinsen would call a 'robust Darwinian naturalism' (and I have called 'an evolutionary success theory'). The question, I have argued, is not whether it commits a 'fallacy', but whether it is *true*. If it is true, then it is either an *a priori* or an *a posteriori* truth. The relevant model for the former is a philosophical analysis like 'Knowledge is true justified belief'. We do not come upon such truths (pretending that it *is* a truth) simply by doing a bit of quick introspection, or by looking in a dictionary. Smith suggests that one way of proceeding is to gather all our platitudes about knowledge – a platitude being something one comes to treat *as* platitudinous in attaining basic competency with the concept – and then to systematize those platitudes (Smith 1993, 1994). 'True, justified belief' may be the best systematization, or encapsulation, of our epistemic platitudes (though it probably isn't). But it is clear that no description worded centrally in *evolutionary* terms is going to be the best systematization of our moral platitudes. Moral concepts, I assume, preserved their identity criteria throughout the nineteenth century: someone saying 'Slavery is morally wrong' in 1890 was not expressing a different proposition to someone uttering the same sentence in 1810 (otherwise, were the 1810 speaker instead to assert 'Slavery is morally permissible', she would not be in disagreement with the 1890 speaker, in which case we could not say that moral attitudes towards slavery changed over the course of the nineteenth century). If this is true, then, according to the theory under question, it was *a priori* available to pre-Darwinian speakers to systematize their moral platitudes in such a way that *natural selection* centrally figured in that explication. But that is absurd, so robust Darwinian naturalism as an *a priori* thesis is a non-starter.

How will it fare as an *a posteriori* thesis? The model here is 'Water is H_2O'. According to the *a posteriori* naturalist, we can 'find out' that two kind terms, perhaps both in common parlance, are, and always have been, co-referential. See, e.g., Boyd (1988) and Brink (1984). This sounds closer to what the robust Darwinian naturalist will presumably claim: when we consider a term like 'moral rightness', and examine the kind of things to which we apply it (and the kind of things from which we withhold it), and then bring in evolutionary theory, perhaps boosted by detailed empirical confirmation, we might discover that (pretty much) all and only the things to which we apply '. . . is right' instantiate a property, or cluster of properties, which may also be described by the predicate '. . . is a type of thing towards which humans, by the process of natural selection, are disposed to have attitude A'. This is potentially threatening to an evolutionary error theory, for we have agreed that there *is* such a property had (pretty much) by all and only the things to which we apply our predicate '. . . is morally right', so is that not immediately to give the game to the (*a posteriori*) evolutionary *success* theorist?

I think not. The worry with this kind of *a posteriori* theory is that it threatens to achieve far too much. Consider our term 'witch', that was once applied to actual persons. It is possible that all and only the persons to whom we applied 'witch' had a certain property, or cluster of properties – perhaps

724

they were women who tended to be of a certain social class, playing a certain socio-political role, who threatened the patriarchal authorities in a particular way (I'm not suggesting that it's anything so simple – it may be disjunctive and vague). But to locate such a property clearly would not be an *a posteriori* vindication of 'witch discourse'. Similarly, we have a term 'phlogiston' that we used to apply to various phenomena: we could point to any open flame and say 'Look, there's the phlogiston escaping.' In recognizing that there clearly is a property, or cluster of properties, that all and only open flames have, have we thereby rescued phlogiston discourse? The reason that the answer is obviously 'No' is that when seventeenth century speakers used the predicate '…is phlogiston' (or '…is a witch') something more was going on than merely applying it to some objects, withholding it from others. What doomed the predicate to emptiness, despite its ostensive paradigms, was that users of the term (considered collectively) thought and said certain things *about* phlogiston, such as 'It is that stuff stored in bodies', 'It is that stuff that is released during combustion', and these concomitant statements are false. (The analogous claim for witches will concern their supernatural abilities.)

In my opinion the same thing will go for moral discourse. It is not enough to find some property had by all and only the things to which we apply our moral terms. There are also very important things which we endorse *about*, say, morally right actions – such as they are the ones which a person *ought to* perform regardless of his desires, they are the ones that we have overriding *reason* to perform, they are the ones the recognition of which will *motivate* an agent. But, as I argued previously, the kind of property adverted to by the robust Darwinian naturalist does not satisfy such a sense of 'inescapable requirement' (or, at least, it will require a great deal more argument to show that it does – the prospects for which I am very skeptical of). Therefore this Darwinian dispositional property, though very probably existing, does not deserve the name 'moral rightness'.

The naturalist might respond: 'So much the worse for our sense of *categorical imperative*. Why not just admit that this aspect of our moral discourse is faulty, and carry on with a revised naturalist discourse?' Well, when Lavoisier gave us oxygen theory in the late eighteenth century, why couldn't the fans of phlogiston just revise their theory, insisting that they had been talking about *oxygen* all along, concerning which they had held some false beliefs about its being stored and release? (Ditto, *mutatis mutandis*, for witch discourse.) The reason that it was *not* available for them to revise and vindicate phlogiston theory in this manner is that the thesis about phlogiston being stored in bodies and released during combustion was too central to the theory to be negotiable – one might say that the whole point of phlogiston discourse was to refer to a *stored and released* material. By the same token, I believe, the whole point of having a moral discourse is to prescribe and condemn various actions with *categorical* force. We have a moral discourse so that various actions (and omissions) can be demanded when desires (whether self-interested or otherwise) are absent, limited, or fail to motivate. If this were not the case, why did we develop a moral discourse at all? – after all, we've always had a perfectly well-structured vocabulary for discussing the means of satisfying desires and fulfilling ends – even long-term ones. Evidently, the language of hypothetical imperatives was not adequate to the task for which we required moral language.

Let me sum up this section before moving on to a rather different topic. The robust Darwinian naturalist – he who agrees that various moral attitudes are the result of natural selection, but hopes to found upon this a moral *success* theory, a kind of moral realism – fails to accommodate some very central moral beliefs. I think there are several fundamental desiderata that will go unsatisfied, but here I have focused on the notion of a categorical imperative. In doing so, the naturalist does not commit any form of 'fallacy' – he merely presents a false theory. Clearly, there are two vital premises to my position that have only been sketched in a rather brief and dogmatic manner: (i) that our moral discourse *is* centrally committed to categorical imperatives, and (ii) that the robust Darwinian naturalist cannot accommodate these imperatives in his system. Successfully combating either claim would undermine much of what I have said.

One might well wonder what work is now being done by the thesis that our sense of categorical requirement is a *biological adaptation*, for if we can show that moral discourse is centrally committed to thesis T, and that T is philosophically indefensible, then we have our error theory right there, with no mention of evolution. This is, of course, exactly what Mackie and others have tried to do – to establish a moral error theory head-on. But any error theorist owes us an account of *why* we have all been led to such a drastic mistake; the absence of such an explanation is likely to raise doubts that we *are* making a mistake at all (i.e., either doubts that T is erroneous, or doubts that our discourse ever committed itself to T). This, I believe, is where an evolutionary account of the development of moral sentiments plays its role. In other words, we have two theses: one is the error theoretic stance for moral discourse, the other is the claim that morality is largely the product of natural selection. The former, by itself, lacks persuasiveness – it lacks an explanation of where the error came from. The claim that morality is an evolutionary trait – that developing a sense of 'intrinsic requirement' would be beneficial to humans *even if there were no such thing* – fills that gap. But the latter thesis, by itself, is insufficient to establish an error theory. I mentioned earlier the possibility of arguing for an error theory using Ockham's Razor: everything

726

that needs explaining is explained by an evolutionary story concerning how and why we have a disposition to make moral judgments, with no need for an additional theory according to which the judgments are *true*. But it can now be seen that Ockham's Razor won't suffice, for the kind of robust Darwinian naturalism that has been under discussion does not posit any extra *ontology* – it rather points to dispositional properties, the existence of which all parties to the debate should antecedently agree to.[8] So Ruse can plausibly claim that we have evolved to believe in objective requirements, but no investigation of the processes of natural selection, of the course of human evolution – no matter how subtle and empirically well-confirmed – will be sufficient to establish that we are victims of *an illusion*. For that we need philosophical argumentation.

Error, abolition and acceptance

Thus far I have argued that those who hope to find in the (probable) fact that certain attitudes have been naturally selected for a *vindication* and *justification* of moral discourse are backing a worthy but misguided cause. Moral injunctions have an *authority* that evolutionary facts cannot underwrite, and being able to appeal to such an authority is the whole point of having a moral discourse. But this authority may yet be shown to be justified in some other manner – certainly there are well-developed philosophical programmes that seek to substantiate it. My judgment is that none of them will be fruitful. Here is not the place to defend that skepticism, but in the remainder I want to investigate what would follow if we decided that the skepticism *is* well-founded – if it is true that we have evolved to accept an illusion, as Ruse thinks.

Let me open discussion with two quotes from earlier articles in this journal. William Hughes writes: 'if [moral values] are unreal then the only rational position is to seek to eradicate moral and ethical language altogether, and replace it with the language of needs and wants (Hughs 1986: 306). And Peter Woolcock, in a cogent critique of Ruse: 'Once we realise [that there are no moral obligations whatsoever], the rational course would seem to be to train ourselves out of any residual tendencies to obey moral laws where we can get away with breaking them. We should deprogramme ourselves out of any inclination to feel guilt, or to want redemption. Contrary to Ruse's denial Nietzsche and Thrasymachus were right – moral thought is overthrown (Woolcock 1993: 428).

A consequence of Ruse's view is that no statement of the form 'S is under a moral obligation to ϕ' is true. Thus no belief having that content is true. Thus, if a person has evidence of this fact – once 'the cat is out of the bag' (as Woolcock puts it) – to have such a belief is irrational. Thus, if we read Ruse's 1986 book and justifiably believe it (or, for that matter, if we read Mackie's relevant work and justifiably believe it), it becomes *irrational* for us to hold moral beliefs. I do not see that Ruse can avoid this conclusion without revising his basic position. It might seem that Hughes' and Woolcock's 'abolitionism' follows close on the heels of this admission, but this is exactly what I want to resist. Moral discourse may still have an active role to play even for those who have seen the cat out of the bag.

One way to proceed (that I don't favour) would be to argue, seemingly paradoxically, that it may sometimes be rational to be irrational. There are different things that admit of the 'rational/irrational' distinction – actions, beliefs, emotions – and it is far from obvious that all are appraised for rationality according to the same framework. For example, certain phobic emotions are deemed irrational, often on the grounds that they are experienced in the presence of inappropriate beliefs (I know the spider is harmless, but it fills me, nevertheless, with dread). Suppose, however, that a person is in an unusual situation, such that having a phobia is greatly to her advantage (perhaps she is developing a worthwhile and loving relationship with her therapist). Suppose, moreover, that there is some action that she can perform that will encourage the development of that phobia (she goes to see the movie *Arachnophobia*, knowing it will traumatize her). Since that action is to her instrumental advantage, and she knows it, we ought to deem it rational; the phobic emotion of fear, however, remains no less irrational. So is *she* rational or irrational? The correct answer, it seems to me, is that according to one normative framework she is rational, according to another she is not. The 'post-Ruseian' moralist may be in the same situation. His ongoing belief in moral obligation is irrational, yet his having that belief may be to his practical advantage, may serve his ends, and therefore if there are actions he can perform to encourage such beliefs, those actions are rational.[9]

As I say, I do not favour this kind of defence, encouraging, as it apparently does, a kind of schizophrenia, or self-deception, in the agent. Besides, Ruse has not shown that it is to any *individual's* instrumental advantage to have moral beliefs, only that having moral beliefs has enhanced reproductive fitness. A group of humans who find in cooperative actions a 'to-be-done-ness' does relatively well, their society flourishes, and their genes are passed on. But of any individual living in such a society we can see that *his* advantage is to defect on promises when he can get away with it. The fact that things would go badly for him if *everyone* thought this way is nothing to him – it merely means that he has to try to encourage moral beliefs in others. Nor does the fact that the trait of 'being a free-rider' is unlikely to be favoured by

728

natural selection in social creatures alter its being to *his* advantage to get a free ride if he can.

Hume (as in so many things) has some interesting thoughts on free-riders (1983, §IX, part 2). First, he points out that there are important values that the free-rider misses – values that by their very nature cannot be gained through secret defection: the satisfaction of fair dealing, comradeship, open cultural participation, etc. Second, free-riders are certainly epistemically fallible, and possibly weak of will, thus 'while they purpose to cheat with moderation and secrecy, a tempting incident occurs, nature is frail, and they give into the snare; whence they can never extricate themselves, without a total loss of reputation, and the forfeiture of all trust and confidence with mankind'. Those looking to defect secretly are likely to miscalculate, get caught, and pay a serious price; if they are also weak of will then the likelihood increases.

It might be argued that what follows from Hume's observation is that clear-headed calculations of expected self-gain will suffice to regulate cooperative behaviour, with no troublesome *moral duties* or *categorical imperatives* entering into the picture at all. The 'sensible knave' would break a promise if she could be sure of getting away with it, but she is sensible enough to know that she is rarely sure of getting away with it, and the price of detection is too great to risk it. Therefore, in all but unusual cases, enlightened self-interest will serve to underwrite all the prescriptions that we would usually call 'moral'. Moral injunctions may be replaced, after all, by 'the language of needs and wants.'

But I don't think that this would be the correct moral to draw from Hume. If Hume's 'knave' really is sensible, he knows that he is epistemically fallible and vulnerable to weakness of will. He knows that the profits of short-term gain are often tempting. He knows, furthermore, that, being human, he is a creature of habit, so a single successful defection might encourage other riskier defections. It is therefore to his advantage to regulate his day-to-day decision procedures by something other than clear-headed egoism, if only because egoistic calculations – as anyone knows who has ever taken up an exercise programme, or embarked on a diet – do not guarantee correct behaviour. What this knave needs is to place a strong value on certain actions, and a strong disvalue on others. He needs to think of cooperative behaviour not in terms of 'This will, in the long-term, be to my benefit – I just shouldn't risk defecting; someone might be watching'; rather, he needs to think of it as '*This must be done*'. When he takes this step, then he is a knave no more. Of course, employing such a moral concept will not *guarantee* correct behaviour either, but it stands a much better chance.

It might seem that we have argued in a circle: we are back to claiming that having moral beliefs is to the advantage of a standardly situated agent, but we have not dispelled the fact that to believe *p* while having been exposed to evidence that firmly discredits *p* is to be irrational. My way out of this circle (and I offer it as a defense of Ruse) is to deny that getting the regulative benefit from moral concepts requires their figuring in *beliefs*. Think of how we best fend off akrasia when commencing a programme of exercise. I tell myself that I *must* run for an hour every other day (that's just a round number; I don't pretend to achieve anything so impressive!). Of course, it's false that I must run this much and no less: if occasionally I run for fifty-five minutes, or occasionally skip a few days, I'll still achieve my goal of fitness perfectly adequately. But the spirit is weak! – if I start allowing these little lapses, the slippery slope of self-sabotage beckons. What keeps me on track for my goal is a firm and non-negotiable rule: an hour every other day, no less. However, I do not need to *believe* this rule for it to work – if someone questions me, suggesting that there's no harm in occasionally skipping a few days, I am not committed to arguing that this is mistaken – what's important is that I rehearse the rule in my mind, that I allow it to influence my actions, that I let it carry weight with me. I *accept* the rule, but I do not believe it. Indeed, if you were to press me seriously about its truth – in a critical context, *not* when I am actually running – then I would happily express my *disbelief* in it.[10]

There is more that we can do with a false theory than either irrationally believe it or abolish it entirely. As a useful fiction it can still have a practical role in our lives (as, indeed, literary fictions have a practical role in our lives). This, I believe, is an option that is available to us concerning *morality* even after we realize that its central concepts are illusions foisted upon us by natural selection. It remains practically advantageous for any ordinarily situated individual to imbue certain cooperative actions with a sense of 'inescapable to-be-done-ness'. It is *more* advantageous for her to do this (I am suggesting) than merely to believe that the same action ought to be performed because it is in her long-term best interest (though she may well believe this as well); and for her really to *believe* that those actions 'must be done' – after reading and justifiably believing Ruse and Mackie, that is – that too would be practically disadvantageous: to believe things the evidence of whose falsehood is available to us is irrational, and is likely to have serious detrimental consequences if adopted as a doxastic policy.

Wittgenstein (1965) once remarked that our moral discourse seems to consist largely of similes. I am reminded of Bentham's slightly bizarre attempt to analyze the idea behind *obligation*: 'the emblematical, or archetypical image, is that of a man lying down with a heavy body pressing upon him (Bentham 1843: 247), as well as Mackie's talk of obligation being an 'invisible cord' and a demand for payment being an 'immaterial suction-pipe' dredging for

730

the owed money (Mackie 1977: 74). 'But,' Wittgenstein continues, 'a simile must be a simile for *something*. ... [Yet] as soon as we try to drop the simile and simply state the facts which stand behind it, we find there are no such facts. And so, what at first appeared to be a simile now seems to be mere nonsense.' Though Wittgenstein concludes that the 'very essence' of morality is its nonsensicality, he does not advocate its abandonment: it is something he 'cannot help respecting deeply' and he refuses to 'belittle this human tendency'.

Wittgenstein's assessment demands the question: 'But *why* do we participate in this "nonsense"?' and an evolutionary story like that favoured by Ruse begins the answer. But another question beckons for both Wittgenstein and Ruse: 'Surely to see nonsense for what it is requires, on pain of irrationality, its rejection?' The argument of this last section explores one way of replying 'Not necessarily.' The question of what we ought to *do*, once we have come to see that our moral discourse is a philosophically indefensible illusion, is a practical question. A neglected answer is that the discourse may be maintained, accepted, but not believed – that it may have the role of a fiction. There is nothing irrational about fictions (so long as we don't believe them); there is nothing irrational about our allowing them to influence our emotions and decisions, or even thinking them of immense importance. Given that the widespread tendency to resist a moral error theory – to think of it as a *dangerous* doctrine – surely does not arise from the manifest plausibility or lucidity of moral concepts, but rather from a fear of what might *happen* if we abolished them, it seems to me quite likely that the practically optimal course, and therefore the rational course – both for society considered collectively, and for the individual – will be to keep these concepts alive.

Notes

[1] I believe that Mackie, who gave us the term 'error theory', would have disagreed with little in Ruse's overall project. Although the evolutionary aspect of Mackie's theory is underdeveloped, there is little doubt that he saw morality as an essentially biological phenomenon (see Mackie 1977: 113).

[2] This is not to say that evolution has favoured cooperation with *anyone* in *any* circumstances. Of course not. Nor do I maintain that morality can be understood entirely in terms of cooperative actions (and sentiments favouring those actions) – attitudes towards various *self-regarding* actions have quite possibly also been selected for. Also, although the disposition to see certain activities and traits as 'intrinsically required' naturally developed in relation to *cooperative* tendencies, there is no reason why cultural pressures might not come to transfer that sense of requirement to other types of action (e.g., in Catholic priests, to celibacy); thus there will be significant cross-cultural differences among moral systems. What they share, at a minimum, is a sense that some actions 'must be (not) done, regardless of the performer's ends', and these required actions will most *probably* attach to cooperative behaviour. These

are important and complex qualifications, but they are not the subject of the present paper, where I keep things simple for brevity.

[3] I say 'untrue' rather than 'false', since the correct conclusion might be that the abstract singular term 'moral obligatoriness' fails to refer to any property at all (as opposed to referring to a property which nothing actually has), in which case one might, for familiar Strawsonian reasons, hold that 'Moral obligatoriness is had by ϕ' is *neither true nor false* (like 'The present king of France is wise'). Since that sentence, arguably, expresses the same proposition (if any) as 'ϕ is morally obligatory', the latter too would be neither true nor false.

[4] Hume (*Treatise*, Book III, part I, section 1) writes: 'Morals excite passions, and produce or prevent actions. Reason of itself is utterly impotent in this particular. The rules of morality, therefore, are not conclusions of our reason'. Many modern ethicists have agreed with him. Michael Smith, though disagreeing with Hume's apparently noncognitivist conclusion, argues in detail for the thesis that one who makes a moral judgment (assuming she is practically rational) feels *prima facie* motivation. See Smith (1994).

[5] I must say, in fairness to Richards, that he does *not* think that the mere fact that we have, as a product of natural selection, a disposition to favour altruism entails that we ought to be altruistic. He notes that we also have evolved aggressive tendencies, but he doesn't think it follows that we ought to act on them. See Richards (1986: 288, 342). However, I must admit that I do not properly understand Richards' attempt to argue for a principled distinction on this point.

[6] It is important to stress that the sense of 'disposition' under discussion is specific: an inherited trait that regulates the formation of certain attitudes when the agent is exposed to certain environmental cues at a certain point in development. Thus when I claim that Jack 'has the disposition', this is a claim about his genetic package; it does not follow that there are any environmental stimuli that Jack could encounter *now* that would result in his forming the attitudes in question.

[7] In particular, I have never understood why William Frankena's sensible 1939 article did not put an end to the whole business.

[8] Compare the kind of 'non-natural' property that Moore thought is the referent of 'good'. If we had a well-confirmed theory that explained all relevant phenomena by appeal only to our *making judgments* that such non-natural properties exist, then Ockham's Razor should serve to establish an error theory – for in order for those judgments to be *true* we would be required to posit some extra kind of entity in the world (i.e., non-natural properties), but this additional ontology would not explain anything that was not explained by the theory that appealed only to (untrue) judgments.

[9] See my 'Rational Fear of Monsters', *British Journal of Aesthetics* (April, 2000) for further discussion along these lines.

[10] These thoughts are developed at further length in my 'Moral Fictionalism' (forthcoming).

References

Anscombe, G.E.M.: 1958, 'Modern Moral Philosophy', *Philosophy* **33**, 1–19.

Bentham, J.: 1843, 'Essay on Logic', in *Collected Works*, Volume VIII, William Tait, Edinburgh.

Boyd, R.: 1988, 'How to be a Moral Realist', in G. Sayre-McCord (ed.), *Essays in Moral Realism*, Cornell University Press, Ithaca.

Brink, D.: 1984, *Moral Realism and the Foundations of Ethics*, Cambridge University Press, Cambridge.

732

Campbell, J.: 1993, 'A Simple View of Colour', in J. Haldane and C. Wright (eds), *Reality, Representation and Projection*, Oxford University Press, New York.

Foot, P.: 1972, 'Morality as a System of Hypothetical Imperatives', *Philosophical Review* **81**, 305–316.

Frankena, W.: 1939, 'The Naturalistic Fallacy', *Mind* **48**, 464–477.

Frankena, W.: 1973, *Ethics*, Prentice Hall, Englewood Cliffs.

Harman, G.: 1975, 'Reasons', *Critica* **7**, 3–13.

Harman, G.: 1977, *The Nature of Morality*, Oxford University Press, New York.

Hughes, W.: 1986, 'Richard's Defence of Evolutionary Ethics', *Biology and Philosophy* **1**, 306–315.

Hume, D.: 1978, *A Treatise of Human Nature*, Clarendon Press, Oxford.

Hume, D.: 1983, *Enquiry Concerning the Principles of Morals*, Hackett Publishing Company, Indianapolis.

Johnston, M.: 1992, 'How to Speak of the Colors', *Philosophical Studies* **68**, 221–263.

Joyce, R.: 2000, 'Rational Fear of Monsters', *The British Journal of Aesthetics* **40**, 209–224.

Kant, I.: 1993, *Groundwork to the Metaphysics of Morals*, translated by H.J. Paton, Routledge, London.

Mackie, J.: 1977, *Ethics: Inventing Right and Wrong*, Penguin Books, New York.

McDowell, J.: 1985, 'Values and Secondary Qualities', in T. Honderich (ed.), *Morality and Objectivity*, Routledge and Kegan Paul, London.

Moore, G.E.: 1903, *Principia Ethica*, Cambridge University Press, Cambridge.

Putnam, H.: 1981, *Reason, Truth and History*, Cambridge University Press, Cambridge.

Richards, R.J.: 1986, 'A Defence of Evolutionary Ethics', *Biology and Philosophy* **1**, 265–293.

Rottschaefer, W.A.: 1998, *The Biology and Psychology of Moral Agency*, Cambridge University Press, Cambridge.

Rottschaefer, W.A. and Martinsen, D.: 1990, 'Really Taking Darwin Seriously: An Alternative to Michael Ruse's Darwinian Metaethics', *Biology and Philosophy* **5**, 149–173.

Ruse, M.: 1986a, *Taking Darwin Seriously*, Basil Blackwell, Oxford.

Ruse, M.: 1986b, 'Evolutionary Ethics: A Phoenix Risen', *Zygon* **21**, 95–112.

Smith, M.: 1993, 'Objectivity and Moral Realism: On the Significance of the Phenomenology of Moral Experience', in J. Haldane and C. Wright (eds), *Reality Representation & Projection*, Oxford University Press, New York.

Smith, M.: 1994, *The Moral Problem*, Blackwell, Oxford.

Wittgenstein, L.: 1965, 'Lecture on Ethics', *Philosophical Review* **74**, 3–12.

Woolcock, P.: 1993, 'Ruse's Darwinian Meta-Ethics: A Critique', *Biology and Philosophy* **8**, 423–439.

[11]

A DARWINIAN DILEMMA FOR REALIST THEORIES OF VALUE

SHARON STREET

1. INTRODUCTION

Contemporary realist theories of value claim to be compatible with natural science. In this paper, I call this claim into question by arguing that Darwinian considerations pose a dilemma for these theories. The main thrust of my argument is this. Evolutionary forces have played a tremendous role in shaping the content of human evaluative attitudes. The challenge for realist theories of value is to explain the relation between these evolutionary influences on our evaluative attitudes, on the one hand, and the independent evaluative truths that realism posits, on the other. Realism, I argue, can give no satisfactory account of this relation. On the one hand, the realist may claim that there is *no* relation between evolutionary influences on our evaluative attitudes and independent evaluative truths. But this claim leads to the implausible skeptical result that most of our evaluative judgements are off track due to the distorting pressure of Darwinian forces. The realist's other option is to claim that there *is* a relation between evolutionary influences and independent evaluative truths, namely that natural selection favored ancestors who were able to grasp those truths. But this account, I argue, is unacceptable on scientific grounds. Either way, then, realist theories of value prove unable to accommodate the fact that Darwinian forces have deeply influenced the content of human values. After responding to three objections, the third of which leads me to argue against a realist understanding of the disvalue of pain, I conclude by sketching how antirealism is able to sidestep the dilemma I have presented. Antirealist theories of value are able to offer an alternative account of the relation between evolutionary forces and evaluative facts — an account that allows us to reconcile our understanding of evaluative truth with our understanding of the many non-rational causes that have played a role in shaping our evaluative judgements.

2. THE TARGET OF THE ARGUMENT: REALIST THEORIES OF VALUE

The defining claim of realism about value, as I will be understanding it, is that there are at least some evaluative facts or truths that hold independently of all our evaluative attitudes.[1] *Evaluative facts or truths* I understand as facts or truths of the form that X is a normative reason to Y, that one should or ought to X, that X is good, valuable, or worthwhile, that X is morally right or wrong, and so on.[2] *Evaluative attitudes* I understand to include states such as desires, attitudes of approval and disapproval, unreflective evaluative tendencies such as the tendency to experience X as counting in favor of or demanding Y, and consciously or unconsciously held evaluative judgements, such as judgements about what is a reason for what, about what one should or ought to do, about what is good, valuable, or worthwhile, about what is morally right or wrong, and so on.

It is important to note that it is not enough to be a realist to claim that the truth of an evaluative judgement holds independently of one's making *that particular* evaluative judgement. Antirealists can agree with that much. Consider, for example, a constructivist view according to which the truth of "X is a reason for agent A to Y" is a function of whether that judgement would be among A's evaluative judgements in reflective equilibrium. This view is antirealist because it understands truths about what reasons a person has as depending on her evaluative attitudes (in particular, on what those attitudes would be in reflective equilibrium). Yet on this

REALIST THEORIES OF VALUE 111

view, it is quite possible for someone to have a reason independently of whether she thinks she does, for whether she has a reason is not a function of whether she (presently) judges she has it, but rather a function of whether that judgement would be among her evaluative judgements in reflective equilibrium. Antirealists can therefore agree with realists that the truth of a given evaluative judgement holds independently of whether one makes that particular judgement. Where antirealists part ways with realists is in denying that there are evaluative truths which hold independently of *the whole set* of evaluative judgements we make or might make upon reflection, or independently of *the whole set* of other evaluative attitudes we hold or might hold upon reflection.

The kind of independence from our evaluative attitudes that realists endorse is what Russ Shafer-Landau has called *stance-independence*.[3] To illustrate: Realists of course agree that the evaluative truth that "Hitler was morally depraved" depends in part on *Hitler's* evaluative attitudes in the sense that if Hitler had valued peace and universal human rights instead of dictatorial power and genocide, then it would have been false instead of true that he was morally depraved. But given that Hitler *did* value dictatorial power and genocide, value realists think that it is true, independent of all of our (and any of Hitler's other) evaluative attitudes, that Hitler was morally depraved. According to realists, the truth that Hitler was morally depraved holds independently of any stance that we (or Hitler) might take toward that truth, whether now or upon reflection.

There are different brands of realism about value. What unites them is the view that there are evaluative facts or truths that hold independently of all our evaluative attitudes (now keeping in mind the qualification about stance-independence). What separates different kinds of realists from one another is how they construe the nature of these facts or truths. According to what I will call *non-naturalist* versions of value realism, evaluative facts or truths are not reducible to any kind of natural fact, and are not the kinds of things that play a role in causal explanations; instead, they are irreducibly normative facts or truths.[4] This brand of realism has been gaining increasing numbers of adherents in recent years, and it lies squarely within the target of the Darwinian Dilemma.

In contrast to non-naturalist versions of value realism, the position I will call *value naturalism* holds that evaluative facts are identical with or constituted by (certain) natural facts, and that evaluative facts *are* the kinds of things that play a role in causal explanations.[5] According to such views, much as water is identical with H_2O, so evaluative properties are identical with certain natural properties, though we may or may not ever be able to provide a reduction telling exactly which natural properties evaluative properties are identical with (different naturalists taking different views on the possibility of such a reduction).[6] Whereas non-naturalist versions of value realism lie straightforwardly within my target in this paper, it is a more complicated matter whether versions of value naturalism lie within my target. Answering this question requires making a distinction (in section 7) between versions of value naturalism which count as genuinely realist on my understanding and versions which don't; my argument will be that the former, but not the latter, are vulnerable to the Darwinian Dilemma. Before introducing these complexities, however, it is important to get the fundamental dilemma for realism on the table.[7]

3.   A CAVEAT

In his 1990 book *Wise Choices, Apt Feelings*, Allan Gibbard notes that his arguments "should be read as having a conditional form: if the psychological facts are roughly as I speculate, here is what might be said philosophically."[8] I attach a similar caveat to my argument in this paper: if the evolutionary facts are roughly as I speculate, here is what might be said philosophically. I try to rest my arguments on the least controversial, most well-founded evolutionary speculations possible. But they are speculations nonetheless, and they, like some of Gibbard's theorizing in *Wise Choices, Apt Feelings*, fall within a difficult and relatively new subfield of evolutionary biology known as evolutionary psychology.[9] According to this subfield, human cognitive traits are (in some cases) just as susceptible to Darwinian explanation as human physical traits are (in some cases). For example, a cognitive trait such as the widespread human tendency to value the survival of one's offspring may, according to evolutionary psychology, be just as susceptible to evolutionary explanation as physical traits such as our bipedalism or our having opposable thumbs. There are many pitfalls that such evolutionary theorizing must avoid, the most important of which is the mistake of assuming that every observable trait (whether cognitive or physical) is an adaptation resulting from natural selection, as opposed to the result of any number of other complex (non-selective or only partially selective) processes that could have produced it.[10] It is more than I can do here to describe such pitfalls in depth or to defend at length the evolutionary claims that my argument will be based on. Instead, it must suffice to emphasize the hypothetical nature of my arguments, and to say that while I am skeptical of the *details* of the evolutionary picture I offer, I think its *outlines* are certain enough to make it well worth exploring the philosophical implications.[11]

4. FIRST PREMISE: THE INFLUENCE OF EVOLUTIONARY FORCES ON THE CONTENT OF OUR EVALUATIVE JUDGEMENTS

In its first approximation, the opening premise of the Darwinian Dilemma argument is this: the forces of natural selection have had a tremendous influence on the content of human evaluative judgements. This is by no means to deny that all kinds of other forces have also shaped the content of our evaluative judgements. No doubt there have been numerous other influences: some of them were perhaps evolutionary factors other than natural selection – for example, genetic drift;[12] and many other forces were not evolutionary at all, but rather social, cultural, historical, or of some other kind. And then there is the crucial and *sui generis* influence of rational reflection that must also be taken into account – a point I return to in the next section. I am discounting none of these other influences. My claim is simply that one enormous factor in shaping the content of human values has been the forces of natural selection, such that our system of evaluative judgements is thoroughly saturated with evolutionary influence. In this section, I make a brief case in support of this view, starting with a highly simplified and idealized evolutionary picture, then discussing two important complications, and ending with a more refined statement of the first premise.

To begin, note the potentially phenomenal costs and benefits, as measured in the Darwinian currency of reproductive success, of accepting some evaluative judgements rather than others. It is clear, for instance, how fatal to reproductive success it would be to judge that the fact that something would endanger one's survival is a reason to do it, or that the fact that someone is kin is a reason to harm that individual. A creature who accepted such evaluative judgements would run itself off cliffs, seek out its predators, and assail its offspring, resulting in the speedy elimination of it and its evaluative tendencies from the world.[13] In contrast, it is clear how beneficial (in terms of reproductive success) it would be to judge that the fact that something would promote one's survival is a reason in favor of it, or that the fact that something would assist one's offspring is a reason to do it. Different evaluative tendencies, then, can have extremely different effects on a creature's chances of survival and reproduction. In light of this, it is only reasonable to expect there to have been, over the course of our evolutionary history, relentless selective pressure on the content of our evaluative judgements, or rather (as I discuss below) "proto" versions thereof. In particular, we can expect there to have been overwhelming pressure in the direction of making those evaluative judgements which tended to promote reproductive success (such as the judgement that one's life is valuable), and against making those evaluative judgements which tended to decrease reproductive success (such as the judgement that one should attack one's offspring).

The hypothesis that this is indeed very roughly what happened is borne out by the patterns of evaluative judgement that we observe in human beings today. There is, of course, a seemingly unlimited diversity to the evaluative judgements that human beings affirm. Yet even as we note this diversity, we also see deep and striking patterns, across both time and cultures, in many of the most basic evaluative judgements that human beings tend to make. Consider, as a brief sampling, the following judgements about reasons:

(1) The fact that something would promote one's survival is a reason in favor of it.

(2) The fact that something would promote the interests of a family member is a reason to do it.

(3) We have greater obligations to help our own children than we do to help complete strangers.

(4) The fact that someone has treated one well is a reason to treat that person well in return.

(5) The fact that someone is altruistic is a reason to admire, praise, and reward him or her.

(6) The fact that someone has done one deliberate harm is a reason to shun that person or seek his or her punishment.

What explains the widespread human acceptance of such judgements? There are so many other possible judgements about reasons we could make – so why these? Why, for instance, do we view the death of our offspring as a horror, rather than as something to be sought after? Why do we think that altruism with no hope of personal reward is the highest form of virtue, rather than something to be loathed and eliminated? Evolutionary biology offers powerful answers to these questions, very roughly of the form that *these* sorts of judgements about reasons tended to promote survival and reproduction much more effectively than the alternative judgements. The details of how survival and reproduction were promoted will vary depending on the evaluative tendency in question. In the case of judgement (1), for instance, the rough explanation is obvious: creatures who possessed this general evaluative tendency tended to do more to promote their survival than those who, say, had a tendency to view the fact that something would promote their survival as counting *against* it, and so the former tended to survive and reproduce in greater numbers. The explanation of evaluative tendencies in the direction of judgements such as (2) and (3) will be somewhat more complicated, drawing on the evolutionary theory of kin selection.[14] The explanation in the case of evaluative tendencies in the direction of judgements (4), (5), and (6), meanwhile, will appeal to the biological theory of reciprocal altruism.[15]

For the sake of contrast, consider the following possible evaluative judgements:

(1′)   The fact that something would promote one's survival is a reason against it.

(2′)   The fact that something would promote the interests of a family member is a reason not to do it.

(3′)   We have greater obligations to help complete strangers than we do to help our own children.

(4′)   The fact that someone has treated one well is a reason to do that individual harm in return.

(5′)   The fact that someone is altruistic is a reason to dislike, condemn, and punish him or her.

(6′)   The fact that someone has done one deliberate harm is a reason to seek out that person's company and reward him or her.

If judgements like these – ones that would, other things being equal, so clearly decrease rather than increase the reproductive success of those who made them – predominated among our most deeply and widely held evaluative judgements across both time and cultures, then this would constitute powerful evidence that the content of our evaluative judgements had not been greatly influenced by Darwinian selective pressures. But these are not the evaluative

REALIST THEORIES OF VALUE 117

judgements we tend to see; instead, among our most deeply and widely held judgements, we observe many like those on the first list – many with exactly the sort of content one would expect if the content of our evaluative judgements had been heavily influenced by selective pressures. In this way, the observed patterns in the actual content of human evaluative judgements provide evidence in favor of the view that natural selection has had a tremendous influence on that content.

A further piece of evidence in favor of this view is the striking continuity that we observe between many of our own widely held evaluative judgements and the more basic evaluative tendencies of other animals, especially those most closely related to us. It does not seem much of a stretch, for example, to say that chimpanzees, in some primitive, non-linguistic sort of fashion, experience certain things in the world as *calling for* or *counting in favor of* certain reactions on their part. Moreover, the *content* of these evaluative experiences seems to overlap significantly with the content of many of our own evaluative tendencies. Like us, individual chimpanzees seem to experience – at some basic motivational level – actions that would promote their survival or help their offspring as in some way "called for." More strikingly, and again at some basic motivational level, chimpanzees seem to experience the fact that another chimpanzee has helped them, whether by sharing food, grooming them, or supporting their position within the group hierarchy, as "counting in favor of" assisting that other individual in similar ways.[16] While more work is needed to make such claims precise and subject them to thorough scientific testing, they have a strong basic plausibility, such that the conspicuous continuities between the basic evaluative tendencies of our close animal relatives and our own evaluative judgements lend further support to the view that evolutionary forces have played a large role in shaping the content of our evaluative judgements. We may view many of our evaluative judgements as conscious, reflective endorsements of more basic evaluative tendencies that we share with other animals.

Now note two important complications to this rough evolutionary sketch. First of all, the discussion so far might have suggested that things happened this way: *first* our ancestors began making evaluative judgements, and *then* tendencies to make some of these evaluative judgements rather than others were selected for. In other words, first came the capacity to make evaluative judgements, and then followed the selection of their content. But the actual course of evolution certainly did not take place in these two stages, much less in that order. Consider again the list of widely held evaluative judgements I mentioned earlier. Behavioral and motivational tendencies in the direction of at least some of the pairings of circumstance and response on this list presumably arose and became entrenched in our ancestors long before the rise of any capacity for full-fledged evaluative judgement – where I am understanding the capacity for *full-fledged evaluative judgement* to involve not only an unreflective capacity to experience one thing as "demanding" or "counting in favor of" another (a more primitive capacity that other animals such as chimpanzees might share with us), but also a reflective, linguistically-infused capacity to judge that one thing counts in favor of another, and to step back from such judgements and call them into question. Behavioral and motivational tendencies to do what would help one's offspring, for example – behavioral and motivational tendencies of gradually increasing degrees of consciousness and complexity – presumably vastly predated the sophisticated, linguistically-infused capacity to make the reflective judgement that "the fact that something would help one's offspring is a reason to do it." Thus, the capacity for full-fledged evaluative judgement was a relatively late evolutionary add-on, superimposed on top of much more basic behavioral and motivational tendencies.[17]

A second complication is this. In order for evolution by natural selection to take place with respect to a given trait, the trait in question must be genetically heritable. Yet it is implausible to think that the acceptance of a *full-fledged evaluative judgement* with a given content – for example, the acceptance of the judgement that "one ought to help those who help you" – is a genetically heritable trait. That is to say: when individuals in a given population vary with respect to whether or not they make this evaluative judgement (or any other), most if not all of that variation is likely *not* due to genetic differences, but other factors (such as culture or upbringing).[18] In contrast, however, it is plausible to suppose that over the course of much of our evolutionary history, what I have been calling "more basic evaluative tendencies" *were* genetically heritable traits, where a *basic evaluative tendency* may be understood very roughly as an unreflective, non-linguistic, motivational tendency to experience something as "called for" or "demanded" in itself, or to experience one thing as "calling for" or "counting in favor of" something else. We may think of these as "proto" forms of evaluative judgement. A relatively primitive version of such a tendency might be possessed by a bird who experiences some kind of motivational "pull" in the direction of feeding its offspring. A more sophisticated version might be possessed by a chimpanzee who has a motivational and perhaps emotional or proto-emotional experience of certain behaviors as "called for" by certain circumstances (for example, the experience of a threat to its offspring as "demanding" a protective response). It seems plausible to hypothesize that over the course of much of our evolutionary history – perhaps up until relatively recently[19] – when individuals in a given population varied with respect to whether they possessed a given basic evaluative tendency, a significant portion of that variation *was* due to genetic differences. So, for example, when individuals varied with respect to the presence or absence of an unreflective tendency to experience the fact that someone helped them as "counting in favor of" helping the other in return, a significant portion of that variation was attributable to genetic differences.

The upshot of these complications is this. The influence of Darwinian selective pressures on the content of human evaluative judgements is best understood as *indirect*. The most plausible picture is that natural selection has had a

120   SHARON STREET

tremendous *direct* influence on what I have called our "more
basic evaluative tendencies," and that these basic evaluative
tendencies, in their turn, have had a major influence on the
evaluative judgements we affirm. By this latter claim I do not
mean that we automatically or inevitably accept the full-
fledged evaluative judgements that line up in content with our
basic evaluative tendencies. Certainly not. For one thing,
other causal influences can shape our evaluative judgements
in ways that make them stray, perhaps quite far, from align-
ment with our more basic evaluative tendencies.[20] For another thing, we are reflective creatures, and as such are
capable of noticing any given evaluative tendency in ourselves, stepping back from it, and deciding on reflection to
disavow it and fight against it rather than to endorse the content suggested by it. My point here is instead the simple and
plausible one that had the general content of our basic evaluative tendencies been very different, then the general content
of our full-fledged evaluative judgements would also have
been very different, and in loosely corresponding ways.[21]
Imagine, for instance, that we had evolved more along the
lines of lions, so that males in relatively frequent circumstances had a strong unreflective evaluative tendency to experience the killing of offspring that were not his own as
"demanded by the circumstances," and so that females, in
turn, experienced no strong unreflective tendency to "hold it
against" a male when he killed her offspring in such circumstances, on the contrary becoming receptive to his advances
soon afterwards. Or imagine that we had evolved more along
the lines of our close primate relatives the bonobos, so that
we experienced sexual relations with all kinds of different
partners as "called for" in all kinds of different circumstances. Finally, imagine that we had evolved more on the
model of the social insects, perhaps possessing overwhelmingly strong unreflective evaluative tendencies in the direction
of devoting ourselves to the welfare of the entire community,
and only the weakest tendency to look out for our own individual survival, being unreflectively inclined to view that survival as "good" only insofar as it was of some use to the

larger community. Presumably in these and other such cases our system of full-fledged, reflective evaluative judgements would have looked very different as well, and in ways that loosely reflected the basic evaluative tendencies in question. My conclusion: the content of human evaluative judgements has been tremendously influenced − *indirectly* influenced, in the way I have indicated, but nevertheless tremendously influenced − by the forces of natural selection, such that our system of evaluative judgements is saturated with evolutionary influence. The truth of some account very roughly along these lines is all that is required for the Darwinian Dilemma to get off the ground.[22]

5.   FIRST HORN OF THE DILEMMA: DENYING A RELATION

The basic problem for realism is that it needs to take a position on what relation there is, if any, between the selective forces that have influenced the content of our evaluative judgements, on the one hand, and the independent evaluative truths that realism posits, on the other. Realists have two options: they may either assert or deny a relation.

Let us begin with the realist's option of claiming that there is *no* relation. The key point to see about this option is that if one takes it, then the forces of natural selection must be viewed as a purely distorting influence on our evaluative judgements, having pushed us in evaluative directions that have nothing whatsoever to do with the evaluative truth. On this view, allowing our evaluative judgements to be shaped by evolutionary influences is analogous to setting out for Bermuda and letting the course of your boat be determined by the wind and tides: just as the push of the wind and tides on your boat has nothing to do with where you want to go, so the historical push of natural selection on the content of our evaluative judgements has nothing to do with evaluative truth. Of course every now and then, the wind and tides might happen to deposit someone's boat on the shores of Bermuda. Similarly, every now and then, Darwinian pressures might have happened to push us toward accepting an evaluative judgement that accords with one of the realist's independent evaluative truths. But this would be purely a matter of chance, since by hypothesis there is no relation between the forces at work and the "destination" in question, namely evaluative truth.

If we take this point and combine it with the first premise that our evaluative judgements have been tremendously shaped by Darwinian influence, then we are left with the implausible skeptical conclusion that our evaluative judgements are in all likelihood mostly off track, for our system of evaluative judgements is revealed to be utterly saturated and contaminated with illegitimate influence. We should have been evolving towards affirming the independent evaluative truths posited by the realist, but instead it turns out that we have been evolving towards affirming whatever evaluative content tends to promote reproductive success. We have thus been guided by the wrong sort of influence from the very outset of our evaluative history, and so, more likely than not, most of our evaluative judgements have nothing to do with the truth. Of course it's *possible* that as a matter of sheer chance, some large portion of our evaluative judgements ended up true, due to a happy coincidence between the realist's independent evaluative truths and the evaluative directions in which natural selection tended to push us, but this would require a fluke of luck that's not only extremely unlikely, in view of the huge universe of logically possible evaluative judgements and truths, but also astoundingly convenient to the realist. Barring such a coincidence, the only conclusion remaining is that many or most of our evaluative judgements are off track. This is the far-fetched skeptical result that awaits any realist who takes the route of claiming that there is no relation between evolutionary influences on our evaluative judgements and independent evaluative truths.

But the realist may not be ready to abandon this route just yet. Let us grant (sticking with this horn of the dilemma) that the distorting influence of natural selection on the content of our evaluative judgements has been tremendous. One might nevertheless object that to draw a skeptical conclusion from this is unwarranted. For the argument so far ignores the power of a very different kind of influence on our system of evaluative judgements – a kind of influence that one might claim *is* related to the truth and that has also been tremendous – namely, the influence of rational reflection. After all, we are not unthinking beings who simply endorse whatever evaluative tendencies were implanted in us by evolutionary forces. Over the course of human history, endless amounts of reflection have gone on and greatly altered the shape of our evaluative judgements. According to the objection at hand, just as a compass and a little steering can correct for the influence of the wind and tides on the course of one's boat, so rational reflection can correct for the influence of selective pressures on our values.[23]

I accept one important point that this objection makes. Any full explanation of why human beings accept the evaluative judgements we do would need to make reference to the large influence of rational reflection. The view I am suggesting by no means involves thinking of us as automatons who simply endorse whatever evaluative tendencies are implanted in us by evolutionary and other forces. On the contrary, the view I am suggesting acknowledges the point that we are self-conscious and reflective creatures, and in a sense seeks to honor that point about us *better* than alternative views, by asking what reflective creatures like ourselves should conclude when we become conscious of what Kant would call this "bidding from the outside" affecting our judgements. (Here I have in mind Kant's statement in the third section of the *Foundations of the Metaphysics of Morals* that "we cannot conceive of a reason which consciously responds to a bidding from the outside with respect to its judgements."[24]) The very fact of our reflectiveness implies that something must happen – that something must change – when we become conscious of any foreign influence (such as these Darwinian forces) on our evaluative judgements. What that change should be is exactly what I am exploring in this paper.

124   SHARON STREET

Where I think the objection goes wrong, then, is as follows. The objection gains its plausibility by suggesting that
rational reflection provides some means of standing apart
from our evaluative judgements, sorting through them, and
gradually separating out the true ones from the false – as if
with the aid of some uncontaminated tool. But this picture
cannot be right. For what rational reflection about evaluative matters involves, inescapably, is assessing some evaluative judgements in terms of others. Rational reflection must
always proceed from some evaluative standpoint; it must
work from some evaluative premises; it must treat some
evaluative judgements as fixed, if only for the time being, as
the assessment of other evaluative judgements is undertaken.
In rational reflection, one does not stand completely apart
from one's starting fund of evaluative judgements: rather,
one *uses* them, reasons in terms of them, holds some of
them up for examination in light of others. The widespread
consensus that the method of reflective equilibrium, broadly
understood, is our sole means of proceeding in ethics is an
acknowledgment of this fact: ultimately, we can test our
evaluative judgements only by testing their consistency with
our other evaluative judgements, combined of course with
judgements about the (non-evaluative) facts. Thus, if the
fund of evaluative judgements with which human reflection
began was thoroughly contaminated with illegitimate influence – and the objector has offered no reason to doubt *this*
part of the argument – then the tools of rational reflection
were equally contaminated, for the latter are always just a
subset of the former. It follows that all our reflection over
the ages has really just been a process of assessing evaluative judgements that are mostly off the mark in terms of
others that are mostly off the mark. And reflection of *this*
kind isn't going to get one any closer to evaluative truth,
any more than sorting through contaminated materials with
contaminated tools is going to get one closer to purity. So
long as we assume that there is no relation between evolutionary influences and evaluative truth, the appeal to
rational reflection offers no escape from the conclusion that,

in the absence of an incredible coincidence, most of our evaluative judgements are likely to be false.[25]

6. SECOND HORN OF THE DILEMMA: ASSERTING A RELATION

So let us now turn to the realist's other option, which is to claim that there *is* indeed some relation between the workings of natural selection and the independent evaluative truths that he or she posits. I think this is the more plausible route for the realist to take. After all, we think that a lot of our evaluative judgements are true. We also think that the content of many of these same evaluative judgements has been influenced by natural selection. This degree of overlap between the content of evaluative truth and the content of the judgements that natural selection pushed us in the direction of making begs for an explanation. Since it is implausible to think that this overlap is a matter of sheer chance – in other words, that natural selection just happened to push us toward true evaluative judgements rather than false ones – the only conclusion left is that there is indeed some relation between evaluative truths and selective pressures. The critical question is what *kind* of relation. Different metaethical views will give different answers, and we may judge them according to those answers.

The realist has a possible account of the relation that might seem attractive on its face. It is actually quite clear, the realist might say, how we should understand the relation between selective pressures and independent evaluative truths. The answer is this: we may understand these evolutionary causes as having *tracked* the truth; we may understand the relation in question to be a *tracking* relation.[26] The realist might elaborate on this as follows. Surely, he or she might say, it is advantageous to recognize evaluative truths; surely it promotes one's survival (and that of one's offspring) to be able to grasp what one has reason to do, believe, and feel. As Derek Parfit has put the point:[27] it is possible that "just as

cheetahs were selected for their speed, and giraffes for their
long necks, the particular feature for which we were selected
was our ability to respond to reasons and to rational requirements."[28] According to this hypothesis, our ability to recognize evaluative truths, like the cheetah's speed and the
giraffe's long neck, conferred upon us certain advantages that
helped us to flourish and reproduce. Thus, the forces of natural selection that influenced the shape of so many of our evaluative judgements need not and should not be viewed as
distorting or illegitimate at all. For the evaluative judgements
that it proved most selectively advantageous to make are, in
general, precisely those evaluative judgements which are true.

Call this proposal by the realist the *tracking account*. The
first thing to notice about this account is that it puts itself
forward as a scientific explanation.[29] It offers a specific
hypothesis as to how the course of natural selection pro-
ceeded and what explains the widespread presence of some
evaluative judgements rather than others in the human population. In particular, it says that the presence of these judgements is explained by the fact that these judgements are true,
and that the capacity to discern such truths proved advantageous for the purposes of survival and reproduction. So, for
instance, if it is asked why we observe widespread tendencies
to take our own survival and that of our offspring to be valuable, or why we tend to judge that we have special obligations to our children, the tracking account answers that these
judgements are true, and that it promoted reproductive success to be able to grasp such truths.

In putting itself forward as a scientific explanation, the
tracking account renders itself subject to all the usual standards of scientific evaluation, putting itself in direct competition with all other scientific hypotheses as to why human
beings tend to make some evaluative judgements rather than
others. The problem for realism is that the tracking account
fares quite poorly in this competition. Even fairly brief consideration suggests that another evolutionary explanation of
why we tend to make some evaluative judgements rather than
others is available, and that this alternative explanation, or

something roughly like it, is distinctly superior to the tracking account.

According to what I will call the *adaptive link account*, tendencies to make certain kinds of evaluative judgements rather than others contributed to our ancestors' reproductive success not because they constituted perceptions of independent evaluative truths, but rather because they forged adaptive links between our ancestors' circumstances and their responses to those circumstances, getting them to act, feel, and believe in ways that turned out to be reproductively advantageous.[30] To elaborate: As a result of natural selection, there are in living organisms all kinds of mechanisms that serve to link an organism's circumstances with its responses in ways that tend to promote survival and reproduction. A straightforward example of such a mechanism is the automatic reflex response that causes one's hand to withdraw from a hot surface, or the mechanism that causes a Venus's-flytrap to snap shut on an insect. Such mechanisms serve to link certain kinds of circumstances — the presence of a hot surface or the visit of an insect — with adaptive responses — the immediate withdrawal of one's hand or the closing of the flytrap. Judgements about reasons — and the more primitive, "proto" forms of valuing that we observe in many other animals — may be viewed, from the external standpoint of evolutionary biology, as another such mechanism. They are analogous to the reflex mechanism or the flytrap's apparatus in the sense that they also serve to link a given circumstance with a given response in a way that may tend to promote survival and reproduction. Consider, for example, the evaluative judgement that the fact that someone has helped one is a reason to help that individual in return. Just as we may see a reflex mechanism as effecting a pairing between the circumstance of a hot surface and the response of withdrawing one's hand, so we may view this evaluative judgement as effecting a pairing between the circumstance of one's being helped and the response of helping in return. Both of these pairings of circumstance and response, at least if the evolutionary theory of reciprocal altruism is correct about the latter case, are ones that tended

to promote the reproductive success of ancestors who possessed them.[31]

Now of course there are radical differences between the
mechanism of a reflex response and the "mechanism" of an
evaluative judgement. The former is a brute, hard-wired physical mechanism, while the latter is a conscious mental state,
subject to reflection and possible revision in light of that
reflection. But this does not change the fact that there is a
deep analogy between their functional roles. From an evolutionary point of view, each may be seen as having the same
practical point: *to get the organism to respond* to its circumstances in a way that is adaptive.[32] Something like a reflex
mechanism does this through a particular hard-wiring of the
nervous system, while an evaluative judgement – or a more
primitive evaluative experience such as some other animals
are likely to have – does this by having the organism experience a particular response as *called for*, or as *demanded by*,
the circumstance in question. In the latter case, the link between circumstance and response is forged by our taking of
the one thing to be a *reason* counting in favor of the other –
that is, by the experience of normativity or value.[33]

For illustration of the differences between the adaptive link
account and the tracking account, consider a few examples.
Consider, for instance, the judgement that the fact that something would promote one's survival is a reason to do it, the
judgement that the fact that someone is kin is a reason to accord him or her special treatment, and the judgement that the
fact that someone has harmed one is a reason to shun that
person or retaliate. Both the adaptive link account and the
tracking account explain the widespread human tendencies to
make such judgements by saying that making them somehow
contributed to reproductive success in the environment of our
ancestors. According to the tracking account, however, making such evaluative judgements contributed to reproductive
success because they are *true*, and it proved advantageous to
grasp evaluative truths. According to the adaptive link
account, on the other hand, making such judgements contributed to reproductive success not because they were true or

false, but rather because they got our ancestors to respond to their circumstances with behavior that itself promoted reproductive success in fairly obvious ways: as a general matter, it clearly tends to promote reproductive success to do what would promote one's survival, or to accord one's kin special treatment, or to shun those who would harm one.

We now have rough sketches of two competing evolutionary accounts of why we tend to make some evaluative judgements rather than others. For reasons that may already have begun to suggest themselves, I believe that the adaptive link account wins this competition hands down, as judged by all the usual criteria of scientific adequacy. In particular, there are at least three respects in which the adaptive link account is superior to the tracking account: it is more parsimonious; it is much clearer; and it sheds much more light on the explanandum in question, namely that human beings tend to make some evaluative judgements rather than others.

Let me start with the parsimony point. The tracking account obviously posits something extra that the adaptive link account does not, namely independent evaluative truths (since it is precisely these truths that the tracking account invokes to explain why making certain evaluative judgements rather than others conferred advantages in the struggle to survive and reproduce). The adaptive link account, in contrast, makes no reference whatsoever to evaluative truth; rather, it explains the advantage of making certain evaluative judgements directly, by pointing out how they got creatures who made them to act in ways that tended to promote reproductive success. Thus, the adaptive link account explains the widespread presence of certain values in the human population more parsimoniously, without any need to posit a role for evaluative truth.[34]

Second, the adaptive link account is much clearer than the tracking account, which turns out to be rather obscure upon closer examination. As we have seen, according to the tracking account, making certain evaluative judgements rather than others promoted reproductive success *because these judgements were true*. But let's now look at this. How exactly is this supposed to work? Exactly why would it promote an organism's reproductive success to grasp the independent evaluative truths posited by the realist? The realist owes us an answer here. It is not enough to say, "Because they are true." We need to know more about *why* it is advantageous to apprehend such truths before we have been given an adequate explanation.

What makes this point somewhat tricky is that on the face of it, it might seem that *of course* it promotes reproductive success to grasp any kind of truth over any kind of falsehood. Surely, one might think, an organism who is aware of the truth in a given area, whether evaluative or otherwise, will do better than one who isn't. But this line of thought falls apart upon closer examination. First consider truths about a creature's manifest surroundings – for example, that there is a fire raging in front of it, or a predator rushing toward it. It is perfectly clear why it tends to promote reproductive success for a creature to grasp such truths: the fire might burn it to a crisp; the predator might eat it up.[35] But there are many other kinds of truths such that it will confer either no advantage or even a disadvantage for a given kind of creature to be able to grasp them. Take, for instance, truths about the presence or absence of electromagnetic wavelengths of the lowest frequencies. For most organisms, such truths are irrelevant to the undertakings of survival and reproduction; hence having an ability to grasp them would confer no benefit. And then one must also take into account the significant costs associated with developing and maintaining such a sophisticated ability. Since for most organisms, this would be energy and resources spent for no gain in terms of reproductive success, the possession of such an ability would actually be positively *disadvantageous*.

With this in mind, let us look again at the evaluative truths posited by realists. Take first the irreducibly normative truths posited by non-naturalist realists such as Nagel, Dworkin, Scanlon, or Shafer-Landau. A creature obviously can't run into such truths or fall over them or be eaten by them. In what way then would it have promoted the reproductive success of our ancestors to grasp them? The realist owes us an answer here, otherwise his or her alleged explanation of why it promotes reproductive success to make certain judgements in terms of the *truth* of those judgements is no explanation at all. To say that these truths could kill you or maim you, like a predator or fire, would be one kind of answer, since it makes it clear how recognizing them could be advantageous. But such an answer is clearly not available in the case of the independent irreducibly normative truths posited by the non-naturalist realists. In the absence of further clarification, then, the non-naturalist's version of the tracking account is not only less parsimonious but also quite obscure.

Value naturalists would appear to have better prospects on this point than non-naturalist realists. Since value naturalists construe evaluative facts as natural facts with causal powers, it is much more comprehensible how grasping such facts could have had an impact on reproductive success. I return to this issue in the following section. For the time being, note the following. The naturalist's proposed version of the tracking account, so far, is this: making some evaluative judgements rather than others tended to promote reproductive success because those judgements constituted perceptions of evaluative facts, which just are a certain kind of natural fact. At least so far, this isn't much of an explanation either. What kinds of natural facts are we talking about, and exactly why did it promote reproductive success to grasp them? The naturalist can certainly try to develop answers to these questions, but at least on the face of things, the prospects appear dim. Take the widespread judgement that one should care for one's offspring, for example. Exactly what natural fact or facts does the evaluative fact that one should care for one's offspring reduce to, or irreducibly supervene upon, and why would perceiving the natural fact or facts in question have promoted our ancestors' reproductive success? It seems unattractive to get into such complexities when one can just say, as the adaptive link account does, that ancestors who judged that they should care for their offspring met with greater reproductive success simply because *they tended to care for their offspring* — and so left more of them.

I've argued that the adaptive link account is both more parsimonious and clearer than the tracking account. My third and final point is that the adaptive link account does a much better job at actually illuminating the phenomenon that is to be explained, namely why there are widespread tendencies among human beings to make some evaluative judgements rather than others. To return to our original questions, why do we tend to judge that our survival is valuable, rather than worthless? Why do we tend to judge that we have special obligations to care for our children, rather than strangers or distant relatives? Why do we tend to view the killing of other human beings as a much more serious matter than the killing of plants or other animals? The adaptive link account has very good answers to such questions, of the general form that ancestors who made evaluative judgements of these kinds, and who as a result tended to respond to their circumstances in the ways demanded by these judgements, did better in terms of reproductive success than their counterparts. It is quite clear why creatures who judged their survival to be valuable would do much better than those who did not, and so on. Now compare the tracking account's explanation. It tries to answer these same questions by saying that these judgements are *true*: that survival *is* valuable, that we *do* have special obligations to care for our children, that the killing of human beings *is* more serious than the killing of plants or other animals. Such answers do not shed much light. In particular, the tracking account fails to answer three questions.

First, how does the tracking account explain the remarkable coincidence that so many of the truths it posits turn out to be exactly the same judgements that forge adaptive links between circumstance and response — the very same judgements we would expect to see if our judgements had been selected on those grounds alone, regardless of their truth? The tracking account has no answer to this question that does not run right back into the parsimony and clarity problems just discussed.

REALIST THEORIES OF VALUE 133

Second, what does the tracking account have to say about our observed predispositions to make other evaluative judgements which (we may decide on reflection) are *not* true? For instance, we observe in human beings a deep tendency to think that the fact that someone is in an "out-group" of some kind is a reason to accord him or her lesser treatment than those in the "in-group." The adaptive link account offers a promising explanation of this, namely that having this evaluative tendency tended to promote reproductive success because those who possessed it tended to shower their assistance on those with a higher degree of genetic relatedness, or on those most able or likely to reciprocate. The tracking account's preferred explanation, however, falls flat, since in this case it is not plausible to answer that this evaluative predisposition developed because it is *true* that the fact that someone is in an "out-group" is a reason to accord him or her lesser treatment than those in the "in-group." More and more, many of us are coming to think that this is not true. The tracking account is thus left with nothing in the way of an explanation as to why we observe such deep tendencies to make the contrary judgement.[36]

Finally, consider the question of all those normative judgements that human beings *could* make but don't. As I have noted, the universe of logically possible evaluative judgements is huge, and we must think of all the possible evaluative judgements that we *don't* see – from the judgement that infanticide is laudable, to the judgement that plants are more valuable than human beings, to the judgement that the fact that something is purple is a reason to scream at it. Here again the adaptive link account has something potentially informative to point out, namely, that such judgements – or evaluative tendencies in these general sorts of directions – forge links between circumstance and response that would have been useless or quite maladaptive as judged in terms of reproductive success. The tracking account has nothing comparably informative to say. It can just stand by and insist that such judgements are false – reaffirming our convictions but adding nothing to our understanding of why we have them.

134   SHARON STREET

To sum up, the set of evaluative judgements that human beings tend to affirm appears to be a disparate mishmash, ranging across all kinds of unrelated spheres and reflecting all kinds of unrelated values – some self-interested, others family-related, still others concerning how we should treat non-relatives and other forms of life, and so on. The power of the adaptive link account is that it exposes much of this seeming unrelatedness as an illusion; it illuminates a striking, previously hidden unity behind many of our most basic evaluative judgements, namely that they forge links between circumstance and response that would have been likely to promote reproductive success in the environments of our ancestors. The tracking account has no comparable explanatory power. Its appeal to the truth and falsity of the judgements in question sheds no light on why we observe the specific *content* that we do in human evaluative judgements; in the end, it merely reiterates the point that we *do* believe or disbelieve these things. When we couple this final point with the points about the parsimony and clarity of the adaptive link account as compared to the tracking account, it is clear which explanation we should prefer. The tracking account is untenable.

One last point remains in order to close off the Darwinian Dilemma. The tracking account was the most obvious and natural account for the realist to give of the relation between selective pressures on our evaluative judgements and the independent evaluative truths that he or she posits. In the wake of the tracking account's failure, one might think that the realist still has the option of developing some alternative account of this relation. But this is not so. Rather, insofar as realism asserts any relation at all between selective pressures on our evaluative judgements and evaluative truths, the position is forced to give a tracking account of this relation. The reason for this stems from the very nature of realism itself. The essence of the realist position is its claim that there are evaluative truths that hold independently of all of our evaluative attitudes. But because it views these evaluative truths as ultimately independent of our evaluative attitudes, the only way for realism *both* to accept that those attitudes have been deeply influenced by evolutionary causes *and* to avoid seeing these causes as distorting is for it to claim that these causes actually in some way *tracked* the alleged independent truths. There is no other way to go. To abandon the tracking account — in other words, to abandon the view that selective pressures pushed us *toward* the acceptance of the independent evaluative truths — is just to adopt the view that selective pressures either pushed us *away from* or pushed us in ways that *bear no relation to* these evaluative truths. And to take *this* view is just to land oneself back in the first horn of the dilemma, in which one claims that there is no relation between selective pressures on our evaluative judgements and the posited independent truths. Realism about value, then, has no escape: it is forced to accept either the tracking account of the relation or else the view that there is no relation at all, and both of these options are unacceptable.[37]

7. FIRST OBJECTION: AN OBJECTION BY THE VALUE NATURALIST

At this point, an important objection remains open to the value naturalist, whose position I touched on only quickly in the argument of the previous section.[38] As we have seen, according to the value naturalist, evaluative facts are identical with (certain) natural facts. As also mentioned earlier, some value naturalists take the position that we may never be able to provide a reduction specifying exactly *which* natural facts evaluative facts are identical with, but let us set this point aside for the moment and assume for the sake of argument that it is agreed upon by all that evaluative facts are identical with such-and-such ordinary natural facts. Since these ordinary natural facts are in the same general category as facts about fires, predators, cliffs, and so on, presumably there is going to be a plausible evolutionary account available as to why we were selected to be able to track them, just as I myself have supposed there is a plausible evolutionary account available as to why we were selected to be able to track facts about fires, predators, cliffs, and so on.[39] There may even be a good evolutionary account of why these natural facts are ones that we take a particularly strong interest in. It thus might seem that we have the outlines of a perfectly good answer to my question regarding the relation between evolutionary pressures on our evaluative judgements and independent evaluative truths. In particular, the relation is this: in ways roughly analogous to the ways in which we were selected to be able to track, with our non-evaluative judgements, facts about such things as fires, predators, and cliffs, so we were also selected to be able to track, with our evaluative judgements, *evaluative* facts, which are just identical with such-and-such natural facts.

This response, I will argue, ultimately just puts off a level the difficulties raised for realism by the Darwinian Dilemma. But first I need to distinguish between versions of value naturalism which count as genuinely realist in my taxonomy and those which don't. My taxonomy, while of course not the only legitimate understanding of realism, is far from *ad hoc*.[40] Rather, it zeroes in on the important question: does the view in question understand evaluative truths as holding, in a fully robust way, independently of all our evaluative attitudes?

Suppose the value naturalist takes the following view. Given that we have the evaluative attitudes we do, evaluative facts are identical with natural facts N. But if we had possessed a completely different set of evaluative attitudes, the evaluative facts would have been identical with the very different natural facts M. Such a view does not count as genuinely realist in my taxonomy, for such a view makes it dependent on our evaluative attitudes *which* natural facts evaluative facts are identical with. On such a view, there is an important sense in which we need only alter our evaluative attitudes in order to change the evaluative facts, for by altering our evaluative attitudes we change which natural facts the evaluative facts are identical with. Views of this kind count as antirealist in my taxonomy, and as such are not a target of my argument; instead they escape the Darwinian Dilemma in the way I discuss in section 10.

Peter Railton's account of individual non-moral good is an example of such a view.[41] According to Railton's proposal, roughly understood, an individual's non-moral good is identical to what that person would desire to desire under conditions of full information. Suppose, then, that what Ann would desire to desire under conditions of full information is (in part) her own longevity. In that case, her individual non-moral good is identical (in part) to her own longevity. But now suppose that Ann undergoes a significant change in her evaluative attitudes, such that it is no longer true of her that under conditions of full information she would desire to desire her own longevity. In that case, her individual non-moral good is no longer identical to her longevity, but is instead identical to something else (whatever it is that she'd now desire to desire under conditions of full information). There is an important sense in which Ann need only alter her evaluative attitudes in order to change the evaluative facts, for by altering her evaluative attitudes she changes which natural facts the evaluative facts are identical with. Railton's proposal therefore counts as antirealist in my taxonomy.[42]

In order to count as genuinely realist, then, a version of value naturalism must take the view that *which* natural facts evaluative facts are identical with is independent of our evaluative attitudes. For ease of expression, let us put the point this way: in order to count as realist, a version of value naturalism must take the view that facts about *natural-normative identities* (in other words, facts about exactly *which* natural facts evaluative facts are identical with) are independent of our evaluative attitudes. On the kind of view I have in mind, evaluative facts are identical with natural facts N, and even if our evaluative attitudes had been entirely different, perhaps not tracking those evaluative/natural facts N at all, but instead tracking some very different natural facts M, the evaluative facts *still* would have been identical with natural facts N, and *not* natural facts M. On this sort of view, for example, Ann's individual non-moral good might be identical (in part) with her longevity, and even if Ann's evaluative attitudes were entirely different — such that she possessed no

138   SHARON STREET

concern whatsoever for her longevity, and would fail to be concerned with it even if fully informed, and so on – her individual good would nevertheless *still* be identical (in part) with her longevity.

There is one artful way of achieving this sort of view that I also wish to rule out as genuinely realist, and that is by means of the move of "rigidifying."[43] Consider, for instance, a view which says that *which* natural facts evaluative facts are identical with is fixed in some way by our *actual* evaluative attitudes (in other words, by *our* attitudes, here and now). And suppose that our actual attitudes determine it that the evaluative facts are identical with natural facts N. On such a view, even if we had had entirely different evaluative attitudes, it *still* would have been the case that the evaluative facts are identical with natural facts N, since those are the ones picked out by our *actual* evaluative attitudes. Such a view is not genuinely realist in my taxonomy, however, for on such a view, there is no robust sense in which other creatures (including other possible versions of ourselves) would be making a *mistake* or *missing anything* if their evaluative attitudes tracked natural facts M, say, instead of natural facts N. For those other creatures could also pull the rigidifying move. And the upshot is that when we say "The good is identical to N" and they say "The good is identical to M," we will not be *disagreeing* with each other, with one of us correct and the other incorrect about which natural facts the good is identical to, but rather simply talking past each other, with the reference of our word "good" fixed by *our* actual evaluative attitudes, and the reference of their word "good" fixed by *their* actual evaluative attitudes.[44] We can of course go on using the word "good" in our sense, according to which we're right to think that the good is identical to N, and they can go on using the word "good" in their sense, according to which they're right to think that the good is identical to M, but there is, on such a view, no standard independent of all of our and their evaluative attitudes determining whose sense of the word "good" is right or better; neither of us can properly accuse the other of having made a mistake.[45] For this reason, views that achieve "independence from our attitudes" by way of the rigidifying move do not count as genuinely realist in my taxonomy.

In sum, what I will call *genuinely realist versions of value naturalism* hold that *which* natural facts evaluative facts are identical with is independent of all our evaluative attitudes, and they do not achieve this result by means of the rigidifying move. When it comes to the case just sketched, a genuinely realist version of value naturalism will hold that even if the two communities' uses of the word "good" track different natural properties, the communities are nevertheless (at least potentially) using the word "good" in the same sense — genuinely disagreeing with one another about the correct natural-normative identity — and that there is a fact of the matter about which (if either) of us is right that obtains independently of all of our and their evaluative attitudes.[46]

Genuinely realist versions of value naturalism are vulnerable to the Darwinian Dilemma. To see this, the first thing to note is the following. How, according to these views, do we figure out the correct natural-normative identities? We may assume that the answer is the one given by value naturalists such as Nicholas Sturgeon and David Brink. Sturgeon writes that if a full account of which natural facts evaluative facts are identical with is to be had, then this account "will have to be derived from our best moral theory, together with our best theory of the rest of the world."[47] And Brink agrees that "determination of just which natural facts and properties constitute which moral facts and properties is a matter of substantive moral theory."[48] These theorists do not propose some completely new approach to substantive moral theory; on the contrary, they think we should proceed in roughly the way we currently do proceed — starting with our existing fund of evaluative judgements, giving more weight to those evaluative judgements which strike us as correct if anything is (for instance, the judgement that Hitler was morally depraved[49]), and then working to bring our evaluative judgements into the greatest possible

coherence with each other and with our best scientific picture of the rest of the world.

It's at this point that the Darwinian Dilemma kicks in. The
genuinely realist value naturalist posits that there are *independent facts about natural-normative identities*. But the value
naturalist also holds that in trying to figure out what those
identities are, we will have to rely very heavily on our existing
evaluative judgements. Yet, as we have seen, those evaluative
judgements have been tremendously influenced by Darwinian
selective pressures. And so the question arises: what is the
relation between evolutionary influences on our evaluative
judgements, on the one hand, and the independent truths
about natural-normative identities posited by the realist, on
the other? In trying to figure out which natural facts evaluative facts are identical with, we have no option but to rely on
our existing fund of evaluative judgements: I judge that Hitler
was morally depraved, for instance, and in doing so steer toward the view that the evaluative fact of someone's being
morally depraved is roughly identical to her having a psychology of such-and-such a character (naturalistically
described) – a psychology that is like Hitler's in certain relevant ways (exactly *which* ways is to be determined by relying
on further evaluative judgements of mine). But in relying on
these and other evaluative judgements, I rely on judgements
that are saturated with evolutionary influence. What then is
the relation between that influence and the independent truths
I'm seeking to uncover – these independent truths about natural-normative identities?

As before, the realist has two options: he or she may either
assert or deny a relation. Suppose that the realist denies that
there is any relation. As before, a highly skeptical result follows. If there is no relation whatsoever between evolutionary
influences on our evaluative judgements and independent
truths about natural-normative identities, then all our hypothesizing about natural-normative identities is hopelessly contaminated with illegitimate influence. Due to the distorting pressure
of Darwinian forces, we are, for all we know, tracking natural
facts M with our evaluative judgements, whereas we ought to

be tracking (say) the entirely different set of natural facts N, the ones which are *really* identical with evaluative facts.

Suppose, on the other hand, the realist value naturalist claims that there *is* a relation between evolutionary pressures on our evaluative judgements and independent truths about natural-normative identities. Here, as before, the realist's only option for spelling out this relation is some version of a tracking account, according to which we were somehow selected to be able to track with our evaluative judgements independent facts about natural-normative identities. But if the tracking account failed as a scientific explanation when it came to arguing that we were selected to track independent evaluative truths, then it will fail even more seriously when it comes to arguing that we were selected to track independent facts about natural-normative identities. For it is even more obscure how tracking something as esoteric as independent facts about natural-normative identities could ever have promoted reproductive success in the environment of our ancestors. The adaptive link account again wins out: the best explanation of why human beings tend to make some evaluative judgements rather than others is not that these judgements constituted an awareness (however imperfect) of independent facts about natural-normative identities, but rather that the relevant evaluative tendencies forged links between our ancestors' circumstances and their responses which tended to promote reproductive success.

I conclude that any genuinely realist version of value naturalism runs headlong into the same basic dilemma I've been sketching. To the extent that a view insists on there being evaluative facts which hold independently of all our evaluative attitudes, it is impossible to reconcile that view with a recognition of the role that Darwinian forces have played in shaping the content of our values. Once we become fully conscious of this powerful "bidding from the outside" with respect to our evaluative judgements, I suggest, our response must be to adjust our metaethical view so as to become antirealists.

142   SHARON STREET

8.   SECOND OBJECTION: THE BYPRODUCT HYPOTHESIS

While the first objection belongs to value naturalists, a second objection might be voiced by realists generally. Confronted with the Darwinian Dilemma, the realist may suggest the following alternative evolutionary hypothesis. Perhaps the human ability to grasp independent evaluative truths was not itself selected for, but is instead the byproduct or outgrowth of some other capacity which *does* have an explanation in terms of natural selection (or else some other, non-selective evolutionary explanation). Many human capacities, after all, are like this: our ability to do astrophysics, for instance, was surely not itself directly selected for, but is instead the byproduct or outgrowth of other capacities which likely do have an explanation in terms of natural selection. Perhaps in some similar fashion our ability to grasp independent evaluative truths has emerged as a byproduct or outgrowth of some other capacity – call it *capacity C*.

This objection has not been properly developed until the realist has explained exactly what capacity C is, how it evolved, and what relation it bears to the capacity to grasp independent evaluative truths. However these details might be filled out, though, the Darwinian Dilemma arises again for such a proposal – this time with regard to capacity C. In particular, the question for the realist becomes this: what relation, if any, does the realist claim obtained between the evolution of capacity C and the independent evaluative truths that he or she posits?

Suppose the realist answers "no relation." Suppose, in other words, that the realist claims that the capacity to grasp independent evaluative truths arose as a complete fluke, as the wholly incidental byproduct of some other capacity C that was selected for on the basis of factors that had nothing whatsoever to do with the task of grasping evaluative truths. If the realist takes this route, then the coincidence point is triggered again: how incredible (not to mention how extraordinarily convenient for the realist) that, as a matter of sheer coincidence, a capacity happened to arise (as the entirely incidental byproduct of some totally unrelated capacity C) which operates to grasp precisely the sort of independent truths postulated by the realist.

To this charge of an implausible coincidence, the realist might protest that this sort of thing happens all the time in evolution – in other words, that one trait arises as the completely incidental byproduct of selection for some other trait. While it is quite true that this happens, the suggestion that this is what happened in the case of our ability to grasp independent evaluative truths is very implausible given the nature of the trait in question. The task of grasping independent evaluative truths presumably requires a highly specialized, sophisticated capacity, one specifically attuned to the evaluative truths in question. The capacity at issue is not a simple, brute sort of feature – not, presumably, if we have any reasonable chance of grasping the truths posited by the realist. But the more complicated and uniquely specialized a faculty is, the less plausible it is to hypothesize that it could have arisen as a sheer fluke, as the purely incidental byproduct of some unrelated capacity that was selected for on other grounds entirely. It is completely implausible, for instance, to suggest that the human eye in its present developed form emerged as the purely incidental byproduct of selection for some other, unrelated capacity.[50] I suggest that it is no more plausible to claim that the sophisticated ability to grasp independent evaluative truths emerged as such a byproduct.

The realist's other option is to maintain that there *is* some relation between the evolution of capacity C and the independent evaluative truths that he or she posits. According to this proposal, it is no fluke that the ability to grasp evaluative truths emerged as a byproduct of capacity C, because there is some relation between capacity C and the capacity to grasp evaluative truths. But now the challenge for the realist is to explain what this relation is. And it's hard to see how the realist can say anything except that capacity C, whatever it may be, involves at least some *basic* sort of ability to grasp independent evaluative truths, of which our present-day ability to grasp evaluative truths is a refined extension, in much

144   SHARON STREET

the same way that our present-day ability to do astrophysics is presumably a refined extension of more basic abilities to discover and model the physical features of the world around us.[51] But at this point the realist has to give some account of how this more basic sort of ability to grasp independent evaluative truths arose. And given what has to be the complexity and specialization of even this more basic ability (a point of comparison is the complexity and specialization of the more basic abilities on which the ability to do astrophysics is based), it is implausible to suggest that the emergence of this more basic ability was a mere fluke. The only alternative to saying that the emergence of this ability was a fluke is to claim that we were in some way selected to track the independent evaluative truths posited by the realist, yet this proposal, for the reasons I've already given, is scientifically unacceptable. The byproduct hypothesis, while it pushes matters off a step by hypothesizing an intervening capacity or set of capacities, does not permit escape from the Darwinian Dilemma for the realist about value.

9.   THIRD OBJECTION: THE BADNESS OF PAIN AS AN ALLEGED INDEPENDENT TRUTH ABOUT VALUE

The case of physical pain – for instance, in the various forms associated with burns, cuts, bruises, broken bones, nausea, and headaches – serves as one of the strongest temptations toward realism about value. Realists frequently appeal to the case of pain when defending their views,[52] and when presented with the Darwinian Dilemma, another such appeal may seem attractive. One possibility is for the realist to argue along the following lines. There are obvious evolutionary explanations of why we tend to feel physical pain when we do: roughly, we tend to feel it in conjunction with bodily conditions or events that diminish reproductive success, such as a cut to the skin or a blow to the head. Pain itself, moreover, due to its very nature, is bad independently of whatever evaluative attitudes we might hold. Together these points provide at least a rough answer to the question of what the relation is between evolutionary pressures and independent evaluative truths: in short, evolutionary pressures led us to feel pain under such-and-such kinds of circumstances, and that experience is, of its very nature, bad independently of all our evaluative attitudes, its badness therefore demanding a realist construal. Taking a slightly different tack, the realist might also argue: It is presumably no mystery from an evolutionary point of view why we're able to "track" pain, and pain itself is an evaluative fact, bad independently of all our evaluative attitudes. So it's fairly clear how we were selected to track (at least these) independent evaluative facts.

While such ideas have some intuitive appeal, evolutionary considerations – including one more application of the Darwinian Dilemma – help us to see how the badness of pain does in fact depend on our evaluative attitudes, revealing a realist understanding of its badness to be mistaken. For the purposes of the ensuing discussion, let us focus on the following evaluative claim: someone's pain counts as a reason for *that person* to do what would avoid, lessen, or stop it.[53] It is no doubt plausible to think that this evaluative truth holds independently of all our evaluative attitudes, just as a realist about value claims. Of course many realists would maintain in addition that someone's pain also counts as a reason for *other people* to do what would avoid, lessen, or stop it, and that this evaluative truth too holds independently of all our evaluative attitudes. I take it, however, that realism about the badness of pain *for the person whose pain it is* a more formidable target than realism about the badness of pain *for people whose pain it isn't*, and so I'll focus on the former sort of realism. If I can raise questions about realism's viability in this toughest, most basic case, then questions about the viability of the latter sort of realism will follow *a fortiori*. In the remainder of this discussion, then, when I talk about the "badness of pain" or say that "pain is bad," I'm using such expressions as a shorthand way of talking about the badness of pain *for the person whose pain it is*.

146   SHARON STREET

To start out, we need to clarify what is meant by "pain." Consider, for the sake of argument, the following proposal:

Pain is a sensation such that the creature having the sensation unreflectively takes that sensation to count in favor of doing whatever would avoid, lessen, or stop it.

Let me now call attention to several points about this definition.

First, the definition draws on the notion of unreflective valuing that I introduced in section 4. In its most rudimentary form of all, such valuing might involve some primitive conscious experience of a motivational "push" or "pull" in the direction of a certain behavior; in its more sophisticated forms, such valuing might involve some emotional or proto-emotional – yet still non-reflective and non-linguistic – experience of a behavior as "demanded" or "counted in favor of" by the circumstances – an experience such as a chimpanzee or grizzly bear might be capable of. Thus, according to the definition of pain under consideration, if a creature does not at an *unreflective* level take a given sensation to "count in favor of" doing what would alleviate it – in other words, if a creature has a sensation that it in no way feels motivationally "pushed" or "pulled" to avoid, lessen, or stop – or if, more complexly, the creature feels no distress at the sensation's presence, no relief when it subsides, and so on – then the sensation in question does not count as a pain. Since unreflective valuing is something that many or most animals are capable of, this definition is consistent with the idea that many or most animals can experience pain.

Second, according to the definition of pain at hand, the word "pain" technically refers to the sensation that is the object of the specified negative unreflective evaluative reaction, as opposed to the composite of the sensation plus the unreflective evaluative reaction. This is analogous to the way in which the expression "Juliet's beloved" refers to Romeo, as opposed to the composite of Romeo plus Juliet's love for him. Yet this does not imply that the unreflective evaluative reaction is not a necessary element of the pain experience. On the contrary, according to the definition at hand, just as Juliet's love for Romeo is what makes Romeo her beloved, so a creature's negative unreflective evaluative reaction to a sensation is (at least part of) what makes that sensation a pain.

Third, accepting this definition of pain involves accepting that there are two elements involved in the experience of pain: a sensation plus an unreflective evaluative reaction to that sensation.[54] But it should be noted that accepting the definition does *not* involve commitment to the idea that it will always or even often be possible to *separate* these two elements in a creature's experience of pain. On the contrary, one might think that in many or perhaps all cases of pain, the sensation and the unreflective evaluative reaction to it are merged inextricably into a single, unified experience – such that it is impossible to have one element of the pain experience without the other, or even to be able to tell the two elements apart when one examines one's pain introspectively. But none of this means that no theoretical analysis of pain into these two elements is possible. Compare the moment when Juliet sees that Romeo is dead. It would be impossible for most of us ever to separate the sight of our beloved's dead body from our evaluative reaction to it, but this does not mean that there is no distinction to be drawn between the two elements of the experience. The experience of pain may well be similar: it may often be impossible, as a practical or introspective matter – but not as a theoretical matter – to separate the sensation that is involved from one's unreflective taking of that sensation to be bad.

Fourth, as it so happens, it appears that it *is* sometimes possible to separate out the two elements involved in the experience of pain. For example, patients who have been suffering from terrible pain sometimes report that after receiving certain drugs or undergoing certain surgeries they feel the same sensation as before and yet it no longer bothers them.[55] Such cases lend support to the idea that there are indeed two distinct elements involved in the experience of pain. Note, however, that according to the definition we're considering, the moment such a separation occurs in practice – in other words, the moment a pain sensation ceases to be the object of a negative unreflective evaluative reaction – it thereby ceases to be a pain, and becomes just another sensation.

Understood as a statement of a *necessary* condition of a sensation's being a pain (and that is how I will be understanding it), I think the proposed definition of pain is a plausible one.[56] Nevertheless, my argument does not depend on one's accepting it. Rather, I'll offer another argument in the form of a dilemma (calling this dilemma the *Pain Dilemma* to distinguish it from the Darwinian Dilemma). I'll argue that the realist about the badness of pain runs into trouble no matter whether he or she accepts or rejects this statement of a necessary condition of a sensation's being a pain.

Suppose, to begin with, that the definition of pain I have just sketched is rejected by the realist. In that case, pain is understood as a sensation such that the creature having it does *not* necessarily unreflectively experience that sensation as counting in favor of whatever would avoid, lessen, or stop it. Instead, pain is given some other definition, according to which there is no inconsistency in supposing that an individual or even an entire species could unreflectively take pain sensations to count in favor of doing whatever would *bring them about and intensify them* rather than whatever would stop or lessen them. On this view, it is perfectly possible, as a conceptual matter, that instead of disliking pain the way we all happen to do, we could naturally enjoy it and be inclined to seek it out, unreflectively experiencing it as counting in favor of what would cause it – just the way we feel about the sensations associated with a massage, for example.

I take it that if pain is understood in such a way, then realism about its status as a reason to do what would avoid, lessen, or stop it is no longer very tempting. Take, for example, the case of a patient who is having a pain sensation (defined however the person opting for the first horn of the Pain Dilemma would like). And suppose that thanks to medication, the person is no longer bothered by this sensation in the slightest, feeling no motivation whatsoever to avoid, lessen, or stop it, no experience of the sensation as something to be gotten rid of. (Someone opting for the first horn of the Pain Dilemma, by definition, must admit that this is possible.) Suppose further that with the help of some new miracle drug, the patient comes positively to *enjoy* the sensation in question – which is to say that he unreflectively feels inclined to treat the sensation as counting in favor of whatever would bring it about or intensify it. (Again, someone opting for the first horn of the Pain Dilemma must agree that this is possible.)

A realist about the badness of pain so understood would have to say that even in such a case, the *sensation itself* still provides the patient with reason to do whatever would avoid, lessen, or stop it, and that this person is making a *mistake* if he goes ahead and endorses his unreflective tendency to think that the sensation counts in favor of doing what would bring it about or intensify it. Such a position is not very plausible. Of course it might be the case that this patient has *other* reasons to take actions that would, as a matter of course, *happen* to stop the sensation in question; indeed, this is likely to be the case if the sensation in question is being caused by an underlying bodily condition that is bad for the person in other respects – for example, because the bodily condition is a hindrance to the pursuit of the person's other ends. But it does not seem plausible to insist that in such a case the *sensation itself* constitutes a reason for the person to do whatever would stop it. On the contrary, if the bodily condition responsible for the sensation is (for example) a broken leg, then it seems that our patient would be perfectly correct to reason as follows: "The fact that allowing my leg to heal would end this sensation that I've come to enjoy (thanks to the miracle drug) is a small thing that counts *against* letting it heal, but all things considered I should go ahead and let it heal anyway, since after all I need to be able to walk to pursue most things that are important to me." Thus, if pain is understood as a sensation such that it is perfectly conceivable that we could unreflectively be inclined to view it as counting in favor of what would bring it about or intensify it, then realism about its status as a reason to do what would avoid, lessen, or stop it is unattractive.

Suppose, however, there is remaining doubt and some are still tempted by a realist position on the badness of pain so understood. It's at this point that the Darwinian Dilemma arises again for the realist. To see how, suppose again that pain is given some definition according to which it is *not* a necessary feature of pain that we unreflectively experience it as counting in favor of what would avoid, lessen, or stop it. In that case, the following becomes a legitimate scientific question: given that it is perfectly conceivable that we all could have ended up taking pain sensations to count in favor of what would cause them and intensify them rather than in favor of what would lessen them and stop them, what explains the fact that such a huge percentage of us so consistently do the latter? Here, as in earlier cases, there is a powerful evolutionary answer. I've left it open how the person opting for the first horn of the Pain Dilemma is defining pain (so long as that definition makes no reference to the idea that pain is a sensation that we unreflectively take to count in favor of what would stop it). But if the proposed definition is to be plausible at all, then it will pick out (predominantly, one assumes) sensations associated with the sorts of bodily conditions that we normally consider painful, such as cuts, burns, bruises, broken bones, and so on. And it is of course no mystery whatsoever, from an evolutionary point of view, why we and the other animals came to take the sensations associated with bodily conditions such as these to count in favor of what would avoid, lessen, or stop them rather than in favor of what would bring about and intensify them. One need only imagine the reproductive prospects of a creature who relished and sought after the sensations of its bones breaking and its tissues tearing; just think how many descendants such a creature would leave in comparison to those who happened to abhor and avoid such sensations.

As in earlier cases, the realist faces a problem when confronted with such an explanation. For once again we see that there is a striking coincidence between the content of the independent evaluative truth posited by the realist, on the one hand, and the content that evolutionary theory would lead us to expect, on the other. The realist tells us that it is an independent evaluative truth that pain sensations (however he or she defines them) are bad, and yet this is precisely what evolutionary theory would have predicted that we come to think. And once again the realist is unable to give any good account of this coincidence. To insist that the coincidence is *mere* coincidence is implausible. The realist's alternative, here as in earlier cases, is to defend some sort of tracking account, according to which we were selected to be able to discern independent evaluative truths, among them the truth that these pain sensations (however the realist is defining them) are bad. Yet here as in earlier cases, the tracking account is scientifically unacceptable. In order to explain why we came to think that these sensations are bad, we need make no reference whatsoever to the *fact* that they are bad; we need only point out how it tended to promote reproductive success to *take* them to be bad (due to their connection with bodily conditions that tended to diminish reproductive success).

The realist, then, is forced to the other horn of the Pain Dilemma. To salvage realism about the badness of pain, he or she is forced to understand pain as a sensation such that the creature who has it unreflectively takes that sensation to count in favor of whatever would avoid, lessen, or stop it. But now notice what this means. In order to salvage his or her view of pain as bad independently of our evaluative attitudes, the realist must admit that pain's badness depends on its being a sensation such that the creature who has it is unreflectively inclined to *take* it to be bad. But this, in turn, is just to admit that its badness depends in an important sense on our evaluative attitudes — in particular, on our being unreflectively inclined to take it to be bad. Pain may well be bad, in other words, but if it is so, its badness hinges crucially on our unreflective evaluative attitudes toward the sensation which pain is. The realist is thus forced to recognize the role of our evaluative attitudes in determining the disvalue of pain. Though initially plausible, it is a mistake to say that pain is bad independently of our evaluative attitudes. Pain, if it is plausibly to be construed as bad independently of our other evaluative attitudes, must be understood as a sensation such that we have a certain evaluative attitude toward it – and it's that evaluative attitude which (at least in part) *makes* the sensation bad.

I conclude that appeals to pain are not a promising avenue for the realist who wishes to escape the Darwinian Dilemma. Appeals to pleasure are no more promising, for there exists an analogous argument against realism about pleasure's status as a reason to do what would bring it about. This argument would center around, and propose an analogous dilemma with regard to, the following definition of pleasure: pleasure is a sensation such that the creature having it unreflectively takes that sensation to count in favor of doing whatever would bring it about, intensify it, or make it continue.

10. HOW ANTIREALISM SIDESTEPS THE DARWINIAN DILEMMA

Let me now sketch how antirealist views on the nature of value sidestep the dilemma for realism that I have described in this paper. Antirealist views understand evaluative facts or truths to be a function of our evaluative attitudes, with different versions of antirealism understanding the exact nature of this function in different ways. For instance, according to the constructivist view mentioned in section 2, the truth of the evaluative judgement that "X is a reason for agent A to Y" is a function of A's evaluative attitudes – in particular, of whether that judgement would be among A's evaluative judgements in reflective equilibrium. Such a view, as I pointed out earlier, leaves room for the possibility of evaluative error. If, for example, A thinks that the fact that someone is a member of some "out-group" is a reason for him to accord that person lesser treatment, then A's judgement is mistaken if that judgement would not be among his evaluative judgements in reflective equilibrium. It is not my purpose to develop or defend such a view here. The point is to give one example of an antirealist view, and to emphasize that antirealist views can leave room for the possibility of evaluative error, even though the standards determining what counts as an error are understood ultimately to be "set" by our own evaluative attitudes.

What then does an antirealist say about the relation between evaluative truths and the evolutionary influences that have shaped our evaluative judgements? First of all, the antirealist opts for what I have said is the more plausible horn of the Darwinian Dilemma, arguing that *of course* there is some relation at work here – of course it is no coincidence that there is such a striking overlap between the content of evaluative truths and the content that natural selection would have tended to push us toward. Of course it's no coincidence that, say, breaking one's bones *is* bad and that's also exactly what evolutionary theory would have predicted we think. But whereas the realist is forced to offer the scientifically unacceptable tracking explanation of this overlap, the antirealist is able to give a very different account.

According to the antirealist, the relation between evolutionary influences and evaluative truth works like this. Each of us begins with a vast and complicated set of evaluative attitudes. We take the breaking of our bones to be bad, we take our children's lives to be valuable, we take ourselves to have reason to help those who help us, and so on. Our holding of each of these evaluative attitudes is assumed by the antirealist to have some sort of causal explanation, just like anything else in the world. And the antirealist grants without hesitation that one major factor in explaining why human beings tend to hold some evaluative attitudes rather than others is the influence of Darwinian selective pressures. In particular, the antirealist has no problem whatsoever with the adaptive link account, if something along those lines turns out to be the best explanation. These and other questions about the best causal explanations of our evaluative attitudes are left in the hands of scientists. Whatever explanation the natural and social scientists ultimately arrive at is granted, and then evaluative truth is understood as a function of the evaluative attitudes we have, however we originally came to have them. Take the constructivist view I've been mentioning as an example. What exactly is the relation between selective pressures and evaluative truth on this view? It may be put this way: evaluative truth is a function of how all the evaluative judgements that selective pressures (along with all kinds of other causes) have imparted to us stand up to scrutiny in terms of each other; it is a function of what would emerge from those evaluative judgements in reflective equilibrium.

Where the realist's tracking account and the antirealist's account divide, then, is over the *direction of dependence* that they take to be involved in the relation between evaluative truths and the evolutionary causes which influenced the content of our evaluative judgements. The realist understands the *evaluative truths* to be prior, in the sense that evolutionary causes are understood to have selected us to track those independent truths. The antirealist, on the other hand, understands the *evolutionary causes* to be prior, in the sense that these causes (along with many others) gave us our starting fund of evaluative attitudes, and evaluative truth is understood to be a function of those attitudes. Both accounts offer an explanation of why it is no coincidence that there is significant overlap between evaluative truths and the kinds of evaluative judgements that natural selection would have pushed us in the direction of. The difference is that the antirealist account of the overlap is consistent with science. Antirealism explains the overlap not with any scientific hypothesis such as the tracking account, but rather with the metaethical hypothesis that value is something that arises as a function of the evaluative attitudes of valuing creatures – attitudes the content of which happened to be shaped by natural selection. The breaking of our bones *is* bad, in other words, and we're well aware of this. But the explanation is not that it is true independently of our attitudes that the breaking of our bones is bad and we were selected to be able to notice this; the explanation is rather that we were selected to *take* the breaking of our bones to be bad, and this evaluative judgement withstands scrutiny from the standpoint of our other evaluative judgements (to speak, for example, in the voice of the constructivist antirealist).[57]

REALIST THEORIES OF VALUE 155

11. CONCLUSION

By understanding evaluative truth as ultimately prior to our evaluative judgements, realism about value puts itself in the awkward position of having to view every causal influence on our evaluative judgements as either a tracking cause or a distorting cause. In the end, this is a difficult position to be in no matter what kind of causal influence is at issue. I have focused on the case of Darwinian influences on our evaluative judgements because I think it raises the problem for realism in a particularly acute form. In principle, however, an analogous dilemma could be constructed using any kind of causal influence on the content of our evaluative judgements. For the argument to work, two conditions must hold. First, the causal influence in question must be extensive enough to yield a skeptical conclusion if the realist goes the route of viewing those causes as distorting. Second, it must be possible to defeat whatever version of the tracking account is put forward with a scientifically better explanation.

At the end of the day, then, the dilemma at hand is not distinctly Darwinian, but much larger. Ultimately, the fact that there are *any* good scientific explanations of our evaluative judgements is a problem for the realist about value. It is a problem because realism must either view the causes described by these explanations as distorting, choosing the path that leads to normative skepticism or the claim of an incredible coincidence, or else it must enter into the game of scientific explanation, claiming that the truths it posits actually play a role in the explanation in question. The problem with this latter option, in turn, is that they don't. The best causal accounts of our evaluative judgements, whether Darwinian or otherwise, make no reference to the realist's independent evaluative truths.

Consider again the old dilemma whether things are valuable because we value them or whether we value them because they are valuable. The right answer, according to the view I've been suggesting, is somewhere in between. Before life began, nothing was valuable. But then life arose and began to value – not because it was recognizing anything, but because creatures who valued (certain things in particular) tended to survive. In this broadest sense, valuing was (and still is) prior to value. That is why antirealism about value is right. But I've emphasized that antirealist views can make room for the possibility of evaluative error, such that we can be wrong about any given evaluative judgement even as we recognize that the standards for such errors are ultimately "set" by our own evaluative attitudes. Because of this, talk of normative perception still makes sense. Now that there are creatures like us with marvelously complicated systems of valuings up and running, it is quite possible to come to value something because one recognizes that it has a value independent of oneself – not in the realist's sense, but in an antirealist's more modest sense. Thus, although valuing ultimately came first, value grew to be able to stand partly on its own. It grew to achieve its own, limited sort of priority over valuing – a priority that we can understand while at the same time being fully conscious of great biddings from the outside.[58]

NOTES

[1] More broadly, realism about value may be understood as the view that there are *mind-independent* evaluative facts or truths. I focus on independence from our *evaluative attitudes* because it is independence from this type of mental state that is the main point of contention between realists and antirealists about value.

[2] My target in this paper is realism about *practical reasons*, or reasons for action, as opposed to *epistemic reasons*, or reasons for belief. While I actually think the Darwinian Dilemma can be extended to apply against realism about epistemic reasons, that topic is more than I'll be able to pursue here. Throughout the paper, I use the word "reason" in the sense of a normative reason – in other words, in the sense of a consideration that *counts in favor of*, or *justifies*, some action.

[3] See Shafer-Landau (2003), p. 15. Shafer-Landau borrows the term from Ronald Milo.

[4] Important statements of this view include Nagel (1986), especially chapter 8, Dworkin (1996), and Shafer-Landau (2003). Scanlon's (1998, chapter 1) view on the nature of reasons is also plausibly read along these

lines. Many of these authors (though not Shafer-Landau) might resist the label "non-naturalist," due to its potential connotations of mysterious "extra" properties in the world, but so long as we keep in mind these authors' insistence that their view involves positing no such properties, the label is useful enough to be adopted.

⁵ Key statements of this view include Sturgeon (1985), Railton (1986), Boyd (1988), and Brink (1989). As Brink notes, value naturalism can be construed as claiming either that evaluative facts are *identical* with natural facts or that evaluative facts are *constituted* by natural facts; Brink argues that value naturalism should be construed as making the constitutive claim – see Brink (1989), section 6.5 and pp. 176–177. For brevity's sake, I gloss over this distinction in what follows and talk simply in terms of identity, not constitution.

⁶ Railton, for example, thinks that such reductions will be forthcoming, and sketches what they might look like in Railton (1986), whereas non-reductionist naturalists such as Sturgeon, Boyd, and Brink think that such reductions may not be, and need not be, forthcoming.

⁷ In addition to naturalist and non-naturalist versions of realism, there is one other very different brand of realism that should be mentioned, namely the quasi-realism of Simon Blackburn (1984, 1993, 1998) and Allan Gibbard (1990, 2003). Their views occupy an uneasy position with regard to the realism/antirealism debate as I am understanding it. There is not space to address their positions here, so for the purposes of this paper I set quasi-realism entirely to one side, and focus exclusively on "non-quasi" brands of realism.

⁸ Gibbard (1990), p. 30.

⁹ For introductions to the field of evolutionary psychology, see Barkow et al. (1992) and Buss (1999).

¹⁰ See Gould and Lewontin (1979). For a more recent overview, see Pigliucci and Kaplan (2000).

¹¹ Cf. Gibbard (1990), p. 30.

¹² Genetic drift is the random fluctuations of gene frequencies within a population – see Avers (1989). Later in this section, I argue that natural selection's influence on our evaluative judgements is best understood as having been *indirect*. A similar point would apply to the influence of other evolutionary forces such as genetic drift.

¹³ This assumes that other things are equal – for example, that the effects of these evaluative judgements on the creature's behavior are not cancelled out by other evaluative judgements that the creature makes. The statement in the text also assumes that a creature is motivated to act in accordance with its evaluative judgements, other things being equal. I do not offer an explicit defense of this internalist assumption in this paper, but I take it to be supported by the plausibility of the overall picture that emerges, and by the hypothesis, argued for in section 6, that the function

of evaluative judgements from an evolutionary point of view is not to "track" independent evaluative truths, but rather to *get us to respond* to our circumstances in ways that are adaptive.

[14]   On this theory, see Hamilton (1963, 1964); Sober and Wilson (1998), chapter 2; and Buss (1999), chapters 7–8.

[15]   On the theory of reciprocal altruism, see Trivers (1971); Axelrod (1984); Sober and Wilson (1998), chapter 2; and Buss (1999), chapter 9. It is important to note what the explanandum is in these sorts of evolutionary explanations. The explanandum is *not* particular attitudes held by particular individuals – for example, your or my or George W. Bush's judgement that the fact that something would help a family member is a reason to do it. Such individual-level facts are not appropriate objects of evolutionary explanation. What *are* appropriate objects of evolutionary explanation are population-level facts about patterns of variation in a given trait across a population – and the widespread presence of certain basic evaluative tendencies in the human population are such objects. For further discussion of how population-level and not individual-level facts are appropriate objects of explanations in terms of natural selection, see Sober (1984), chapter 5.

[16]   See, for instance, de Waal (1996).

[17]   I am indebted to Peter Godfrey-Smith for helpful comments regarding these points.

[18]   For discussion of the concept of genetic heritability, see Block and Dworkin (1974).

[19]   For discussion of how the genetic heritability of a trait can vary over time, see Block and Dworkin (1974), p. 41.

[20]   Indeed, it is likely that we were selected above all else to be extremely flexible when it comes to our evaluative judgements – not locked into any particular set of them but rather able to acquire and adjust them in response to the conditions in which we find ourselves. In suggesting that we possess basic evaluative tendencies, then, I am simply suggesting that when it comes to certain core issues such as our individual survival, the treatment of our offspring, and reciprocal relations with others, there are likely to be strong predispositions in the direction of making some evaluative judgements rather than others, for instance (referring back to my earlier lists) judgements (1) through (6) as opposed to judgements (1′) through (6′).

[21]   This counterfactual claim is all I need for the purposes of my argument. While one might inquire into the exact causal process by which basic evaluative tendencies have influenced the content of human evaluative judgements, it is not necessary for me to enter into such questions here.

[22]   In the remainder of the paper, I will often speak loosely about the influence of natural selection on our evaluative judgements, without reiter-

ating the complications I have discussed in this section. These complications should nevertheless be kept in mind throughout.

[23] I owe this last way of putting the point to Paul Boghossian.

[24] Kant (1959 [1785]), p. 448.

[25] If one holds that the assessment of (non-evaluative) factual judgements also proceeds via reflective equilibrium, one might wonder why the points in this paragraph don't apply equally well to rational reflection about scientific matters (for example). The key difference is that in the scientific case, our "starting fund" of (non-evaluative) factual judgements need not be viewed as mostly "off track." For further discussion, see note 35 below.

[26] I borrow the term "tracking" from Robert Nozick, who uses it in similar contexts in Nozick (1981).

[27] In correspondence.

[28] Nozick suggests something very similar when he writes that "it seems reasonable to assume there has been some evolutionary advantage in acting for (rational) reasons. The capacity to do so, once it appeared, would have been selected for. Organisms able and prone to act for (rational) reasons gained some increased efficiency in leaving great-grand progeny" (1981), p. 337.

[29] This brings out the interesting way in which non-naturalist versions of value realism, in spite of their insistence that values are *not* the kinds of things that play a role in causal explanations, are ultimately forced (unless they opt for the first horn of the dilemma) to take a stand on certain matters of scientific explanation – in particular, on questions about why human beings tend to make some evaluative judgements rather than others, and on the origins of our capacity to grasp independent evaluative truths. Indeed, as I'll try to show, these realists are forced (again, unless they opt for the first horn) to posit a causal role for evaluative truths in the course of our species' evolution.

[30] For closely related points, see Blackburn, who writes that an evaluative attitude's "function is to mediate the move from features of a situation to a reaction" (1993), p. 168; and Gibbard, who writes that the "biological function [of normative judgements] is to govern our actions, beliefs, and emotions" (1990), p. 110.

[31] In order for a mechanism which *effects a pairing* between the circumstance of a hot surface and the response of withdrawing one's hand to be adaptive, there must of course be a means of *detecting* the presence of a hot surface. Similarly, in order for a "mechanism" which *effects a pairing* between the circumstance of one's being helped and the response of helping in return to be adaptive, there must be a means of *detecting* or *tracking* circumstances in which one is helped. In proposing the adaptive link account, what I mean to be focusing in on are the mechanisms which *effect the pairing* between (perceived) circumstance and response, and *not*

160   SHARON STREET

the mechanisms which do the (separate) job of *tracking circumstances*. While in the case of an automatic reflex mechanism, it may be hard to pull these mechanisms apart, the two jobs are nevertheless theoretically distinct, and the "mechanisms" clearly do come apart in the case of (non-evaluative) factual judgement versus evaluative judgement. Our capacity for (non-evaluative) factual judgement does the job of *tracking circumstances* (tracking, among innumerable other things, which individuals have helped us), whereas our capacity for evaluative judgement does the job of *effecting pairings of (perceived) circumstance and response* (getting us, among many other things, to respond to those who have helped us with help in return).

[32] I do not mean to be offering a full explanation of why we have a capacity to make evaluative judgements. Among other things, I say nothing to address the question: why did we evolve this "normative capacity" as a means of forging links between circumstance and response instead of, for instance, having such links forged solely by brute reflex mechanisms? The answer presumably has to do with the incredible flexibility and plasticity of the former capacity as opposed to reflex mechanisms, but this is not a question that I need to enter into for the purposes of my argument.

[33] Here I have suggested that it's a certain kind of *conscious experience* – for example, the conscious experience of the fact that someone has helped you as "counting in favor of" helping in return – that does the work of forging adaptive links between circumstance and response, and which was selected for. But a qualification is needed here, for it may be that this *conscious experience* was not itself directly selected for, but is rather an incidental byproduct of underlying information-processing and behavior-control systems which *were* selected for. If this is so, it does not pose any problem for my argument, since the only point I need for my argument is that the content of our evaluative judgements has been greatly affected by the influence of natural selection. This point still holds even if what was selected for are certain information-processing and behavior-control systems, which in turn give rise, as an incidental byproduct, to conscious experiences – here, in particular, of some things as "counting in favor of" other things.

[34] For related discussion, see Blackburn (1993), p. 69 and Gibbard (1990), pp. 107–108.

[35] It is points like this which explain why the Darwinian Dilemma doesn't go through against realism about non-evaluative facts such as facts about fires, predators, cliffs, and so on. In short, the difference is that in the case of such non-evaluative facts, unlike in the case of evaluative facts, the tracking account prevails as the best explanation of our capacity to make the relevant sort of judgement. In order to explain why it proved advantageous to form judgements about the presence of fires, predators, and cliffs, one will need to posit in one's best explanation that

there *were indeed* fires, predators, and cliffs, which it proved quite useful to be aware of, given that one could be burned by them, eaten by them, or could plummet over them. For related discussion, see Gibbard (1990), chapter 6 and (2003), pp. 253–258.

[36] Someone drawn to the tracking account might argue that when it comes to our *corrected* evaluative judgements – for example, our judgement that membership in an "out-group" is no reason to accord a person lesser treatment – the tracking account provides a better explanation. But this is not so. It is perfectly compatible with the adaptive link account that we come to reject some of our basic evaluative tendencies on the basis of other evaluative judgements we hold.

[37] There is one other option available here: to posit that evaluative truths are in some way a *function* of our evaluative attitudes. This is exactly the way to go, in my view. But to make this move – to accept that evaluative truths are ultimately a function of our evaluative attitudes – is just to abandon value realism and embrace antirealism.

[38] I am indebted to Nishi Shah for very helpful comments and discussion regarding this objection and the material throughout this section.

[39] See note 35 above.

[40] Thanks to Dale Jamieson for pressing me to be explicit about this.

[41] See Railton (1986).

[42] One might object that while it depends on Ann's evaluative attitudes that her good is identical (in part) to her *longevity*, it presumably does not depend on her evaluative attitudes that her good is identical to *what she would desire to desire under conditions of full information*. This latter identity holds independently of her evaluative attitudes. Does that mean that Railton's view is realist after all on my taxonomy? The answer is no, since a view which *identifies* evaluative facts with facts about our evaluative attitudes (identifying them in particular with what those attitudes pick out as valuable under certain conditions) cannot properly be said to hold that evaluative facts are *independent* of our evaluative attitudes – any more than a view which identifies water with H_2O can properly be said to hold that facts about water are independent of facts about H_2O.

[43] For discussion of the "rigidifying" move, see Darwall et al. (1992), pp. 162–163.

[44] For similar points, see Hare (1952), pp. 148–149; Horgan and Timmons (1991, 1992); and Smith (1994), pp. 32–35.

[45] This assumes that each community has correctly identified the natural properties tracked by its own evaluative attitudes.

[46] Brink (2001), section 7 seems to take such a view.

[47] Sturgeon (1985), p. 59.

[48] Brink (1989), pp. 177–178.

[49] See Sturgeon (1985).

[50] Of course it may be that the very first *rudiments* of the human eye emerged as the purely incidental byproduct of selection for some other capacity, and then these rudiments conferred advantages of their own and began to be selected for more directly. And the realist might claim that something similar could have happened in the case of our ability to grasp independent evaluative truths: first came an extraordinarily rudimentary form of this ability, and then the ability began to be selected for more directly. The realist may certainly pursue this proposal. But if he or she does so, then he or she has opted for the second horn of the dilemma, claiming that there *was* indeed a relation between the operation of selective pressures on the relevant capacity and the independent evaluative truths he or she posits.

[51] For related discussion, see Gibbard (2003), pp. 265–267.

[52] One prominent such appeal is the discussion of pain in Nagel (1986), pp. 156–162.

[53] More precisely: someone's pain counts as a *pro tanto* reason for that person to do what would avoid, lessen, or stop it, where a *pro tanto* reason is a reason that is good as far as it goes, but which may be outweighed by other considerations. (So, for example, the fact that it will be painful is a good reason *not* to go to the dentist, so far as that reason goes, but it may well be the case that this reason is counterbalanced and ultimately outweighed by good reasons in favor of going.)

[54] My treatment of pain has been influenced by Korsgaard (1996), lecture 4, particularly when it comes to the idea that there are two elements involved in the experience of pain. One of the most important differences between Korsgaard's treatment of pain and mine, however, is this. Korsgaard takes the position that pain itself never provides a reason for its sufferer to do what would alleviate it; it is rather merely the perception of some *other* reason that the sufferer has. In contrast, on my view, pain itself (at least virtually always) *does* provide a reason for its sufferer to do what would alleviate it. I take this to be the more intuitively plausible position.

[55] For references, see Hall's (1989) discussion, to which I am indebted.

[56] One might well doubt whether the definition states a sufficient condition of a sensation's being a pain. One might, for instance, be concerned about the way in which the definition would apparently count sensations like the taste of something rancid, the smell of rotten eggs, or the sound of fingernails on a chalkboard as pains, since we have unreflective negative evaluative reactions to these sensations too. One might also think that a complete definition would say more about how pain involves not only an unreflective negative evaluative reaction to a *sensation*, but also to the *bodily condition* of which that sensation is (at least in many cases) a perception. With regard to this latter point, however, the definition at

hand already addresses it, at least up to a point. For note that the definition may be understood as positing *two* unreflective evaluative reactions: first, an unreflective taking of the *sensation* in question to be bad, and second, an unreflective taking of *whatever would avoid, lessen, or stop the sensation* to be good. Assuming that an underlying bodily condition is what is causing the sensation in question, and that the elimination of that bodily condition would stop it, the second unreflective evaluative element involves taking that bodily condition to be bad and the elimination of it to be good. So, for example, if one is experiencing pain due to a broken leg, part of what this involves, according to the definition, is the unreflective taking of whatever would stop this sensation to be good, which in turn involves the unreflective taking of the healing of one's leg to be good (and the unreflective taking of its current broken condition to be bad). Thanks to Hilla Jacobson for discussion of related issues.

[57] In objection to this section's argument, someone might try to replicate against antirealism the move I made in section 7 against realist versions of value naturalism. In particular, an objector might charge that the antirealist, in arriving at his or her view on the way in which evaluative truth is a function of our evaluative attitudes, must rely heavily on our substantive evaluative judgements (regarding practical reasons). Since those judgements are contaminated with evolutionary influence (and since the antirealist presumably wishes to say that his or her metaethical view is true independently of our evaluative attitudes), the objector might argue that the Darwinian Dilemma threatens antirealism as much as it does realism. There is not space to address this objection in depth here, but in brief my reply is that in arriving at his or her metaethical view, the antirealist does *not* need to rely on our substantive evaluative judgements (regarding practical reasons). This may be seen by imagining an alien investigator who (1) quite recognizably possesses evaluative concepts; (2) accepts evaluative judgements (regarding practical reasons) with *entirely different* substantive content than our own; and who nevertheless (3) arrives at the same metaethical view as the human antirealist; and (4) does so based on the exact same considerations. Examples of such considerations might include the Darwinian Dilemma itself (which the alien investigator could accept), or the types of considerations that David Lewis (1989) offers in favor of his analysis of value. (Lewis's proposal counts as antirealist in my taxonomy – see his remark that if we had been disposed to value seasickness and petty sleaze, then in one good sense, though not the only sense, of "value," "it would have been true for us to say 'seasickness and petty sleaze are values'" [1989], p. 133.) I assume here for the sake of argument that the alien investigator would share our judgements regarding *epistemic reasons*; this assumption is complicated by the fact that (as

164 SHARON STREET

mentioned in note 2) I believe the Darwinian Dilemma can be extended to apply against realism about epistemic reasons. A full discussion of this objection would address such complications.

[58] For their comments on earlier versions of this paper, I am indebted to Melissa Barry, Paul Boghossian, Hartry Field, Patricia Kitcher, Philip Kitcher, Christine M. Korsgaard, Thomas Nagel, Derek Parfit, T.M. Scanlon, Nishi Shah, Michael Strevens, and audiences at Amherst College, Columbia University, and Duke University. I am also indebted to Paul Bloomfield and Earl Conee for their comments on this paper at the 2005 Pacific Division APA.

REFERENCES

Avers, C.J. (1989): *Process and Pattern in Evolution*, New York: Oxford University Press.

Axelrod, R. (1984): *The Evolution of Cooperation*, New York: Basic Books.

Barkow, J.H., Cosmides, L. and Tooby, J. (1992): *The Adapted Mind: Evolutionary Psychology and the Generation of Culture*, New York: Oxford University Press.

Blackburn, S. (1984): *Spreading the Word: Groundings in the Philosophy of Language*, Oxford: Oxford University Press.

Blackburn, S. (1993): *Essays in Quasi-Realism*, New York: Oxford University Press.

Blackburn, S. (1998): *Ruling Passions*, Oxford: Oxford University Press.

Block, N.J. and Dworkin, G. (1974): 'IQ, Heritability, and Inequality, Part 2', *Philosophy and Public Affairs* 4, 40–99.

Boyd, R. (1988): 'How to Be a Moral Realist', in G. Sayre-McCord (ed.), *Essays on Moral Realism* (pp. 181–228), Ithaca: Cornell University Press.

Brink, D.O. (1989): *Moral Realism and the Foundations of Ethics*, Cambridge: Cambridge University Press.

Brink, D.O. (2001): 'Realism, Naturalism, and Moral Semantics', *Social Philosophy and Policy* 18, 154–176.

Buss, D.M. (1999): *Evolutionary Psychology: The New Science of Mind*, Boston: Allyn and Bacon.

Darwall, S., Gibbard, A. and Railton, P. (1992): 'Toward Fin de Siècle Ethics: Some Trends', *Philosophical Review* 101, 115–189.

Dworkin, R. (1996): 'Objectivity and Truth: You'd Better Believe It', *Philosophy and Public Affairs* 25, 87–139.

Gibbard, A. (1990): *Wise Choices, Apt Feelings*, Cambridge: Harvard University Press.

Gibbard, A. (2003): *Thinking How to Live*, Cambridge: Harvard University Press.

Gould, S.J. and Lewontin, R.C. (1979): 'The Spandrels of San Marco and the Panglossian Paradigm: A Critique of the Adaptationist Programme', *Proceedings of the Royal Society of London, Series B, Biological Sciences* 205, 581–598.

Hall, R.J. (1989): 'Are Pains Necessarily Unpleasant?', *Philosophy and Phenomenological Research* 49, 643–659.

Hamilton, W.D. (1963): 'The Evolution of Altruistic Behavior', *American Naturalist* 97, 354–356.

Hamilton, W.D. (1964): 'The Genetical Evolution of Social Behavior, I and II', *Journal of Theoretical Biology* 7, 1–52.

Hare, R.M. (1952): *The Language of Morals*, Oxford: Oxford University Press.

Horgan, T. and Timmons, M. (1991): 'New Wave Moral Realism Meets Moral Twin Earth', *Journal of Philosophical Research* 16, 447–465.

Horgan, T. and Timmons, M. (1992): 'Troubles for New Wave Moral Semantics: The Open Question Argument Revived', *Philosophical Papers* 21, 153–175.

Kant, I. (1959 [1785]): *Foundations of the Metaphysics of Morals*, L.W. Beck (trans.), New York: Macmillan.

Korsgaard, C.M. (1996): *The Sources of Normativity*, Cambridge: Cambridge University Press.

Lewis, D. (1989): 'Dispositional Theories of Value', *Aristotelian Society Supplementary Volume* 63, 113–137.

Nagel, T. (1986): *The View from Nowhere*, Oxford: Oxford University Press.

Nozick, R. (1981): *Philosophical Explanations*, Cambridge: Harvard University Press.

Pigliucci, M. and Kaplan, J. (2000): 'The Fall and Rise of Dr. Pangloss: Adaptationism and the *Spandrels* Paper 20 Years Later', *Trends in Ecology and Evolution* 15, 66–70.

Railton, P. (1986): 'Moral Realism', *Philosophical Review* 95, 163–207.

Scanlon, T.M. (1998): *What We Owe to Each Other*, Cambridge: Harvard University Press.

Shafer-Landau, R. (2003): *Moral Realism: A Defence*, Oxford: Oxford University Press.

Smith, M. (1994): *The Moral Problem*, Oxford: Blackwell.

Sober, E. (1984): *The Nature of Selection: Evolutionary Theory in Philosophical Focus*, Cambridge: MIT Press.

Sober, E. and Wilson, D.S. (1998): *Unto Others: The Evolution and Psychology of Unselfish Behavior*, Cambridge: Harvard University Press.

Sturgeon, N. (1985): 'Moral Explanations', in D. Copp and D. Zimmerman (eds.), *Morality, Reason and Truth* (pp. 49–78), Totowa, NJ: Rowman and Allanheld.

166 SHARON STREET

Trivers, R. (1971): 'The Evolution of Reciprocal Altruism', *Quarterly Review of Biology* 46, 35–57.
de Waal, F. (1996): *Good Natured: The Origins of Right and Wrong in Humans and Other Animals*, Cambridge: Harvard University Press.

Department of Philosophy
New York University
Silver Center Room 503
100 Washington Square East
New York, NY 10003
USA
E-mail: sharon.street@nyu.edu

Part V
Normative Ethics

[12]

A Defense of Evolutionary Ethics*

ROBERT J. RICHARDS

Committee on the Conceptual Foundations of Science
The University of Chicago, IL 60637
U.S.A.

ABSTRACT: From Charles Darwin to Edward Wilson, evolutionary biologists have attempted to construct systems of evolutionary ethics. These attempts have been roundly criticized, most often for having committed the naturalistic fallacy. In this essay, I review the history of previous efforts at formulating an evolutionary ethics, focusing on the proposals of Darwin and Wilson. I then advance and defend a proposal of my own. In the last part of the essay, I try to demonstrate that my revised version of evolutionary ethics: (1) does not commit the naturalistic fallacy as it is usually understood; (2) does, admittedly, derive values from facts; but (3) does not commit any fallacy in doing so.

KEY WORDS: Altruism, C. Darwin, ethics, evolution, naturalistic fallacy, sociobiology.

I

Introduction

"The most obvious, and most immediate, and most important result of the *Origin of Species* was to effect a separation between truth in moral science and truth in natural science," so concluded the historian of science Susan Cannon (1978, p. 276). Darwin had demolished, in Cannon's view, the truth complex that joined natural science, religion, and morality in the Nineteenth Century. He had shown, in Cannon's terms, "whatever it is, 'nature' isn't any good" (1978, p. 276). Those who attempt to rivet together again ethics and science must, therefore, produce a structure that can bear no critical weight. Indeed, most contemporary philosophers suspect that the original complex cracked decisively because of intrinsic logical flaws, so that any effort at reconstruction must necessarily fail. G. E. Moore believed those making such an attempt would perpetrate the "naturalistic fallacy," and he judged Herbert Spencer the most egregious offender. Spencer uncritically transformed scientific assertions of fact into moral imperatives. He and his tribe, according to Moore, fallaciously maintained that evolution, "while it shews us the direction in which we *are* developing, thereby and for that reason shews us the direction in which we *ought* to develop" (1903, p. 46).

Those who commit the fallacy must, it is often assumed, subvert morality altogether. Consider the self-justificatory rapacity of the Rocke-

266   ROBERT J. RICHARDS

fellers and Morgans at the beginning of this century, men who read
Spencer as the prophet of profit and preached the moral commandments
of Social Darwinism. Marshall Sahlins warns us, in his *The Use and Abuse
of Biology,* against the most recent consequence of the fallacy, the ethical
and social preachments of sociobiology. This evolutionary theory of
society, he finds, illegitimately perpetuates Western moral and cultural
hegemony. Its parentage betrays it. It came aborning through the narrow
gates of Nineteenth-Century *laissez faire* economics:

> Conceived in the image of the market system, the nature thus culturally figured has
> been in turn used to explain the human social order, and vice versa, in an endless
> reciprocal interchange between social Darwinism and natural capitalism. Sociobiology
> . . . is only the latest phase in this cycle (1976, p. xv).

An immaculately conceived nature would remain silent, but a Malthusian
nature urges us to easy virtue.

The fallacy might even be thought to have a more sinister outcome.
Ernst Haeckel, Darwin's champion in Germany, produced out of evolutionary theory moral criteria for evaluating human "Lebenswerth." In his
book *Die Lebenswunder* of 1904, he seems to have prepared instruments
for Teutonic horror:

> Although the significant differences in mental life and cultural conditions between the
> higher and lower races of men is generally well known, nonetheless their respective
> *Lebenswerth* is usually misunderstood. That which raises men so high over the animals
> — including those to which they are closely related — and that which gives their life
> infinite worth is culture and the higher evolution of reason that makes men capable of
> culture. This, however, is for the most part only the property of the higher races of
> men; among the lower races it is only imperfectly developed — or not at all. Natural
> men (e.g., Indian Vedas or Australian negroes) are closer in respect of psychology to
> the higher vertebrates (e.g., apes and dogs) than to highly civilized Europeans. Thus
> their individual *Lebenswerth* must be judged completely differently (1904, pp. 449—
> 50).

Here is science brought to justify the ideology and racism of German
culture in the early part of this century: sinning against logic appears to
have terrible moral consequences.

But was the fault of the American industrialists and German mandarins
in their logic or in themselves? Must an evolutionary ethics commit the
naturalistic fallacy, and is it a fallacy after all? These are questions I wish
here to consider.

Social Indeterminacy of Evolutionary Theory

Historians, such as Richard Hofstadter, have documented the efforts of
the great capitalists, at the turn of the century, to justify their practices by
appeal to popular evolutionary ideas. John D. Rockefeller, for instance,

declared in a Sunday sermon that "the growth of a large business is merely a survival of the fittest." Warming to his subject, he went on:

> The American Beauty rose can be produced in the splendor and fragrance which bring cheer to its beholder only by sacrificing the early buds which grow up around it. This is not an evil tendency in business. It is merely the working-out of a law of nature and a law of God (quoted in Hofstadter 1955, p. 45).

More recently, however, other historians have shown how American progressives (Banister 1979) and European socialists (Jones 1980) made use of evolutionary conceptions to advance their political and moral programs. For instance, Enrico Ferri, an Italian Marxist writing at about the same time as Rockefeller, sought to demonstrate that "Marxian socialism . . . is only the practical and fruitful complement in social life of that modern scientific revolution, which . . . has triumphed in our days, thanks to the labours of Charles Darwin and Herbert Spencer" (1909, p. xi). Several important German socialists also found support for their political agenda in Darwin: Eduard Bernstein argued that biological evolution had socialism as a natural consequence (1890—91); and August Bebel's *Die Frau und der Sozialismus* (1879) derived the doctrine of feminine liberation from Darwin's conception. Rudolf Virchow had forecast such political uses of evolutionary theory when he warned the Association of German Scientists in 1877 that Darwinism logically led to socialism (Kelly 1981, pp. 59—60).

While Virchow might have been a brilliant medical scientist, and even a shrewd politician, his sight dimmed when inspecting the finer lines of logical relationship: he failed to recognize that the presumed logical consequence of evolutionary theory required special tacit premises imported from Marxist ideology. Add different social postulates, of the kind Rockefeller dispensed along with his dimes, and evolutionary theory will demonstrate the natural virtues of big business. Though, as I will maintain, evolutionary theory is not compatible with every social and moral philosophy, it can accommodate a broad range of historically representative doctrines. Thus, in order for evolutionary theory to yield determinate conclusions about appropriate practice, it requires a mediating social theory to specify the units and relationships of concern. It is therefore impossible to examine the "real" social implications of evolutionary theory without the staining fluids of political and social values. The historical facts thus stand forth: an evolutionary approach to the moral and social environment does not inevitably support a particular ideology.

Those apprehensive about the dangers of the naturalistic fallacy may object, of course, that just this level of indeterminacy — the apparent ability to give witness to opposed moral and social convictions — shows the liability of any wedding of morals and evolutionary theory. But such objection ignores two historical facts: first, that moral barbarians have frequently defended heinous behavior by claiming that it was enjoined by holy writ and saintly example — so no judgment about the viability of an ethical system can be made simply on the basis of the policies that it has been called upon to support; and, second, that several logically different systems have traveled under the name "evolutionary ethics" — so one cannot condemn all such systems simply because of the liabilities of one of another. In other words, we must examine particular systems of evolutionary ethics to determine whether they embody any fallacy and to discover what kinds of acts they sanction.

Elsewhere I have described the moral systems of several evolutionary theorists and have attempted to assess the logic of those systems (Richards, in press), so I won't rehearse all that here. Rather, I will draw on those systems to develop the outline of an evolutionary conception of morals, one, I believe, that escapes the usual objections to this approach. In what follows, I will first sketch Darwin's theory of morals, which provides the essential structure for the system I wish to advance, and compare it to a recent and vigorously decried descendant, the ethical ideas formulated by Edward Wilson in his books *Sociobiology* (1975) and *On Human Nature* (1978). Next I will describe my own revised version of an evolutionary ethics. Then I will consider the most pressing objections brought against an ethics based on evolutionary theory. Finally, I will show how the proposal I have in mind escapes these objections.

II

Darwin's Moral Theory

In the *Descent of Man* (1871), Darwin urged that the moral sense — the motive feeling which fueled intentions to perform altruistic acts and which caused pain when duty was ignored — be considered a species of social instinct (1871, chs. 2, 3, 5; Richards 1982; in press, ch. 5). He conceived social instincts as the bonds forming animal groups into social wholes. Social instincts comprised behaviors that nurtured offspring, secured their welfare, produced cooperation among kin, and organized the clan into a functional unit. The principal mechanism of their evolution, in Darwin's view, was community selection: that kind of natural selection operating at levels of organization higher than the individual. The degree to which social instincts welded together a society out of its striving members depended on the species and its special conditions. Community selection worked most effectively among the social insects, but Darwin thought its power was in evidence among all socially dependent animals, including that most socially advanced creature, man.

In the *Descent*, Darwin elaborated a conception of morals that he first

A DEFENSE OF EVOLUTIONARY ETHICS 269

outlined in the late 1830s (Richards 1982). He erected a model depicting four over-lapping stages in the evolution of the moral sense. In the first, well-developed social instincts would evolve to bind proto-men into social groups, that is, into units that might continue to undergo community selection. During the second stage, creatures would develop sufficient intelligence to recall past instances of unrequited social instincts. The primitive anthropoid that abandoned its young because of a momentarily stronger urge to migrate might, upon brutish recollection of its hungry offspring, feel again the sting of unfulfilled social instinct. This, Darwin contended, would be the beginning of conscience. The third stage in the evolution of the moral sense would arrive when social groups became linguistically competent, so that the needs of individuals and their societies could be codified in language and easily communicated. In the fourth stage, individuals would acquire habits of socially approved behavior that would direct the moral instincts into appropriate channels — they would learn how to help their neighbors and advance the welfare of their group. So what began as crude instinct in our predecessors, responding to obvious perceptual cues, would become, in Darwin's construction, a moral motive under the guidance of social custom and intelligent decision. As the moral sense evolved, so did a distinctively human creature.

Under prodding from his cousin Hensleigh Wedgwood, Darwin expanded certain features of his theory in the second edition of the *Descent*. He made clear that during the ontogenesis of conscience, individuals learned to avoid the nagging persistence of unfulfilled social instinct by implicitly formulating rules about appropriate conduct. These rules would take into account not only the general urgings of instinct, but also the particular ways a given society might sanction their satisfaction. Such rules, Darwin thought, would put a rational edge on conscience and, in time, would become the publicly expressed canons of morality. With the training of each generation's young, these moral rules would recede into the very bones of social habits and customs. Darwin, as a child of his scientific time, also believed that such rational principles, first induced from instinctive reactions, might be transformed into habits, and then infiltrate the hereditary substance to augment and reform the biological legacy of succeeding generations.

Darwin's theory of moral sense was taken by some of his reviewers to be but a species of utilitarianism, one that gave scientific approbation to the morality of selfishness (Richards in press, ch. 5). Darwin took exception to such judgments. He thought his theory completely distinct from that of Bentham and Mill. Individuals, he emphasized, acted instinctively to avoid vice and seek virtue without any rational calculations of benefit. Pleasure may be our sovereign mistress, as Bentham painted her, but some human actions, Darwin insisted, were indifferent to her allure. Pleasure was neither the usual motive nor the end of moral acts. Rather, moral behavior, arising from community selection, was ultimately directed to the vigor and health of the group, not to the pleasures of its individual members. This meant, according to Darwin, that the criterion of morality — that highest principle by which we judge our behavior in a cool hour — was not the general happiness, but the general good, which he interpreted as the welfare and survival of the group. This was no crude utilitarian theory of morality dressed in biological guise. It cast moral acts as intrinsically altruistic.

Darwin, of course, noticed that men sometimes adopted the moral patterns of their culture for somewhat lower motives: implicitly they formed contracts to respect the person and property of others, provided they received the same consideration; they acted, in our terms, as reciprocal altruists. Darwin also observed that his fellow creatures glowed or smarted under the judgments of their peers; accordingly, they might betimes practice virtue in response to public praise rather than to the inner voice of austere duty. Yet men did harken to that voice, which they understood to be authoritative, if not always coercive.

From the beginning of his formulation of a moral theory, in the late 1830s, Darwin recognized a chief competitive advantage of his approach. He could explain what other moralists merely assumed: he could explain how the moral criterion and the moral sense were linked. Sir James Mackintosh, from whom Darwin borrowed the basic framework of his moral conception, declared that the *moral sense* for right conduct had to be distinguished from the *criterion* of moral behavior. We instinctively perceive murder as vile, but in a cool moment of rational evaluation, we can also weigh the disutility of murder. When a man jumps into the river to save a drowning child, he acts impulsively and without deliberation, while those safely on shore may rationally evaluate his behavior according to the criterion of virtuous behavior. Mackintosh had no satisfactory account of the usual coincidence between motive and criterion. He could not easily explain why impulsive actions might yet be what moral deliberation would recommend. Darwin believed he could succeed where Mackintosh failed; he could provide a perfectly natural explanation of the linkage between the moral motive and the moral criterion. Under the aegis of community selection, men in social groups evolved sets of instinctive responses to preserve the welfare of the community. This common feature of acting for the community welfare would then become, for intelligent creatures who reacted favorably to the display of such moral impulses, an inductively derived but dispositionally encouraged general principle of appropriate behavior. What served nature as the criterion for selecting behavior became the standard of choice for her creatures as well.

A DEFENSE OF EVOLUTIONARY ETHICS 271

Wilson's Moral Theory

In his book *On Human Nature* (1978) Edward Wilson elaborated a moral theory that he had earlier sketched in the concluding chapter of his massive *Sociobiology* (1975). Though Wilson's proposals bear strong resemblance to Darwin's own, the similarity appears to stem more from the logic of the interaction of evolutionary theory and morals than from an intimate knowledge of his predecessor's ethical views. Wilson, like Darwin, portrays the moral sense as the product of natural selection operating on the group. In light of subsequent developments in evolutionary theory, however, he more carefully specifies the unit of selection as the kin, the immediate and the more remote. The altruism evinced by lower animals for their offspring and immediate relatives can be explained, then, by employing the Hamiltonian version — i.e., kin selection — of Darwin's original concept of community selection. Also like Darwin, Wilson suggests that the forms of altruistic behavior are constrained by the cultural traditions of particular societies. But unlike Darwin, Wilson regards this "hard-core" altruism, as he calls it, to be insufficient, even detrimental to the organization of societies larger than kin groups, since such altruism does not reach beyond blood relatives. As a necessary compromise between individual and group welfare, men have adopted implicit social contracts; they have become reciprocal altruists.

Wilson calls this latter kind of altruism, which Darwin also recognized, "soft-core," since it is both genetically and psychologically selfish: individuals agree mutually to adhere to moral rules in order that they might secure the greatest amount of happiness possible. Though Wilson deems soft-core altruism as basically a learned pattern of behavior, he conceives it as "shaped by powerful emotional controls of the kind intuitively expected to occur in its hardest forms" (1978, p. 162). He appears to believe that the "deep structure" of moral rules, whether hard-core or soft, express a genetically determined disposition to employ rules of the moral form. In any case, the existence of such rules ultimately can only have a biological explanation, for "morality has no other demonstrable ultimate function" than "to keep human genetic material intact" (1978, p. 167).

Wilson's theory has recently received vigorous defense from Michael Ruse (1984). Ruse endorses Wilson's evaluation of the ethical as well as the biological merits of soft-core altruism:

> Humans help relatives without hope or expectation of the ethical return. Humans help nonrelatives insofar as and only insofar as they anticipate some return. This may not be an anticipation of immediate return, but only a fool or a saint (categories often linked) would do something absolutely for nothing (p. 171).

Ruse argues that principles of reciprocal altruism have become inbred in the human species and manifest themselves to our consciousness in the

272   ROBERT J. RICHARDS

form of feelings. The common conditions of human evolution mean that
most men share feelings of right and wrong. Nonetheless, ethical standards, according to Ruse, are relative to our evolutionary history. He
believes we cannot justify moral norms through other means: "All the
justification that can be given for ethics lies in our evolution" (1984, p.
177).[1]

A Theory of Evolutionary Ethics

The theory I wish to advocate is based on Darwin's original conception
and has some similarities to Wilson's proposal. It is theory, however,
which augments Darwin's and differs in certain respects from Wilson's.
For convenience I will refer to it as the revised version (or RV for short).
RV has two distinguishable parts, a speculative theory of human evolution
and a more distinctively moral theory based on it. Evolutionary thinkers
attempting to account for human mental, behavioral, and, indeed, anat-
omical traits usually spin just-so stories, projective accounts that have
more or less theoretical and empirical support. Some will judge the
evidence I suggest for my own tale too insubstantial to bear much critical
weight. My concern, however, will not be to argue the truth of the
empirical assertions, but to show that if those assertions are true they
adequately justify the second part of RV, the moral theory. My aim, then,
is fundamentally logical and conceptual: to demonstrate that an ethics
based on presumed facts of biological evolution need commit no sin of
moral logic, rather can be justified by using those facts and the theory
articulating them.

RV supposes that a moral sense has evolved in the human group.
"Moral sense" names a set of innate dispositions that, in appropriate
circumstances, move the individual to act in specific ways for the good of
the community. The human animal has been selected to provide for the
welfare of its own offspring (e.g., by specific acts of nurture and protection); to defend the weak; to aid others in distress; and generally to
respond to the needs of community members. The individual must learn
to recognize, for instance, what constitutes more subtle forms of need and
what specific responses might alleviate distress. But, so RV proposes, once
different needs are recognized, feelings of sympathy and urges to remedial
action will naturally follow. These specific sympathetic responses and
pricks to action together constitute the core of the altruistic attitude. The
mechanism of the initial evolution of this attitude I take to be kin
selection, aided, perhaps, by group selection on small communities.[2]
Accordingly, altruistic motives will be strongest when behavior is directed
toward immediate relatives. (Parents, after all, are apt to sacrifice considerably more for the welfare of their children than for complete
strangers.) Since natural selection has imparted no way for men or animals

A DEFENSE OF EVOLUTIONARY ETHICS 273

to perceive blood kin straight off, a variety of perceptual cues have become indicators of kin. In animals it might be smells, sounds, or coloring that serve as the imprintable signs of one's relatives. With men, extended association during childhood seems to be a strong indicator. Maynard Smith, who has taken some exception to the evolutionary interpretation of ethics, yet admits his mind was changed about the incest taboo (1978). The reasons he offers are: (1) the deleterious consequences of inbreeding; (2) the evidence that even higher animals avoid inbreeding; and (3) the phenomenon of kubbutzim children not forming sexual relations. Children of the kubutz appear to recognize each other as "kin," and so are disposed to act for the common good by shunning sex with each other.

On the basis of such considerations, RV supposes that early human societies consisted principally of extended kin groups, of clans. Such clans would be in competition with others in the geographical area, and so natural selection might operate on them to promote a great variety of altruistic impulses, all having the ultimate purpose of serving the community good.

Men are cultural animals. Their perceptions of the meaning of behaviors, their recognition of "brothers," their judgments of what acts would be beneficial in a situation — all of these are interpreted according to the traditions established in the history of particular groups. Hence, it is no objection to an evolutionary ethics that in certain tribes — whose kin systems only loosely recapitulate biological relations — the natives may treat with extreme altruism those who are only cultural but not biological kin.[3] In a biological sense, this may be a mistake; but on average the cultural depiction of kin will serve nature's ends.

RV insists, building on Darwin's and Wilson's theories, that the moral attitude will be informed by an evolving intelligence and cultural tradition. Nature demands we protect our brother, but we must learn who our brother is. During human history, evolving cultural traditions may translate "community member" as "red Sioux," "black Mau Mau," or "White Englishman," and the "community good" as "sacrificing to the gods," "killing usurping colonials," or "constructing factories." But as men become wiser and old fears and superstitions fade, they may come to see their brother in every human being and to discover what really does foster the good of all men.

RV departs from Wilson's sociobiological ethics and Ruse's defense of it, since they regard reciprocal altruism as the chief sort, and "keeping the genetic material intact" (Wilson 1978, p. 167) as the ultimate justification. Reciprocal altruism, as a matter of fact, may operate more widely than the authentic kind; it may even be more beneficial to the long-term survival of human groups. But this does not elevate it to the status of the highest kind of morality, though Wilson and Ruse suggest it does. And while the evolution of authentic altruistic motives may serve to perpetuate genetic

274   ROBERT J. RICHARDS

stock, that only justifies altruistic behavior in an empirical sense, not a moral sense. That is, the biological function of altruism may be understood (and thus justified) as a consequence of natural selection, but so may aggressive and murderous impulses. Authentic altruism requires a moral justification. Such justification, as I will undertake below, will show it morally superior to contract altruism.

The general character of RV may now be a little clearer. Its further features can be elaborated in a consideration of the principal objections to evolutionary ethics.

III

Systems of evolutionary ethics, of both the Darwinian and the Wilsonian varieties, have attracted objections of two distinct kinds: those challenging their adequacy as biological theories and those their adequacy as moral theories. Critics focusing on the biological part have complained that complex social behavior does not fall obviously under the direction of any genetic program, indeed, that the conceptual structure of evolutionary biology prohibits the assignment of any behavioral pattern exclusively to the genetic program and certainly not behavior that must be responsive to complex and often highly abstract circumstances (i.e., requiring the ability to interpret a host of subtle social and linguistic signs) (Gould 1977, pp. 251—59; Burian 1978; Lewontin et al. 1984, pp. 265—90). A present-day critic of Darwin's particular account might also urge that the kind of group selection his theory requires has been denied by many recent evolutionary theorists (Williams 1966, pp. 92—124), and that even of those convinced of group selection (e.g., Wilson), a number doubt it has played a significant role in human evolution. And if kin selection, instead of group selection of unrelated individuals, be proposed as the source of altruism in humans, a persistent critic might contend that human altruistic behavior is often extended to non-relatives. Hence kin selection cannot be the source of the ethical attitude (Mattern 1978; Lewontin et al., 1984, p. 261).

Within the biological community, the issues raised by these objections continue to be strenuously debated. So, for instance, some ethologists and sociobiologists would point to very intricate animal behaviors that are, nonetheless, highly heritable (Wilson 1975; Eibl-Eibesfeldt, 1970). And Ernst Mayr has proposed that complex instincts can be classified as exhibiting a relatively more open or a more closed program: the latter remain fairly impervious to shifting environments, while the former respond more sensitively to changing circumstances (1976). Further, different animal species show social hierarchies of amazing complexity (e.g., societies of low-land baboons) and display repertoires of instinctive behaviors whose values are highly context dependent (e.g., the waggle-

A DEFENSE OF EVOLUTIONARY ETHICS 275

dance of the honey bee, which specifies direction and distance of food sources). This suggests the likelihood that instinctual and emotional responses in humans can be triggered by subtle interpretive perception (e.g., the survival responses of fear and flight can be activated by a stranger who points a gun at you in a Chicago back alley). Cross-cultural studies, moreover, have evinced similar patterns of moral development, which could be explained, at least in part, as the result of a biologically based program determining the sequence of moral stages that individuals in conventional environments follow (Wilson 1975, pp. 562—63). Further, recent impressive experiments have shown that group selection may well be a potent force in evolution (Wade 1976; 1977). Finally, some anthropologists have found kin selection to be a powerful explanation of social behavior in primitive tribes (Chagnon and Irons 1979). How these issues will eventually fall out, however, is not immediately my concern, since only developing evolutionary theory can properly arbitrate them. At this time we can say, I believe, that the objections based on a particular construal of evolution seem not to be fatal to an evolutionary ethics — and this admission suffices for my purposes.

Concerning the other class of objections, those directed to the distinctively moral character of evolutionary ethics, resolution does not have to wait, for the issues are factually mundane, though conceptually tangled. Against the moral objections, I will attempt to show that the evolutionary approach to ethics need abrogate no fundamental meta-ethical principles. For the sake of getting to the conceptual difficulties, I will assume that the biological objections concerning group and kin selection and an evolutionary account of complex social behavior have been eliminated. With this assumption, I can then focus on the question of the moral adequacy of an evolutionary ethics.

The objections to the adequacy of the distinctively moral component of evolutionary ethics themselves fall into two classes: objections to the entire framework of evolutionary ethics and objections based on the logic or semantics of the conceptual relations internal to the framework. For convenience I will refer to these as *framework questions* and *internal questions*. Questions concerning the framework and the internal field overlap, since some problems will be transitive — i.e., a faulty key principle may indict a whole framework. The interests of clarity may, however, be served by this distinction. Another helpful distinction is that between ethics as a descriptive discipline and ethics as an imperative discipline. The first will try to give an accurate account of what ethical principles people actually use and their origin: this may be regarded as a part of social anthropology. The latter urges and recommends either the adoption of the principles isolated or that they be considered *the ethically adequate principles*. The former kind of theory will require *empirical justification*, the latter *moral justification*.

276   ROBERT J. RICHARDS

Let me first consider some important internal challenges to both the empirical and moral justification of evolutionary ethics. It has been charged, for instance, that the concept of "altruism" when used to describe a soldier bee sacrificing its life for the nest has a different meaning than the nominally similar concept that describes the action of a human soldier who sacrifices his life for his community (Burian 1978; Mattern 1978; Alper 1978). It would be illegitimate, therefore, to base conclusions about human altruism on the evolutionary principles governing animal altruism. Some critics further maintain that the logic of the concept's role in sociobiology and in any adequate moral system must differ, since the biological usage implies genetic selfishness, while the moral use implies unselfishness.[4] I do not believe these are lethal objections. First, the term "altruism" does not retain a univocal meaning even when used to describe various human actions. Its semantic role in a description of parents' saving for their children's education surely differs from its role in a description of a stranger's jumping into a river to save a drowning child. Nonetheless the many different applications to human behavior and the several applications to animal behavior intend to pick out a common feature, namely that the action is directed to the welfare of the recipient and cost the agent some good for which reciprocation would not normally be expected. Let us call this "action altruism." We might then wish to extend, as sociobiologists are wont to do, the description "altruistic" to the genes that prompt such action, but that would be by causal analogy only (as when we call Tabasco "hot" sauce). Hence the explanation of human or animal "action altruism" by reference to "selfish genes" involves no contradiction; for the concept of "genetic selfishness" is antithetic neither to "action altruism" — since it is not applied to the same category of object — nor to "genetic altruism," for they are implicitly defined to be compatible by sociobiologists. It is only antithetic to clarity of exposition. For the real question at issue in applying the concepts of (action) altruism and (action) selfishness is whether the agent is motivated principally to act for the good of another or himself. Of course, one could, as a matter of linguistic punctiliousness, refrain from describing any animal behavior or genetic substrate as "altruistic." The problem would then cease to be semantic and become again one of the empirical adequacy of evolutionary biology to account for similar patterns of behavior in men and animals.

Though some varieties of utilitarianism denominate behavior morally good if it has certain consequences, the evolutionary ethics that I am advocating regards an action good if it is intentionally performed from a certain kind of motive and can be justified by that motive. I will assume as an empirical postulate that the motive has been established by community or kin selection. The altruistic motive encourages the agent to attended to the needs of others, such needs as either biology or culture (or both) interpret for the agent. Aristotelian-Thomistic ethics, as well as the very different Kantian moral philosophy, holds that action from appropriate motives, not action having desirable consequences, is necessary to render an act moral. The common-sense moral tradition sanctions the same distinction; that tradition prompts us, for example, to judge those Hippocratic physicians who risked their lives during the Athenian plague as moral heros — even though their therapies just as often hastened the deaths of their patients. The Hippocratics acted from altruistic motives — ultimately to advance the community good (i.e., the health and welfare of the group), proximately to do so through certain actions directed, unfortunately, by invincibly defective medical knowledge.

This non-consequentialist feature of RV leads, however, to another important internal objection. It suggests that either animal altruism does not stem from altruistic motives, or that animals are moral creatures (since moral creatures are those who act from moral motives) (Mattern 1978). Yet if animal altruism does not arise from altruistic motives and thus is only nominally similar to human altruism, then there is no reason to postulate community selection as the source of both and we cannot, therefore, use evidence from animal behavior to help establish RV. Thus either the evolutionary explanation of morals is deficient or animals are moral creatures. But no system that renders animals moral creatures is acceptable. Hence the evolutionary explanation is logically deficient.

To answer this objection we must distinguish between altruistic motives and altruistic (or moral) intentions. Though my intention is to write a book about evolutionary theories of mind, my motive may be either money (foolish motive that), prestige, professional advancement, or something of a higher nature. Human beings form intentions to act for reasons (i.e., motives), but animals presumably do not. We may then say that though animals may act from altruistic motives, they can neither form the intention of doing so, nor can they justify their behavior in terms of its motive. Hence they are not moral creatures. Three conditions, then, are necessary and sufficient for denominating an action moral: (1) it is performed from an altruistic (or moral) motive; (2) the agent intended to act from the motive; and (3) the agent could justify his action by appeal to the motive.

The distinction between motives and intentions, while it has the utility of overcoming the objection mentioned, seems warranted for other reasons as well. Motives consist of cognitive representations of goals or goal directed actions coupled with positive attitudes about the goal (e.g., the Hippocratic physician wanted to reinstate a humoral balance so as to effect a cure). Appeal to the agent's motives and his beliefs about the means to attain desired goals (e.g., the physician believed continued purging would produce a balance) provides an explanation of action (e.g., that the physician killed his patient by producing a severe anemia). Intentions, on the other hand, should not be identified with motives or beliefs, though they operate on both. Intentions are conscious acts that recruit motives and beliefs to guide behavior (e.g., the physician, motivated by commitment to the Oath, intended to cure his patient through purging). Intentions alone may not adequately explain action (e.g., the physician killed his patient because he intended to cure him!). Intentions, however, confer moral responsibility, while mere motives only furnish a necessary condition for the ascription of responsibility. To see this, consider Sam, a man who killed his mistress by feeding her spoiled paté. Did he murder her? Before the court, Sam planned to plead that yes, he had the motive (revenge for her infidelity) and yes, he knew spoiled paté would do it, but that in giving her the paté he nonetheless did not intend to kill her. He thought he could explain it by claiming that his wife put him in an hypnotic trance that suppressed his moral scruples. Thus, though he acted on his desire for revenge, he still did not intend to kill his erstwhile lover. Sam's lawyer suggested a better defense. He should plead that though he had the motive and knew that spoiled paté would do her in, yet he did not intend to kill her since he did not know this particular paté was spoiled. The moral of this sordid little example is threefold. First, simply that motives differ from intentions. Second, that for moral responsibility to be attributed, motives must not only be marshalled (as suggested by the second defense), but consciously marshalled (as suggested by the first defense). And finally, that conscious marshalling of motives and beliefs allows a justification of action (or in their absence, an excuse) by the agent (as suggested by both defenses).

The charge that RV would make animals moral creatures is thus over-turned. For we assume that animals, though they may act from altruistic motives, cannot intend to do so. Nor can they justify or defend their behavior by appeal to such motives. Generally we take a moral creature to be one who can intend action and justify it.

In addition to these several objections to specific features of the internal logic and coherence of RV (and other similar systems of evolutionary ethics), one important objection attempts to indict the whole framework by pointing out that the logic of moral discourse implies the agent can act freely. But if evolutionary processes have stampted higher organisms with the need to serve the community good, this suggests that ethical decisions are coerced by irrational forces — that men, like helpless puppets, are jerked about by strands of their DNA. There are, however, four considerations that should defuse the charge that an evolutionary construction of behavior implies the denial of authentic moral choice. First, we may simply observe that the problem of compatibility of moral discourse and scientific discourse (which presumes, generally, that every event, at least at the macroscopic level, has a cause) is hardly unique to evolutionary ethics. Most every ethical system explicitly or implicitly recognizes the validity of causal explanations of human behavior (which

A DEFENSE OF EVOLUTIONARY ETHICS 279

explanatory efforts imply the principle that every event has a cause). Hence, this charge is not really a challenge to an evolutionary ethics, but to the possibility of meaningful ethical discourse quite generally. Nonetheless, let us accept the challenge and move to a second consideration. Though evolutionary processes may have resulted in sets of instinctual urges (e.g., to nurture children, alleviate obvious distress, etc.) that promote the welfare of the community, is this not a goal at which careful ethical deliberation might also arrive? Certainly many moral philosophers have thought so. Moreover, an evolutionary account of why men generally act according to the community good does not invalidate a logically autonomous argument which concludes that this same standard is the ultimate moral standard. The similar case of mathematical reasoning is instructive. Undoubtedly we have been naturally selected for an ability to recognize the quantitative aspects of our environment. Those protomen who failed to perform simple quantitative computations (such as determining the closest tree when the saber-tooth charged) have founded lines of extinct descendants. A mathematician who concedes that this brain has been designed, in part at least, to make quantitative evaluations need not discard his mathematical proofs as invalid, based on a judgment coerced by an irrational force. Nor need the moralist (Fried 1978). Third, the standard of community good must be intelligently applied. Rational deliberation must discover what actions in contingent circumstances lead to enhancing the community welfare. Such choices are not automatic but the result of improvable reason. Finally, the evolutionary perspective indicates that external forces do not conspire to wrench moral acts from a person. Rather, man is ineluctably a moral being. Aristotle believed that men where by nature moral creatures. Darwin demonstrated it.

I wish now to consider one final kind of objection to an evolutionary ethics. It requires special and somewhat more extended treatment, since its force and incision have been thought to deliver the coup de grace to all Darwinizing in morals.

IV

RV Escapes the Usual Form of the Naturalist Fallacy

G. E. Moore first formally charged evolutionary ethicians — particularly Herbert Spencer — with committing the naturalistic fallacy (1903, pp. 46—58). The substance of the charge had been previously leveled against Spencer by both his old friend Thomas Huxley (1893) and his later antagonist Henry Sidgwick (1902, p. 219). Many philosophers subsequently have endorsed the complaint against those who would make the Spencerean turn. Bertrand Russell, for instance, thumped it with characteristic *élan*:

280   ROBERT J. RICHARDS

> If evolutionary ethics were sound, we ought to be entirely indifferent as to what the
> course of evolution may be, since whatever it is is thereby proved to be the best. Yet if
> it should turn out that the Negro or the Chinaman was able to oust the European, we
> should cease to have any admiration for evolution; for as a matter of fact our prefer-
> ence of the European to the Negro is wholly independent of the European's greater
> prowess with the Maxim gun (quoted by Flew 1967, p. 44).

Anthony Flew glosses this passage with the observation that "Russell's
argument is decisive against any attempt to define the ideas of right and
wrong, good and evil, in terms of a neutrally scientific notion of evolution"
(1967, p. 45). He continues in his tract *Evolutionary Ethics* to pin-point
the alleged fallacy:

> For any such move to be sound [i.e., "deducing ethical conclusions directly from
> premises supplied by evolutionary biology"] the prescription in the conclusion must be
> somehow incapsulated in the premises; for, by definition, a valid deduction is one in
> which you could not assert the premises and deny the conclusion without thereby con-
> tradicting yourself (1967, p. 47).

Flew's objection is, of course, that one could jolly well admit all the
declared facts of evolution, but still logically deny any prescriptive
statement purportedly drawn from them.

This objection raises two questions for RV: Does it commit the fallacy
as here expressed? and Is it a fallacy after all? I will endeavor to show that
RV does not commit this supposed fallacy, but that even if at some level it
derives norms from facts, it would yet escape unscathed, since the
"naturalist fallacy" describes no fallacy.

There are two ways in which evolutionary ethics has been thought to
commit the naturalist fallacy.[5] Some versions of evolutionary ethics have
represented the current state of our society as ethically sanctioned, since
whatever has evolved is right. Haeckel believed, for instance, that evolution had produced a higher German culture which could serve as a norm
for judging the moral worth of men of inferior cultures. Other versions of
evolutionary ethics have identified certain long-term trends in evolution,
which they *ipso facto* deem good; Julian Huxley, for example, held efforts
at greater social organization were morally sanctioned by the fact that a
progressive integration has characterized social evolution (1947, p. 136).
But RV (and its parent, Darwin's original moral theory) prescribes neither
of these alternatives. It does not specify a particular social arrangement as
being best; rather, it supposes that men will seek the arrangement that
appears best to enhance the community good. The conception of what
constitutes such an ideal pattern will change through time and over
different cultures. Nor does this theory isolate a particular historical trend
and enshrine that. During long periods in our prehistory, for instance, it
might have been deemed in the community interest to sacrifice virgins,
and this ritual might in fact have contributed to community cohesiveness

and thus have been of continuing evolutionary advantage. But RV does not sanction thereafter the sacrifice of virgins, only acts that, on balance, appear to be conducive to the community good. As the rational capacities of men have evolved, the ineffectiveness of such superstitious behavior has become obvious. The theory maintains that the criterion of morally approved behavior will remain constant, while the conception of what particular acts fall under the criterion will continue to change. RV, therefore, does not derive ethical imperatives from evolutionary facts in the usual way.

But does RV derive ethical norms from evolutionary facts in some way? Unequivocally, yes. But to see that this involves no logically — or morally — fallacious move requires that we first consider more generally the roles of factual propositions in ethics.

Empirical Hypotheses in Ethics

Empirical considerations impinge upon ethical systems both as *framework* assumptions and as *internal* assumptions. In analyzing ethical systems, therefore, framework questions or internal questions may arise. Framework questions, as indicated above, concern the relationship of the ethical system to other conceptual systems and, via those other systems, to the worlds of men and nature. They stimulate such worries as: Can the ethical system be adopted by men in our society? How can such a moral code be justified? Must ethical systems require rational deliberation before an act can be regarded as moral? Internal questions concern the logic of the moral principles and the terms of discourse of a given ethical system. They involve such questions as: Is abortion immoral in this system? What are the principles of a just war in this system? Some apparent internal questions — such as, "What is the justification for fostering the community good?" — are really framework questions — to wit, "How can this system, whose highest principle is 'foster the community good,' be justified?" The empirical ties an ethical framework has to the worlds of men and nature are transitive: they render the internal principles of the system ultimately dependent upon empirical hypotheses and assumptions.

Every ethical system fit for men includes at least three kinds of empirical assumptions (or explicit empirical hypotheses) regarding frameworks and, transitively, internal elements. First, every ethical system recommended for human adoption makes certain framework assumptions about man's nature — i.e., about the kind of creatures men are such that they can heed the commands of the system. Even the austere ethics of Kant supposes human nature to be such, for instance, that intellectual intuitions into the noumenal realm are foreclosed; that behavior is guided by maxims; that human life is finite; that men desire immortality, etc. An evolutionary ethics also forms empirical suppositions about human nature, ones extracted from evolutionary theory and its supporting evidentiary base. Consequently, no objection to RV (or any evolutionary ethics) can be made on the grounds that it requires empirical assumptions — all ethical systems do.

A second level of empirical assumption is required of a system designed for culturally bound human nature: connections must be forged between the moral terms of the system — e.g., "goods," "the highest good," etc. — and the objects, events, and conditions realized in various human societies. What are goods (relative and ultimate) in one society (e.g., secular Western society) may not be in another (e.g., a community of Buddhists monks). In one sense these are internal questions of how individual terms of the system are semantically related to characterizations of a given society's attitudes, observations, and theoretical knowledge (e.g., the virtue of sacrificing virgins, since that act produces life-giving crops; the evil of thermonuclear war, since it will likely destroy all human life; etc). But quickly these become framework questions. So, the question of what a society deems the highest good may become the question of justifying a system whose ultimate moral principle is, for example, "Seek the sensual pleasure of the greatest number of people." Since the interpretation of moral terms will occur during a particular stage of development, it may be that certain acts sanctioned by one society's moral system might be forbidden by ours, yet still be, as far as we are concerned, moral. That is, we may be ready not only to make the analytic statement that "The sacrifice of virgins was moral in Inca society," but also to judge the Inca high priest as a good and moral man for sacrificing virgins. Such judgment, of course, would not relieve us of the obligation to stay, if we could, the priest's hand from plunging in the knife.

A third way in which empirical assumptions enter into framework questions regards the methods of justifying the system and its highest principles. Consider an ethical system that has several moral axioms, of the kind we might find adopted in our own society: e.g., lying is always wrong; abortion is immoral; adultery is bad, etc. If asked to justify these precepts, someone might attempt to show that they conformed to a yet more general moral canon, such as the Golden Rule, the Ten Commandments, the Greatest Happiness Principle, etc. But another common sort of justification might be offered. Appeal might be made to the fact that moral authorities within our society have condemned or praised certain actions. Such an appeal, of course, would be empirical. Yet the justifying argument would meet the usual criterion of validity, if the contending parties implicitly or explicitly agreed on a meta-moral inference principle such as "Conclude as sound ethical injunctions what moral leaders preach." Principles of this kind — comparable to Carnap's "meaning postulates" — implicitly regulate the entailment of propositions within a particular community of discourse.[6] They would include rules that govern use of the standard logical elements (e.g., "and," "or," "if . . . , then," etc.) as well as the other terms of discourse. Thus in a community of analytic philosophers, the rule "From 'a knows x,' conclude 'x'" authorizes arguments of the kind: "Hilary knows we are not brains in vats, so we are not brains in vats." In a particular community, the moral discourse of its members could well be governed by a meta-moral inference principle of the sort mentioned. Such an inference rule would justify the argument from moral authority, because the interlocutors could not assert the premise (e.g., "Moral leaders believe abortion is wrong") and deny the conclusion (e.g., "Abortion is wrong") without contradiction. In this case, then, one would have a perfectly valid argument that derived morally normative conclusions from factual propositions.

The cautious critic, however, might object that this argument does not draw a moral conclusion (e.g., "Abortion is wrong") solely from factual premises (e.g., "Moral leaders believe abortion is wrong"), but also from the meta-moral inference principle, which is not a factual proposition — hence, that I have not shown a moral imperative can be derived from factual premises alone. Moreover, so the critic might continue, the inference principle actually endorses a certain moral action (e.g., shunning abortion) and thus incorporates a moral imperative — consequently that I have assumed a moral injunction rather than deriving it from factual premises. This two pronged objection requires a double defense, one part that examines the role of inference principles, the other that analyzes what such principles enjoin.

The logical structure of every argument has, implicitly at least, three distinguishable parts: (1) one or more premises; (2) a conclusion; and (3) a rule or rules that permit the assertion of the conclusion on the basis of the premises. The inference rule, however, is not 'from which' a conclusion is drawn, but 'by which' it is drawn. If rules were rather to be regarded as among the premises from which the conclusion was drawn, there would be no principle authorizing the move from premises to conclusion and the argument would grind to a halt (as Lewis Carroll's tortoise knew). Hence, the first prong of the objection may be bent aside.

The second prong may also be diverted. An inference principle logically only endorses a conclusion on the basis of the premises — i.e., it enjoins not a moral act (e.g., shunning abortion) but an epistemological act (e.g., accepting the proposition "Abortion should be shunned"). Once we are convinced of the truth of a proposition, we might, of course, act in light of it; but that is an entirely different matter — at least logically. These two considerations, I believe, take the bite out of the objection.

We have just seen how normative conclusions may be drawn from factual premises. This would be an internal justification if the contending parties initially agreed about inference principles. However, they may not agree, and then the problem of justification would become the framework issue of what justifies the inference rule. It would also turn out to be a framework question if the original challenge were not to an inference rule, but to a cardinal principle (e.g., the Greatest Happiness, the Golden Rule, etc.) that was used as the axiom whence the moral theorems of the system were derived. To meet a framework challenge, one must move outside the system in order to avoid a circularly vicious justification. When philosophers take this step, they typically begin to appeal (and ultimately must) to common-sense moral judgments. They produce test cases to determine whether a given principle will yield the same moral conclusions as would commonly be reached by individuals in their society. In short, frameworks, their inference rules, and their principles are usually justified in terms of intuitively clear cases — i.e., in terms of matters of fact. Such justifying arguments, then proceed from what people as a matter of fact believe to conclusions about what principles would yield these matters of fact.

This method of justifying norms is not confined to ethics. It is also used, for example, in establishing "modus ponens" as the chief principle of modern logic: i.e., modus ponens renders the same arguments valid that rational men consider valid. But this strategy for justifying norms utilizes empirical evidence, albeit of a very general sort. Quite simple the strategy recognizes what William James liked to pound home: that no system can validate its own first principles. The first principles of an ethical system can be justified only by appeal to another kind of discourse, an appeal in which factual evidence about common sentiments and beliefs is adduced. (It is at this level of empirical appeal, I believe, that we can dismiss Wilson's suggestion that contract altruism — i.e., "I'll scratch your back, if and only if you'll scratch mine" — is the highest kind. For most men would declare an action non-moral if done only for personal gain. I will discuss the relation of this kind of empirical strategy to RV in a moment.)

The contention that the inference principles or cardinal imperatives of a moral system can ultimately be justified only by referring to common beliefs and practices seems degenerately relativistic. To what beliefs, to what practices, to what men shall we appeal? Should we look to the KKK for enlightenment about race relations? Further, even if the argument were correct about the justification of logical rules by appeal to the practices of rational men, the same seems not to hold for moral rules, because persons differ far less in their criteria of logical soundness. The analogy between logical imperatives and moral imperatives thus appears to wither. These objections are potent, though I believe they infect all attempts to justify moral principles (Gewirth 1982, pp. 43—5). In the case of evolutionary ethics, however, I think the prognosis is good. I will take up the last objection first and then turn to the first to sketch an answer that will be completed in the final section of this essay.

The last objection actually grants my contention that logical inference rules or principle are justified by appeal to beliefs and practices; presumably the objection would then be deflated if a larger consensus were likely in the case of moral justification. The second objection, then, either accepts my analysis of justificatory procedures or it amounts to the first objection that appeal to the beliefs and practices of men fails to determine the reference class and becomes stuck in the moral muck of relativism. My sketchy answer to the second objection, which will be filled in below, is simply that the reference class is moral men (just as in logical justification it is the class of rational men) and that we can count on this being a rather large class because evolution has produced it so (just as it has produced a large class of rational creatures). Indeed, one who cannot comprehend the soundness of basic moral principles, along with one who cannot comprehend the soundness of basic logical principles, we regard as hardly a man. Moreover, we have evolved, so I contend, to recognize and approve of moral behavior when we encounter it (just as we have evolved to recognize and approve of logical behavior). Those protohuman lineages that have not had these traits selected for, have not been selected at all. This does not mean, of course, that every infant slipping fresh out of the womb will respond to others in altruistic ways or be able to formulate maxims of ethical behavior. Cognitive maturity must be reached before the individual can become aware of the signs of human need and bring different kinds of response under a common description — e.g., altruistic or morally good behavior. Likewise, maturity and cultural transmission must complement the urges for logical consistency that nature has instilled. We should not, therefore, be misled by the KKK example. Most Klansmen are probably quite moral people. They simple have unsound beliefs about, among other things, different races, international conspiracies, etc. Our chief disagreement with them will not be with their convictions about heeding the community good, but with their beliefs about what leads to that good.

This brief discussion of justification of ethical principles indicates how the concept of justification must, I believe, be employed. "To justify" means "to demonstrate that a proposition or system of propositions conforms to a set of acceptable rules, a set of acceptable factual propositions, or a set of acceptable practices. The order of justification is from rules to empirical propositions about beliefs and practices. That is, if rules serving as inference principles or the rules serving as premises (e.g., the Golden Rule) of a justifying argument are themselves put to the test, then they must be shown to conform either to still more general rules or to empirical propositions about common beliefs and practices. Barring an infinite regress, this procedure must end in what are regarded as acceptable beliefs or practices. Aristotle, for instance, justified the forms of syllogistic reasoning by showing that they made explicit the patterns employed in argument by rational men. Kant justified the categorical imperative and the postulates of practical reason by demonstrating, to his

286   ROBERT J. RICHARDS

satisfaction, that they were the necessary conditions of common moral experience: that is, he justified normative principles by showing that their application to particular cases reproduced the common moral conclusions of 18th century German burgers and Pietists.

If this is an accurate rendering of the concept of justification, then the justification of first moral principles and inference rules must ultimately lead to an appeal to the beliefs and practices of men, which of course is an empirical appeal. So moral principles ultimately can be justified only by facts. The rebuttal, then, to the charge that at some level evolutionary ethics must attempt to derive its norms from facts is simply that every ethical system must. Consequently, either the naturalistic fallacy is no fallacy, or no ethical system can be justified. But to assert that no ethical system can be justified is just to say that ultimately no reasons can be given for or against an ethical position, that all ethical judgments are nonrational. Such a view sanctions the canonization of Hitler along with St. Francis. Utilizing, therefore, the common rational strategy of appealing to common beliefs and practices to justify philosophical positions, we must reject the idea that the 'naturalistic fallacy' is a fallacy.

The Justification of RV as an Ethical System

RV stipulates that the community welfare is the highest moral good. It supposes that evolution has equipped human beings with a number of social instincts, such as the need to protect offspring, provide for the general well-being of members of the community (including oneself), defend the helpless against aggression, and other dispositions that constitute a moral creature. These constitutionally imbedded directives are instances of the supreme principle of heeding the community welfare. Particular moral maxims, which translate these injunctions into the language and values of a given society, would be justified by an individual's showing that, all things considered, following such maxims would contribute to the community welfare.

To justify the supreme principle, and thus the system, requires a different kind of argument. I wish to remind the reader, however, that I will attempt to justify RV as a moral system *under the supposition that it correctly accounts for all the relevant biological facts.* I will adopt the forensic strategy that several good arguments make a better case than one. I have three justifying arguments.

First Justifying Argument. The first argument is adapted from Alan Gewirth who, I believe, has offered a very compelling approach to deriving an "ought" from an "is." He first specifies what the concept of 'ought' means (i.e., he implicitly indicates the rule governing its deployment in arguments). He suggests that it typically means: "necessitated or required by reasons stemming from some structured context" (1982, p. 108). Thus in the inference "It is lightning, therefore it ought to thunder," the "ought" means, he suggests, "given the occurrence of lightning, it is required or necessary that thunder also occur, this necessity stemming from the law-governed context of physical nature" (1982, p. 108). Here descriptive causal laws provide the major (unexpressed) premise of the derivation of "ought" from "is." The practical sphere of action also presents structured contexts. So, for example, as a member of the University, I ought to prepare my classes adequately. Now Gewirth observes that derivation of a practical ought, such as the one incumbent on a university professor, requires first that one accept the structured context. But then, he contends, only hypothetical 'ought's are produced: e.g., *If* I am a member of the University, then I ought to prepare classes adequately. Since nothing compels me to become a member of the University, I can never be categorically enjoined: "Prepare classes adequately." Gewirth further argues that if one decides to commit oneself to the context, e.g., to university membership, then the derivation of 'ought' will really be from an obligation assumed, that is from one 'ought' to another 'ought.' He attempts to overcome these obstacles by deriving 'ought's from a context that the person cannot avoid, cannot choose to accept or reject. He claims that the generic features of human action impose a context that cannot be escaped and that such a context requires the agent regard as good his freedom and well being. From the recognition that freedom and well-being are necessary conditions of all action, the agent can logically derive, according to Gewirth, the proposition "I have a right to freedom and basic well-being." This 'rights' claim, which indeed implies "I ought to have freedom and well-being," can only be made if the agent must grant the same right to others. Since the claim depends only on what is required for human agency and not on more particular circumstances, Gewirth concludes that every one must logically concede the right to any other human agent.

Gewirth's derivation of 'ought' from 'is' has been criticized by Alisdair MacIntyre among other. MacIntyre simply objects that because I have a need for certain goods does not entail that I have a right to them, i.e., that others are obliged to help me secure them (1981, pp. 64—5). This, I believe, is a sound objection to Gewirth's formulation. Gewirth's core position, however, can be preserved, if we recognize that a generally accepted moral inference principle sanctions the derivation of rights-claims from empirical claims about needs common to all men. Anyone who doubts the validity of such an inference principle need only perform the empirical test mentioned above (i.e., consult the kind of inferences most men actually draw). Yet even if we granted the force of MacIntyre's objection to Gewirth, the evolutionary perspective permits a similar derivation, though without the objectionable detour through human needs.

288   ROBERT J. RICHARDS

Evolution provides the structured context of moral action: it has con-
situted men not only to be moved to act for the community good, but also
to approve, endorse, and encourage others to do so as well. This particular
formation of human nature does not impose an individual need, not
something that will be directly harmful if not satisfied; hence, the question
of a logical transition from an individual (or generic) need to a right does
not arise. Rather, the constructive forces of evolution impose a practical
necessity on each man to promote the community good. We must, we are
obliged to heed this imperative. We might attempt to ignore the demand
of our nature by refusing to act altruistically, but this does not diminish its
reality. The inability of men to harden their consciences completely to
basic principles of morality means that sinners can be redeemed. Hence,
just as the context of physical nature allows us to argue "Since lightening
has struck, thunder ought to follow," so the structured context of human
evolution allows us to argue *"Since each man has evolved to advance the
community good, each ought to act altruistically."*

Two important objections might be lodged at this juncture. First, that
just because evolution has outfitted men with a moral sense of commitment to the community welfare, this fact *ipso sole* does not impose any
obligation. After all, evolution has installed aggressive urges in men, but
they are not morally obliged to act upon them. A careful RVer will
respond as follows. An inborn commitment to the community welfare, on
the one hand, and an aggressive instinct, on the other, are two greatly
different traits. In the first, the particular complex of dispositions and
attitudes produced by evolution (i.e., through kin and group selection in
my version) leads an individual to behave in ways that we can generally
characterize as acting for the community good; in the second, the behavior
cannot be so characterized. Moral 'ought'-propositions are not sanctioned
by the mere fact of evolutionary formation of human nature, but by the
fact of the peculiar formation of human nature we call "moral," which has
been accomplished by evolution. (The evolutionary formation of human
nature according to other familiar biological relations might well sanction
such propositions as "Since he has been constituted an aggressive being by
evolution, he ought to react hostilely when I punch him in the nose.")

The second objection points out what appears to be a logical gap
between the structured context of the evolutionary constitution of man
and an 'ought'-proposition. Even if it is granted that evolution has formed
human nature in a particular way, call it the "moral way" (the exact
meaning of which must yet be explored), yet what justifies concluding that
one 'ought' to act altruistically? What justified the move, of course, is an
inference principle to the effect: "From a particular sort of structured
context, conclude that the activity appropriate to the context ought to
occur." Gewirth, in his attempt to show that moral "ought"s can be derived
from "is"s, depends on such a rule; and significantly, MacIntyre's response

does not challenge it. Indeed, MacIntyre employs another such inference rule, which Gewirth would likely endorse: i.e., "From 'needs'-propositions alone one may not conclude to 'claims'-propositions." All meta-level discussions, all attempts to justify ethical frameworks depend on such inference rules, whose ultimate justification can only be their acceptance by rational and moral creatures.

Second Justifying Argument. The second justifying argument amplifies the first. It recognizes that evolution has formed a part of human nature according to the criterion of the community good (i.e., according to the principles of kin and group selection). This we call the moral part. The justification for the imperative advice to a fellow creature "Act for the community good" is therefore: "Since you are a moral being, constituted so by evolution, you ought act for the community good." To bring a further justification for the imperative would require that the premise of this inference be justified, which would entail furnishing factual evidence as to the validity of evolutionary theory (including RV). And this, of course, would be ultimately to justify the moral imperative by appeal to empirical evidence. The justifying argument, then, amounts to: *the evidence shows that evolution has, as a matter of fact, constructed human beings to act for the community good; but to act for the community good is what we mean by being moral. Since, therefore, human beings are moral beings — an unavoidable condition produced by evolution — each ought act for the community good.* This second justifying argument differs from the first only in stressing: (1) that ultimate justification will require securing the evidentiary base for evolutionary theory and the operations of kin and group selection in forming human nature; and (2) that the logical movement of the justification is from — (a) the empirical evidence and theory of evolution, to (b) man's constitution as an altruist, to (c) identifying being an altruist with being moral, to (d) concluding that since men so constituted are moral, they morally ought promote the community good.

Three points need to be made about this second justifying argument in light of these last remarks, especially those under number (2). To begin, the general conclusion reached — i.e., "Since each human being is a moral being, each ought act for the community good" — does not beg the question of deriving moral imperatives from evolutionary facts. The connection between being human and being moral is contingent, due to the creative hand of evolution: it is because, so I allege, that creatures having a human frame and rational mind also underwent the peculiar processes of kin and group selection that they have been formed 'to regard and advance the community good' and approve of altruism in others. (There is a sense, of course, in which a completely amoral person will be regarded as something less than human.) Having such a set of attitudes and acting on them is what we mean by being moral.[7] Further, given our

notion of what it is to be moral, it is a factual question as to whether
certain activity should be described as "moral behavior."

The second point is an evolutionary Kantian one and refers back to the
previous discussion on the nature of justification. If challenged to justify
altruism as being a moral act in reference to which 'ought'-propositions
can be derived, a defender of RV will respond that the objecter should
consult his own intuitions and those commonly of men. If the evolutionary
scenario of RV is basically correct, then the challenger will admit his own
intuitions confirm that he especially values altruistic acts, that spontaneously he recognizes the authority of the urge to perform them, and that he
would encourage them in others — all of which identifies altruistic
behavior with moral behavior. But if he yet questions the reliability of his
own intuitions or if he fails to make the identification (because his own development has been devastatingly warped by a wicked aunt), then
evidence for evolutionary theory and kin and group selection must be
adduced to show that men generally (with few exceptions) have been
formed to approve, endorse, and encourage altruistic behavior.

The third point glosses the meaning of "ought." In reference to structured contexts, "ought to occur," "ought to be," "ought to act," etc.
typically mean "must occur," "must be," "must act, *provided there is no
interference.*" Structured contexts involve causal processes. Typically
"ought" adds to "must" the idea that perchance some other cause might
disrupt the process (e.g., "Lightening has flashed, so it ought to thunder,
that is, it must thunder, provided that no sudden vacuum in the intervening space is created, that there is an ear around to transduce movement of
air molecules into nerve potentials, etc."). In the context of the evolutionary constitution of human moral behavior, "ought" means that the
person must act altruistically, provided he has assessed the situation
correctly and a surge of jealousy, hatred, greed, etc. does not interfere.
The "must" here is a causal "must"; it means that in ideal conditions — i.e.,
perfectly formed attitudes resulting from evolutionary processes, complete
knowledge of situations, absolute control of the passions, etc. — altruistic
behavior would necessarily occur in the appropriate conditions. When
conditions are less than ideal — when, for example, the severe stress of
war causes an individual to murder innocent civilians — then we might be
warranted in expressing another kind of 'ought'-proposition: e.g., under
conditions of brutalizing war, some soldiers ought to murder non-com-
batants. In such cases, of course, the "ought" is not a moral ought; it is not
a moral ought because the 'ought'-judgment is not formed in recognition
of altruism as the motive for behavior. In moral discourse, expressions of
'ought'-propositions have the additional function of encouraging the agent
to avoid or reject anything that might interfere with the act. The "ought"
derived from the structured context of man's evolutionary formation, then,
will be a moral ought precisely because the activities of abiding the

community good and approving of altruistic behavior constitute what we mean and (if RV is correct) must mean by "being moral."

This second justifying argument recognizes that there are three kinds of instances in which moral imperatives will not be heeded. First, when a person misconstrues the situation (e.g., when a person, without warrant, takes the life of another, because he didn't know the gun was loaded and, therefore, could not have formed the relevant intention). But here, since the person has misunderstood the situation, no moral obligation or fault can be ascribed. The second case occurs when a person does understand the moral requirements of the situation, but refuses to act accordingly. This is analogous to the case when we say thunder ought to have followed lightening, though did not (because of some intervening cause). The person who so refuses to act on a moral obligation will not be able, logically, to justify his action, and will be called a sinner. Finally, there is the case of the person born morally deficient, the sociopath who robs, rapes, and murders without a shadow of guilt. Like the creature born without cerebral hemispheres, the sociopath has been deprived of what we have come to regard as an essential organ of humanity. We do not think of him as a human being in the full sense. RV implies that such an individual, strange as it seems, cannot be held responsible for his actions. He cannot be held morally guilty for his crimes, since he, through no fault of his own, has not been provided the equipment to make moral decisions. This does not mean, of course, that the community should not be protected from him, nor that it should permit his behavior to go unpunished; indeed, community members would have an obligation to defend against the sociopath and inflict the kind of punishment that might restrain unacceptable bahavior.

Third Justifying Argument. The final justifying argument for RV is second order. It shows RV warranted because it grounds other of the key strategies for justifying moral principles. Consider how moral philosophers have attempted to justify the cardinal principles of their systems. Usually they have adopted one of three methods. They might, with G. E. Moore, proclaim that certain activities or principles of behavior are intuitively good, that their moral character is self-evident. But such moralists have no ready answer to the person who might truthfully say, "I just don't see it, sorry." Nor do they have any way to excluding the possibility that a large number of such people exist or will exist. Another strategy is akin to that of Kant, which is to assert that men have some authentic moral experiences, and from these an argument can be made to a general principle in whose light their moral character is intelligible. But this tactic, too, suffers from the liability that men may differ in their judgments of what actions are moral. Finally, there is the method employed by Herbert Spencer. He asks of someone proposing another principle — Spencer's was that of

292   ROBERT J. RICHARDS

greatest happiness — to reason with him. The outcome should be — if Spencer's principle is the correct one — that the interlocutor will find either that actions he regards as authentically moral do not conform to his own principle, but to Spencer's, or that his principle reduces to or is another version of Spencer's principle. But here again, it is quite possible that the interlocutor's principle will cover all the cases of action he describes as moral, but will not be reducible to Spencer's principle. No reason is offered for expecting ultimate agreement in any of these cases.

All three strategies suppose that one can find near-universal consent among men concerning what actions are moral and what principles sanction them. Yet no way of conceptually securing such agreement is provided. And here is where RV obliges: it shows that the pith of every man's nature, the core by which he is constituted a social and moral being, has been created according to the same standard. Each heart must resound to the same moral cord, acting for the common good. It may, of course, occur that some men are born deformed in spirit. There are psychopaths among us. But these, the theory suggests, are to be regarded as less than moral creatures, just as those born severely retarded are thought to be less than rational creatures. But for the vast community of men, they have been stamped by nature as moral beings. RV, therefore, shows that the several strategies used to support an ultimate ethical principle will, in fact, be successful, successfully showing, of course, that the community good is the highest ethical standard. But for RV to render successful several strategies for demonstrating the validity of the highest ethical principle is itself a justification.

In this defense of evolutionary ethics, I have tried to do three things — to demonstrate that if we grant certain empirical propositions, then my revised version (RV) of evolutionary ethics: (1) does not commit the naturalistic fallacy as it is usually formulated; (2) does, admittedly, derive values from facts; but (3) does not commit any fallacy in doing so. The ultimate justification of evolutionary ethics can, however, be accomplished only in the light of advancing evolutionary theory.

NOTES

* I am grateful to my colleagues Alan Gewirth and Robert DiSalle, who tried to warn me off bad arguments.
[1] I am in sympathy with the spirit of Ruse's defense of Wilson, though, as will be indicated below, I take exception to a major conclusion concerning the moral primacy of reciprocal altruism and certain aspects of the justification of the system he advances.
[2] The usual models of group selection assume that individual selection and group selection work at cross-purposes, that, for instance, the individual must pay a high price for altruistic behavior (e.g., bees' disemboweling themselves by stinging enemies; risking one's life to save a drowning child, etc.). But in most familar cases, individuals perform altruistic acts at

THE CHALLENGE OF EVOLUTIONARY ETHICS   293

little practical cost. In a hostile environment, those small tribal groups populated by altruists and co-operators would have a decided advantage. Cheating would not likely become wide spread, since the advantage would be quite small and the possible cost quite high (e.g., ostracism of the individual or death of the tribe). Under such circumstances group selection, especially on tribes laced with relatives, might well become a force to install virtuous behavior. For an analysis of the problematic assumptions of most group selection models, see Wade (1978).

[3] This is largely the objection of Marshall Sahlins to the sociobiology of human behavior (1976).

[4] Playing on the apparent reduction of altruistic behavior to genetic selfishness and then to selfishness simply, Lewontin et al. complain: "by emphasizing that even altruism is the consequence of selection for reproductive selfishness, the general validity of individual selfishness in behaviors is supported ... Sociobiology is yet another attempt to put a natural scientific foundation under Adam Smith" (1984, p. 264).

[5] In an early discussion of evolutionary ethics, Ruse (1979, pp. 199—204) affirmed that any evolutionary ethics must commit the naturalist fallacy, and admitted that the two characteristics mentioned in the text produce the most potent objections to evolutionary ethics: without begging the question, we would have no way of specifying what trends or what aspects of the evolutionary process should constitute the moral standard.

[6] For a consideration of inference principles of the kind mentioned, see Carnap (1956, pp. 222—32), Sellars (1948), and McCawley (1981, p. 46).

[7] Gewirth (1982, pp. 82—3) endorses the following criteria as establishing a motive as moral: the agent takes it as prescriptive; he universalizes it; he regards it as over-riding and authoritative; and it is formed of principles that denominate actions right simply because of their effect on other persons. These criteria are certainly met in altruistic behavior described by RV.

[13]

Is Darwin Right?

Keith Sutherland and Jordan Hughes

Look in the *Oxford English Dictionary* and you will find at least three different meanings of the word 'right'. In the first (adjectival) sense of the word — 'correct' — the answer to the question 'is Darwin right', is clearly 'yes'. The views advanced by Larry Arnhart in *Darwinian Natural Right*[1] are also generally right,[2] at least from the perspective of this review. The book tackles a very broad and difficult subject and meets the many obstacles with admirable clarity, breadth, and scholarship. The fact that we disagree with Arnhart in some of the particulars is hardly surprising, given the scope of his undertaking. Not only is Arnhart's synthesis of Aristotle, Hume and Darwin well supported by the textual citations, but the many *controversial* claims are both interesting and provocative. The book's challenging central thesis warrants the attention of all ethicists, political theorists and evolutionary psychologists.

A second (substantive) meaning of 'right' provides Arnhart with his title. According to the *OED,* 'right' in this second sense is 'the standard of permitted and forbidden action within a certain sphere; law; a rule or canon'. The dictionary goes on to elaborate: 'that which is consonant with equity or the light of nature; and which is morally just or due' — and this is where the controversial issues begin. Is morality derived from reason (abstract principles like 'equity') or *pace* G.E. Moore, 'the light of nature'? Does any empirical science of ethics necessarily commit what Moore called the 'naturalistic fallacy'? Can ethics properly begin with *descriptive* premises and arrive at *prescriptive* conclusions? According to most standard interpretations of a famous passage written by David Hume, any such derivation of 'ought' from 'is' is

Correspondence: Keith Sutherland, Imprint Academic, PO Box 1, Thorverton EX5 5YX, UK.
    Email: keith@imprint.co.uk.
Jordan Hughes, Cognitive Neuroscience Lab, Department of Cognitive Science, 0515,
    University of California, San Diego, La Jolla, CA 92093. Email: jordan@ucsd.edu

[1] Review of Larry Arnhart, *Darwinian Natural Right: The Biological Ethics of Human Nature* (SUNY Series in Philosophy and Biology), May 1998, ISBN 0791436942, $24.95. Unless otherwise attributed, page numbers throughout this essay refer to this edition. The authors are grateful to James Blair and Paul Rogers for comments on an earlier draft of this essay. Due to space constraints the appendices (on religion and Hobbes) are only available on the web: http://www.imprint.co.uk/arnhart

[2] Unless you live in Kansas and definitely not 'right-on' in the sense of 'politically correct'.

illicit.[3] So widely held is this interpretation of Hume's dictum that R.M. Hare dubbed it 'Hume's Law' and (unfortunately, in our opinion) the name has stuck. Arnhart rejects these constraints. He thinks that morality is an empirical rather than an abstract philosophical subject area and we will devote the greater part of this review essay to a discussion of that claim.

A third (adjectival) meaning of 'right' is derived from right-handedness and has come to be associated with politicians of a conservative disposition. Critics of sociobiology and evolutionary psychology like Stephen Jay Gould, Richard Lewontin and Steven Rose are (or were) 'more or less Marxist' (Brown, 2000, p. 58). Is a Darwinian approach to ethics of most interest to the political right? Arnhart has elsewhere (2000) suggested this might well be the case but not everyone would agree, including the editor of a special issue of this very journal published earlier this year.[4]

Darwinian Natural Right

Bring me my Bow of burning gold:
Bring me my Arrows of desire: William Blake

The theme of Arnhart's book is 'the good is the desirable', and he attempts to find common ground between Aristotle's idea of 'natural right', Hume's idea of the natural moral sense and Darwin's idea of the moral sense as shaped by natural selection. Arnhart describes this as an 'empiricist' or naturalistic view of ethics, in sharp contrast with the 'transcendentalist' or Kantian view of ethics in which morality is viewed as an autonomous realm of rationality separated from natural desires.

Although it is often believed that the current wave of 'evolutionary psychology' is just 1960s sociobiology with a make-over, the debate over ethics and evolution is sharply divided on this topic. Prominent evolutionary psychologists like Richard Dawkins, David Buss, Steven Pinker and George C. Williams all claim that the moral realm of values transcends (or is at odds with) the natural realm of facts.

> Be warned that if you wish, as I do, to build a society in which individuals cooperate generously and unselfishly towards a common good, you can expect little help from biological nature. Let us try to teach generosity and altruism, because we are born selfish (Dawkins, 1976, p. 3).

This denial that evolution can account for ethics belongs to a tradition of thought that goes back to T.H. Huxley's lecture *Evolution and Ethics* (1894). Huxley believed that human nature is essentially evil — a product of a nasty and unsympathetic natural world, which he famously described as 'red in tooth and claw' — and that ethical behaviour is only the result of cultural influences. Morality, he argued, is a human invention explicitly devised to control and combat selfish and competitive tendencies generated by the evolutionary process. By depicting morality in this way, Huxley was advocating that the search for the origins of morality should be de-coupled from evolution and conducted outside of biology.[5]

---

[3] A close relative of the 'naturalistic fallacy', this pattern of argument often goes by the same name.

[4] Katz (2000). See also Dickens (2000); Singer (2000).

[5] This viewpoint has been described as 'Calvinist sociobiology' by Franz de Waal, because it 'presumes an implicit acceptance of the doctrine of original sin and the natural depravity of human beings' (de Waal, 1996, pp. 13–20).

By contrast, sociobiologist E.O. Wilson's insistence that ethics is rooted in natural moral sentiments shaped by natural selection belongs to the tradition that goes back to Darwin's theory of the 'moral sense' in *The Descent of Man*. Huxley adopted a transcendental dualism derived ultimately from Kant, while Darwin adopted an empiricist naturalism derived from Hume and, before him, Aristotle.

Darwin's ethical naturalism was developed by Edward Westermarck who argued that ethics can be explained as a cultural expression of moral emotions rooted in human nature. Westermarck's theory of the incest taboo was one example of how a Darwinian could explain an ethical norm as rooted in natural emotions shaped by natural selection. It is not surprising therefore that Wilson uses Westermarck's account of the incest taboo to illustrate how Darwinian biology could explain the 'epigenetic rules' of ethics. Although evolutionary psychologists have generally accepted Westermarck's theory of incest avoidance, they would mostly say that the specifically moral content of the incest taboo as a norm transcends biological nature.

Arnhart claims that if Darwinism really is a comprehensive account of human behaviour, then Darwinians must defend a naturalistic account of ethics such as that developed by Darwin, Westermarck and Wilson, in contrast to the dualistic explanation favoured by most evolutionary psychologists.[6] This leads him to formulate ten propositions (pp. 6–7):

1. The good is the desirable, because all animals capable of voluntary movement pursue the satisfaction of their desires as guided by their information about the world.

2. Only human beings, however, can pursue happiness as a deliberate conception of the fullest satisfaction of their desires over a whole life, because only they have the cognitive capacities for reason and language that allow them to formulate a plan of life, so that they can judge present actions in the light of past experience and future expectations.

3. Human beings are by nature social and political animals, because the species-specific behavioural repertoire of *Homo sapiens* includes inborn desires and cognitive capacities that are fulfilled in social and political life.

4. The fulfilment of these natural potentials requires social learning and moral habituation; and although the specific content of this learning and habituation will vary according to the social and physical circumstances of each human group, the natural repertoire of desires and cognitive capacities will structure this variability.

5. We can judge divergent ways of life by how well they nurture the natural desires and cognitive capacities of human beings in different circumstances, but deciding what should be done in particular cases requires prudential judgments that respect the social practices of the group.

6. Rather than identifying morality with altruistic selflessness, we should see that human beings are moved by self-love, and as social animals they are moved to love others with whom they are bonded as extensions of themselves.

7. Two of the primary forms of human sociality are the familial bond between parents and children and the conjugal bond between husband and wife.

---

[6] Readers of this journal will be aware that Steven Pinker also adopts the closet dualist view that consciousness is likewise beyond naturalistic explanation, using an argument derived from Colin McGinn.

8. Human beings have a natural moral sense that emerges as a joint product of moral emotions such as sympathy and anger and moral principles such as kinship and reciprocity.
9. Modern Darwinian biology supports this understanding of the ethical and social nature of human beings by showing how it could have arisen by natural selection through evolutionary history.
10. Consequently, a Darwinian understanding of human nature supports a modern version of Aristotelian natural right.

Objections

This is all highly controversial. The immediate objection is that Arnhart's theory contravenes the prime orthodoxy of the social sciences: human cultures are quintessentially diverse and definitions of 'the good' are at the best arbitrary and, at the worst, a reflection of the distribution of power within a society. According to the 'standard social science model' (SSSM) there is no such thing as 'human nature' and even basic appetites are moulded by cultural forces (social constructivism).

It is not possible to understand how such a profoundly unbiological (and counter-intuitive) point of view became ubiquitous in the social sciences without a brief review of the intellectual history of the twentieth century. Although it was Herbert Spencer, a conservative liberal, who first coined the term 'sociology', Spencer's organic, universalist and biological view of society — immensely popular in late-Victorian England and America — soon went into sharp decline. This was partly on account of increasing ethnographic evidence for the sheer diversity of human cultures and partly on account of the growing influence of the view of Spencer's contemporary Karl Marx that there was no such thing as human nature, only human history. Despite the rearguard efforts of the likes of Pitirim Sorokin at Harvard, within a few generations Marx's viewpoint prevailed in most social science faculties and remained firmly ensconced until the fall of the Berlin Wall. Social Darwinism and eugenics also became tarred by the Nazi brush[7] so it is unsurprising that relativism and social constructivism became the only show in town (MacIntyre, 1982).

The other influence on the social sciences was the new behaviourism emanating from psychology departments at the beginning of the twentieth century. Under the influence of Watson and Skinner psychology eschewed any reference to innatism through the adoption of a radical learning paradigm. The language of conditioning and reinforcement left no place for any talk of universals like a biological 'human nature'. The so-called 'cognitive revolution' of the second half of the century altered this not one iota. The widespread adoption of the computer model, and the spread of structuralism and functionalism — with the associated notion that the 'substrate' was of no particular importance — left no place for a natural history approach to human cognition.[8]

Given this political and intellectual climate, it is unsurprising that the first wave of biological naturalism to hit the social sciences — the sociobiology of the 1960s —

---

[7] It is more than coincidental that Gould, Lewontin and Rose, the three most vocal critics of sociobiology, are all Jewish (Brown, 2000, p. 58).

[8] Lévi-Strauss's *La Pensée Sauvage* (1966) is a prime example of the static ahistorical approach favoured by the structuralists.

should have met with such fierce resistance.[9] But the climate had changed dramatically by the 1990s. The political meltdown of Eastern Europe and the victory of market capitalism meant that only a few diehards were left to stem the rising tide of naturalism in the social sciences. And the belated realization that the 'problem of consciousness' was still ignored by science led the Churchlands to start a campaign for the inclusion of neurobiology within the cognitive sciences. Although all sorts of Cartesian snares have been laid to entrap these poachers and intruders, the technology-driven growth of cognitive neuroscience will ensure that the social sciences of the twenty-first century will no longer escape the strictures of naturalism.

The 'Naturalistic Fallacy'

Given the widespread view that the problem of *consciousness* will eventually yield to naturalistic explanation, why is it then that *ethics* — one of the fruits of human consciousness — is still seen as the domain of philosophers, rather than social scientists? The philosopher G.E. Moore advanced one of the most influential expressions of this view in his famous book *Principia Ethica* (1903). Moore contended that moral philosophers, and particularly the Utilitarians, were logically confused. According to Moore's famous 'open question argument', the word 'good' cannot be defined in terms of natural qualities, as it always makes sense to ask (i.e. it is 'an open question') whether anything possessing natural qualities is *good*. Moore claimed that goodness is a simple, unanalysable, non-natural quality and he charged that any attempt to reduce goodness to any more basic concept, whether natural (e.g. pleasure, for Utilitarians) or *non*-natural (e.g. God's nature, for religious ethicists) committed what he called the 'naturalistic fallacy'. This term has gained a somewhat broader meaning than Moore originally had in mind; it has become applied generally to any attempt to derive values from facts.[10]

Evolutionary psychologists by and large accept Moore's argument, as documented by Ullica Segerstrale's account of the 1996 convention of the Human Behavior and Evolution Society. The keynote address, which deeply disturbed many in the audience, was given by E.O. Wilson. Segerstrale writes (2000, p. 363):

> What baffled and annoyed many — including the conference organizers — was Wilson's insistence that there was only one science, and that the humanities should learn from science, because science prescribes the correct values for us! It seemed that after a twenty-year pause, Wilson had again shamelessly reiterated in public that controversial point from his first chapter in *Sociobiology* — the extrapolation from *is* to *ought* — and with a new vehemence.

After the conference, the local organizer — John Beckstrom — wrote a letter to Wilson expressing the displeasure of the HBES membership:

> I hope I misunderstood you. Among other things, you seemed to be (1) promoting the use of evolutionary history of human behavior in establishing values for the humanities and society in general and (2) denigrating philosophers who point out the follies of the

[9] A group of Maoist student protesters assaulted Wilson at a scientific meeting uttering the cry 'E.O. Wilson, you can't hide / We charge you with genocide' (Brown, 2000, p. 72).

[10] The 'fact–value' distinction was first given its familiar modern form by Max Weber (1904) with regard to the newly emerging social sciences. However, within Anglo-American philosophy, the most influential discussions in ethics have predominantly responded to the related concerns addressed by Hume and Moore.

68   K. SUTHERLAND AND J. HUGHES

Naturalistic Fallacy. In other words, you seemed to be advocating normative uses of sociobiology. If you were, I would have to oppose vigorously your position and I expect many in attendance with whom I later discussed your speech would do likewise.

The irony is that many, if not most, contemporary moral philosophers subscribe neither to Moore's 'open question argument', nor to the celebrated 'fallacy'. Gilbert Harman, in his 1977 book *The Nature of Morality,* stated

> ... as it stands the open question argument is invalid. An analogous argument could be used on someone who was ignorant of the chemical composition of water to 'prove' to him that water is not H_2O. This person will agree that it is not an open question whether water is water but it is an open question, at least for him, whether water is H_2O. Since this argument would not show that water is not H_2O, the open question argument in ethics cannot be used as it stands ... (in Cahn & Markie, 1998, pp. 545–6).

Oliver Curry (2000) offers a similar argument: 'If one approaches ethics with the view that ethical properties are natural properties (what else could they be?),[11] then it is not an error to deduce ethical conclusions from natural premises'. And Bernard Williams calls the 'naturalistic fallacy' a 'spectacular misnomer':

> It is hard to think of any other widely used phrase in the history of philosophy that is such a spectacular misnomer. In the first place, it is not clear why those criticized were committing a fallacy (which is a mistake in inference) as opposed to making what in Moore's view was an error, or else simply redefining a word. More important, the phrase appropriated to a misconceived purpose the useful word 'naturalism.' A naturalistic view of ethics was previously contrasted with a supernaturalistic view, and it meant a view according to which ethics was to be understood in worldly terms, without reference to God or any transcendental authority. It meant the kind of ethical view that stems from the general attitude that man is part of nature. ... What causes even more confusion is that not everyone who, according to Moore, committed this [so-called] fallacy was also a naturalist in the broad and useful sense. Some of the most conspicuous offenders were antinaturalist in the broad and useful sense, such as those who defined goodness in terms of what is commanded or willed by God (Williams, 1985, pp.121–2).

Or, as three other prominent ethicists stated in a recent article:

> It has been known for the last fifty years that Moore discovered no *fallacy* at all. Moreover, Moore's accident-prone deployment of his famous 'open question argument' in defending his claims made appeal to a now defunct intuitionistic Platonism.... To grant Moore all of the resources he deploys or assumes in his official presentation of the open question argument would suffice to bring the whole enterprise of conceptual analysis to a standstill and show nothing about Good in particular. (Darwall *et al.,* 1997).

Why, then, are most social scientists so afraid of this nonexistent 'fallacy'? Understandably, nobody wants to be guilty of committing a *fallacy*, a relatively elementary but devastating error in logic. This is probably why the 'spectacular misnomer', as Williams calls it, is uncannily resistant to the demise it so heartily deserves.

---

[11] However to Moore it is just as much a mistake to say that there can be a behavioural definition of 'the good' as it is to claim that the phenomenal property of 'yellowness' (the Liberal Party had not at the time been eclipsed by the party of the Red Flag) can be completely specified by the wavelength of light and the principles of neurophysiology (Moore, 1903, p. 10). In this sense his remarks on ethics can be seen to prefigure the debate on visual qualia that is so familiar to readers of this journal. Perhaps it is no coincidence that prominent Darwinians such as Daniel Dennett and Nicholas Humphrey are disinclined to accept that the problem of qualia is beyond the scope of naturalistic explanation, but we cannot say whether these two authors would agree that ethics is equally amenable to reductive naturalism.

REVIEW ESSAY: IS DARWIN RIGHT? 69

The bottom line is that the term 'naturalistic fallacy' is *itself* fallacious. Ethics is about human behaviour and, as is true of any other aspect of human behaviour, it is fully amenable to empirical analysis. As long as the spectres of the 'naturalistic fallacy' and the 'is–ought gap' continue to haunt the discussions of social scientists, we will be stymied in our efforts to advance a scientific understanding of those aspects of human behaviour concerned with morals and other kinds of values.[12]

As the resources of cognitive neuroscience continue to expand, it becomes increasingly apparent that *all* human behaviour, including those kinds that are commonly characterized as having significant moral dimensions, arises from the complex information-processing capacities of the human brain. The philosophers and social scientists who continue to preach the doctrine of an unbridgeable gulf forever separating facts and values have simply not devoted sufficient attention to the burgeoning literature investigating features of human neurocognition that are deeply interconnected with moral judgment and social rationality.

David Hume's famous dictum, usually interpreted as prohibiting any derivation of an 'ought' from an 'is', is generally assumed to support Moore's argument against the 'naturalistic fallacy'. If this were true then it would be devastating to Arnhart's case. However, as Arnhart points out (p. 70):

> The common interpretation of Hume as having separated *is* and *ought* depends on only one paragraph in his *Treatise of Human Nature* (1888, 469–70). Some Hume scholars have shown that if one considers carefully both the textual and historical contexts of this paragraph, one sees that the common interpretation is wrong (Buckle 1991, 282–84; Capaldi 1966, 1989; Martin 1991). The textual context makes clear that Hume's claim is that moral distinctions are derived not from pure reason alone but from a moral sense. The historical context makes clear that Hume is restating Francis Hutcheson's criticisms of some early modern rationalists such as Samuel Clarke and William Wollaston, who believed that moral distinctions could be derived from abstract reasoning about structures in the universe that were completely independent of human nature.

The famous paragraph from Hume (1739/1985, p. 521) reads as follows:

> In every system of morality, which I have hitherto met with . . . the author proceeds for some time in the ordinary ways of reasoning, and establishes the being of a God, or makes observations concerning human affairs; when of a sudden I am surpriz'd to find, that instead of the usual copulations of propositions, is and is not, I meet with no proposition that is not connected with ought or ought not. . . . [A]s this ought, or ought not, expresses some new relation or affirmation, 'tis neccesary that shou'd be observ'd and explain'd; and at the same time that a reason should be given for what seems altogether inconceivable, how this new relation can be a deduction from others, which are entirely different from it. . . . [T]his small attention wou'd subvert all the vulgar systems of morality.

But Hume's text is open to a variety of interpretations, for example (Curry, 2000):

> 'Ought', thought Hume, was merely short-hand for saying 'is what our moral sense dictates in a given circumstance'[13] Moral judgments, then, are 'factual judgments about the species-typical pattern of moral sentiments in specified circumstances' and they are 'accurate when they correctly report what our moral sentiments would be in a given set of circumstances' (Arnhart, 1995, p. 389).

---

[12] One of this review's authors (Hughes) defends a fully scientific approach to ethics in his paper 'A Normative Science Manifesto' (2000).

[13] 'Ought-propositions cannot ever be deduced from is-propositions. But the reason for this is that sentences expressing ought-propositions are paraphrases of certain sentences expressing is-propositions, and paraphrasing is not deducing' (Yalden-Thomson, 1978).

Curry's take on Hume's intentions is entirely consistent with the interpretation suggested by Arnhart, who says, 'If we accept the common view of Hume as having argued that we cannot infer what *ought* to be from what *is* the case, then it would seem that he contradicts himself by deriving morality from the natural inclinations of human beings. . . . [T]he dichotomy between *is* and *ought* falsely attributed to Hume was actually first formulated by Immanuel Kant' (p. 73).

However, this is a decidedly non-standard interpretation of Hume's intentions in the famous passage. Many philosophers would claim that excessive conceptual contortions are required to hold Hume to a position that actually *denies* the apparent import of the passage. Hume practised irony, but he generally said what he meant. As Arnhart points out, several philosophers have contested the standard interpretation of the passage, however, others have argued persuasively in the opposite direction (for papers on both sides, see Hudson, 1969). Most would agree with Arnhart's perspective that Hume's ethical views are fundamentally consistent with Darwinian naturalism, but that does not necessarily mean that the ordinary understanding of Hume's dictum is mistaken. Barry Stroud, a prominent Hume scholar, offers a slightly different analysis of the same passage.

> Although many claims have been made both for and about this passage, Hume seems primarily concerned to re-emphasize the point that it is because of the special character of moral judgments that they cannot be 'perceiv'd by reason'. We undoubtedly make transitions from beliefs about the way things are to the judgment that things ought to be a certain way. That is to say, we observe actions and discover by reasoning some of their other characteristics and their consequences, and then we immediately and quite naturally arrive at moral judgments or conclusions. But if we understand the peculiar nature of these 'conclusions' — if we recognize their 'active' or motivational force — we see that the transitions by which they are reached are not ones that reason determines us to make. Once we come to have certain beliefs about the way things are, then, because of natural human dispositions we come to feel certain sentiments which we express in moral judgments.
>
> Hume takes himself to have explained the only way in which such transitions can occur. Because of the 'active' power of moral judgments, we arrive at them from other beliefs only by the interposition of a feeling or preference, since feeling or preference must be present for action to take place. Given what he takes to be the undeniability of those facts, Hume expresses the conviction that anyone else who tries to explain how we arrive at moral judgments will come to agree with him. ... That is why some 'small attention' to the question he raises would subvert all the vulgar systems of morality. He sees his 'subversive' answer to it as the only possible answer (Stroud, 1977, pp. 187–8).

Stroud's view has the virtue of being both naturalistic (as Hume certainly intended) and also consistent with the general consensus about Hume's intentions, *vis-à-vis* morality and the passions. According to Stroud's interpretation, Hume *did* argue that one cannot derive an 'ought' from an 'is', because the transition requires the involvement of the passions, not reason. According to that perspective, Hume *accepts* the fact-value dichotomy and assumes that *moral sentiments* (i.e. passions) are necessary to bridge the gulf. That much is clearly a naturalistic view, and is compatible both with Darwin and with Arnhart's interpretation of Darwinian ethics.

In contrast with both Arnhart and Curry, this approach lets Hume off the 'naturalistic fallacy' hook, but does not deny that he viewed the transition from 'is' to 'ought' as illicit. Instead, it recognizes that, for Hume, the necessary (and generally missing) steps in that transition involve *moral sentiments*, not rational deductions. For Hume, you *cannot* get from statements of fact to statements of value without including reference to the passions. In other words, Hume really *did* believe that you cannot derive an 'ought' from an 'is'. He thought that human nature was organized in such a way that the source of the moral sentiments was not to be found in the intellect, but in the passions (roughly emotion and motivation).

Arnhart and other ethical naturalists could employ an alternative approach to resolving the apparent contradiction between Hume's naturalism and his famous 'is–ought' dichotomy. Modern approaches to ethical theory (unknown to Hume) frequently divide the subject into substantive (or normative) ethics and metaethics. The former category includes practical moral discussions of what one should or should not do in various situations; the latter debates more theoretical questions such as how one might justify or explain one's normative convictions. By clearly recognizing this distinction, we can explain how Hume, as a bona fide ethical naturalist, could argue coherently against deriving 'ought' from 'is'. The apparent inconsistency in Hume's position evaporates when one sees that the focus of the famous passage is not metaethics, but substantive ethics. In this paragraph, Hume was simply not addressing questions about whether or not human moral systems can be fully founded in nature. Undoubtedly he would concur with the Darwinian perspective, because he believed that moral sentiments are central to ethics and are part of human nature.

> Given Hume's view that a desire or aversion is involved in the production of every action, his conception of morality has the consequence that a desire or aversion is somehow involved in the making of every moral judgment. But for Hume desires or aversions are themselves feelings or sentiments, and that is why he says you can never find the vice until you look into your own breast and find a 'sentiment of disapprobation' towards the action. That sentiment, of course, is not discovered by reasoning, but by being felt. It is a sentiment that occurs in the mind whenever we observe or contemplate actions or characters that have certain characteristics, and if we never got such feelings there would be no such thing as morality. Morality is thus based on feeling or sentiment, not reason (Stroud, 1977, p. 179).

This view of Hume's approach to morality is consistent with both Darwin and with Arnhart's main thesis. Emotions and motivation evolved as constituent parts of human biology. Identifying his views with Hume's, Arnhart says, 'a reason for action is a complex psychological state in which the conative component (a desire) plays the primary role, and the cognitive component (a belief) plays the secondary role' (pp. 19–20). Hume would certainly have endorsed a Darwinian approach to naturalistic ethics; for him the passions and the moral sentiments are a natural part of human psychology. However, he was not (in the famous paragraph) addressing the question of whether moral principles could ever be derived from non-moral premises. Rather, he was lamenting the common tendency in 'the vulgar systems of morality' to slide from statements of fact to moral assertions, without acknowledging the addition of a moral component (which, for Hume, could only originate in the passions).

An ethical system can affirm the 'is–ought' gap and still be fully naturalistic, so long as the distinction between substantive ethics and metaethics is clearly understood and honoured. Hume's system was one such. Metaethically, he had no problem with founding 'oughts' in human nature. With respect to substantive ethics, however, he claimed that one cannot properly derive an 'ought' from an 'is'.

Desire and its Discontents

Of course, Hume might have been wrong — indeed we believe that in this matter he probably *was* wrong. We think that his theory of human nature suffers from many of the basic flaws that have plagued folk psychology for centuries. His views on moral motivation relied on an artificial dichotomy between passion and reason. Consequently, he drastically oversimplified the psychology of motivation and his account of the connections among reason, emotion, and action were schematic to the point of caricature. His psychology was seriously infected with residual dualism; it would have been nearly impossible for him to avoid that. The fundamental distinction between reason and passion, so basic to his moral system, has been shown to be a false dichotomy (Damasio, 1994; 1999; Rolls, 1999; Panksepp, 1998). In order to develop an adequate account of the relationships between action and reason, we require a much more finely variegated analysis of the microstructure of motivation than anything to be found in Hume's psychological theory.

Arnhart's reliance upon Hume's inadequate, folk-psychological theories of moral motivation leads to one of the most serious problems with his account. The first and most foundational of Arnhart's 'ten propositions' equates the good with the desirable:

1. The good is the desirable, because all animals capable of voluntary movement pursue the satisfaction of their desires as guided by their information about the world (p. 6).

This view is clearly consistent with the priority that Hume attached to the moral sentiments; however, it suffers from the same flaws. 'Desire' and 'desirability' are relatively abstract theoretical concepts. Folk psychological theories generally give 'desire' a central role in the mental economy, but the term has little other than heuristic value as a descriptor of neurocognitive information processing. Consider the complex interactions among perception, memory, attention, emotion, and motor control that are implicitly assumed whenever we talk about acting upon some desire or other. Furthermore, many of the 'twenty natural desires' Arnhart describes incorporate complex combinations of individual and social components, as he freely admits (pp. 29–36). We don't doubt that these 'desires' are deeply rooted in the biology of human nature, nor that they evolved according to Darwinian principles. Our concern is ontological; Arnhart treats desires as natural kinds, fundamentally real psychological entities. We think that his approach accords insufficient attention to neurobiological realism.

The problems associated with Arnhart's accepting a basically folk psychological account of beliefs and desires could be resolved via relatively painless revisions by thickening the theoretical commitments to cognitive neuroscience. The term 'desire' can be understood as a theoretical cipher or variable, an abstract placeholder to be fleshed out with the appropriate neurobiological particulars in each specific case. To do so would require significant revisions to the rather anaemic model of motivation and would seriously undercut the apparently happy three-way marriage that Arnhart portrays among Darwin, Hume, and Aristotle. A truly Darwinian ethics must make neurobiological realism central to its psychological commitments. The timeworn, artificial dichotomies between reason and passion, essential to Hume's moral psychology, are empirically inadequate. Nevertheless, Arnhart's basic naturalistic project could be preserved, even if Hume's contributions were significantly curtailed.

One of the most widely-discussed books on this topic is Antonio Damasio's *Descartes' Error* (1994). Damasio examined the clinical records of the unfortunate Phineas Gage, who suffered a massive injury to the pre-frontal area of his brain and underwent serious personality changes as a consequence. Although the damage was in an area that is normally associated with emotion, Damasio discovered that, in contemporary patients with similar lesions, the ability to take rational decisions regarding social and ethical behaviour was severely impaired by the damage to their social-emotional brain. 'Descartes' [and Kant's] error', according to Damasio, was to separate our rational souls from their physical and social context.

Of course it was Freud who overturned the viewpoint, common since antiquity, according to which the rational is a completely independent force. 'Evidently, it is a force that is stronger in some people than in others, and, when it comes to action, it is less often frustrated in some agents than in others. But the old idea was that there is no interfering with its inner working' (Pears, 1984). Although the Freudian view of irrationality, with its posit of the dynamic unconscious and other such sub-agents is now dismissed by psychologists, nevertheless after Freud there was no escaping the fact that reason had descended from the Platonic realms and was now incarnate. Damasio's work has contributed to the reclamation of rationality down from the Platonic realm to embodied, culture-specific behaviour, no longer in a position to act as a 'god's eye' arbiter for the onward march of civilisation. Modern neurology tells us that reason is embodied and cannot be separated from social and affective processes.

Arnhart devotes a chapter to the study of psychopathy that draws heavily on Damasio's discovery of the interdependence of reason and emotion. Central to his case is the growing evidence that psychopathy is characterized by an *absence* of emotion — a 'poverty of desire' — hence the fact that psychopaths typically push experience to the limit in order to feel anything at all. Psychopaths appear to lack empathy for suffering in their victims precisely on account of their own lack of feelings.

He draws heavily on Ann Rule's 1989 study of the serial killer Ted Bundy. Not only did Bundy lack the natural moral emotions of sympathy, guilt or shame, he was in fact an 'emotional robot, programmed by himself to reflect the responses that he has found society demands' (Rule, 1989, p. 403). Psychopathy is four to seven times more prevalent among men than among women, and Arnhart speculates that it 'might be an extreme manifestation of typically masculine traits' (p. 215). He goes on to suggest an adaptationist account of psychopathy in terms of an 'evolutionary niche for Machiavellians' (p. 219).

Arnhart argues that it is precisely because psychopaths cannot develop the moral emotions that they are incapable of ethical behaviour. Since they lack the moral sentiments that make moral persuasion possible, our only appeal with such people is force and fear. Arnhart rejects the standard liberal argument that such people cannot be blamed for their behaviour so punishment is inappropriate. To Arnhart it is the moral emotions that make us human, and society has to defend itself from 'dysfunctional deviations from human normalcy' (p. 229). The law of Moses required that such people be taken to the gates of the city and stoned to death; the State of Florida used a slightly more high-tech approach with Ted Bundy, but the outcome was the same.

Given the growing evidence from cognitive neuroscience regarding the seamless link between rationality, the moral emotions and the development and integrity of the pre-frontal cortex, it's hard to understand why so many philosophers, remaining under the spell of Kant, still argue that morality requires a purely rational logic of universal rules:

> Psychopaths show us that that cannot be true. There is no evidence that psychopaths have any deficit in their capacity for abstract rationality or pure logic. Their immorality comes not from any defect of abstract reason, but from their emotional poverty. They cannot be moral, because they lack the social emotions — such as sympathy, guilt, and shame — that sustain moral conduct (p. 229).

If anything the cool rational strategy advocated by Kant 'has far more to do with the way patients with prefrontal damage go about deciding than with how normals usually operate' (Damasio, 1994, p. 172).[14]

* * *

The centrality of 'desirability' to Arnhart's project presents another — and deeper — problem. Even if Arnhart were to fully update his theoretical taxonomy to more accurately reflect the complex interactions among 'beliefs' and 'desires', he would still face a very serious challenge that arises from equating 'the good' with 'the desirable'. Aside from all of the terminological problems that spring from any attempt to treat 'the desirable' as a single, homogeneous psychologically salient construct, the proposed equation gives very short shrift to the issue of moral conflicts.

Arnhart gestures towards this problem, but he fails to resolve it. Nor could he have done any better, given his basic commitment to what he takes to be the fundamental naturalistic equation between goodness and desirability. In one of the passages where he struggles with this problem, he writes:

> What human beings happen to desire at any moment is not always desirable, therefore, insofar as it does not always promote their flourishing. The common experience of regretting what we have done reminds us that we often mistakenly desire what is not truly desirable: for example, we might discover that satisfying some present desire impedes the satisfaction of some future desire; or we might find that in pursuing some narrowly selfish desires, we have failed to cultivate those social bonds of affection and cooperation that we need to satisfy our social desires. Learning how to manage our desires over a complete life in a manner that is appropriate for our individual and social circumstances requires proper habituation and prudent reflection (p. 82).

Earlier, he lists 'four sources of moral disagreement': (1) fallible beliefs about circumstances; (2) fallible beliefs about desires; (3) variable circumstances; and (4) variable desires. He is not attempting to sweep these problems under the rug. He does recognize and acknowledge the *fact* of moral disagreements and conflicts. However, he does not seem to identify the severity of the problem for his basic normative

[14] However a recent paper by Blair and Cipolotti (2000) suggests that Damasio and Arnhart are wrong to claim that psychopathy is characterized by an absence of emotion and that there are important differences between patients with orbitofrontal trauma who present symptoms of 'acquired sociopathy' and developmental psychopaths. Psychopaths show skin conductance responses to very threatening or disgusting images. However the authors acknowledge that developmental psychopaths do have a specific emotional impairment, often connected with a dysfunction within the neuro-cognitive system that responds to the sad and frightened faces of others.

principle, the foundational equation between goodness and desirability. Under each of the four headings he points to problems associated with the variable contents of 'desirability': (1): 'we have differing views of the relevant circumstances'; (2): 'we are often unsure about what we truly desire; (3): 'although the pattern of natural human desires is universal, satisfying those desires in different individual and social circumstances requires different patterns of conduct appropriate to the circumstances; (4): 'there is both normal and abnormal variation in human desires'. The common element among all of these sources of disagreement is the fact (as he acknowledges) that 'desirability' is *very* far from uniform across persons, times, and cultures. The upshot is that some versions of the desirable are incommensurable with others, resulting in conflicts that cannot be reconciled by any universal moral code, able to go beyond the intrinsically conflicting versions of what is desirable. He comes closest to admitting the depth of this problem in Chapter Six, 'Man and Woman'.

> But if I am right about this, if human beings are not bound together by a universal senti-
> ment of disinterested humanitarianism, then deep conflicts of interest between individu-
> als or between groups can create moral tragedies in which there is no universal moral
> principle or sentiment to resolve the conflict ... The only alternative, which I do not
> regard as a realistic alternative, is to invoke some transcendental norm of impartial jus-
> tice (such as Christian charity) that is beyond the order of nature (p. 149).

If the good is the desirable, then, even according to his own analysis, there will be multiple and incommensurable instances of the good. The deep problem that his equation cannot resolve enters at exactly this point. When goods collide, how are the collisions to be reconciled? In the natural world, the way that such conflicts are typically resolved boils down to the efficient use of force and dominance. Obviously, humans too often employ just such means, but any overriding commitments that we have to the Aristotelian principle of *eudaimonia* (well-being) will lead us to reject the rule of force as an ultimate arbiter of moral conflicts.

Humanity's dual evolutionary heritage instils in us behavioural mandates that descend from both *biology* and *culture*. This fact constantly confronts us with contradictory requirements; primeval urges compete with social and cultural constraints, frequently forcing us into profound conflicts of interest. One of the sad ironies of human life is the fact that our species has been so successful—in reproductive terms — that we now face massive ecological and social burdens forced by the exponentially increasing human population. Biological evolution has bequeathed us a neurocognitive legacy that, all too often, is radically incompatible with the contingencies of human life on earth in the twenty-first century. If we have nothing other than the inherently subjective measure of 'desirability' to serve as the ultimate criterion for value, then we have no principled basis upon which to answer the unavoidable question 'whose desires?'.

Morality requires exactly such an answer. By what means are we to understand the *proper* way to assess conflicting claims about whose version of desirability is 'best'? Arnhart's basic moral principle, that the good is the desirable, is simply inadequate as an arbiter among profoundly incommensurable desires. During his discussion of the common objections to a Darwinian ethical system, Arnhart notes that 'While a naturalist explanation of morality would assume a moral universalism founded on the unity of human nature, a culturalist explanation of morality would assume a moral relativism founded on the diversity of cultural traditions' (p. 9). He rejects moral relativism, but does not explain how we are to determine which version of 'the desirable' is best. If we are to avoid the pitfalls of relativism, we require objective, quantifiable criteria by which we can adjudicate among competing moralities. Such criteria are not to found in the idea that 'the good is the desirable', because we can never establish a single, ultimate standard of desirability against which competing claims can be measured. What might seem to be self-evidently desirable for one person, immersed in one particular sociocultural, economic, and political situation, will inevitably strike many others as equally self-evidently *wrong*. Desirability is always, in the final analysis, a profoundly subjective, and hence relative matter. Who decides which version is right, and how are they to decide?

One way that Arnhart might escape this problem is to expand his reliance upon Aristotle's conception of *eudaimonia* (well-being, flourishing). He nods in that direction several times, but the centrality of his emphasis on desire interferes with his fully embracing *eudaimonia* as a foundation for ethics, as Aristotle did explicitly. As we noted earlier, Arnhart starts his book by summarizing ten basic propositions of 'Darwinian natural right'. Proposition 5 states 'We can judge divergent ways of life by how well they nurture the natural desires and cognitive capacities of human beings in different circumstances ...' (p. 6). Implicit within this idea is the suggestion that we can assess competing moral systems by weighing their relative contributions to human flourishing. This is a promising option and is fully consistent with the basic naturalistic tenor of the project. The problem is that, for such an adjudicatory procedure to be possible, it would have to ultimately appeal to considerations beyond desirability. It might be considered virtually truistic to note that conditions that support human flourishing are desirable. However (and this is the crux of the matter), determinations regarding which conditions promote human flourishing are *not* ultimately based upon which conditions are desirable. Rather, they depend upon measures of health and well-being that apply universally to human beings in virtue of their common biological structures (Hughes, 2000).

The incommensurability of conflicting versions of the good is a fundamentally political issue. If it cannot be resolved in a consensual manner then we can confidently expect the Leviathan to raise its sleeping head as the influence of religion and other traditional forms of arbitration and consensus building declines.

Political Entailments

What are the political implications, if any, of the view that the ultimate source of morality is human biology? The application of our scientific understanding of neurology and psychology to society is fraught with perils — indeed Antonio Damasio is quick to caution his readers against following the earlier example of the Social Dawinists in drawing political inferences from science. He argues that science is descriptive and politics is prescriptive and that the two should be kept quite separate.

As remarked earlier the three most vocal critics of sociobiology — Stephen Jay Gould, Richard Lewontin and Steven Rose — all share a leftist political orientation. Steven and Hilary Rose describe evolutionary psychology as 'transparently part of a right-wing libertarian attack on collectivity, above all the welfare state' (Rose and Rose, 2000). And in many respects one has to sympathize with their critique —

ethical and political philosophy has to take issue with all forms of single factor reductionism: in the words of R.M. McShea (1986):

> If we hold fast to and develop the modest but powerful notion of man as a part of nature we will be preserved from such stilted and dangerous nonsense as is involved in seeing him as crucially Rational, as cosmically important, as driven immanently toward survival, creative labour, Freedom, pleasure, or some mystic fulfilment of his Spirit or of History.

Political philosophy is 'structurally normative' (*ibid*, p. 210) and our theories in this field have always been shaped by our overall cosmology. James Rutherford makes a similar point:

> A new Darwinism, which recognizes both natural and cultural evolution, rejects the false exclusionary dichotomies of nature versus nurture, fact versus value, and nature versus free will (Arnhart, 1995). If facts are nor related to values, for example, the phrase 'political science' is an oxymoron (Rutherford, 1999, p. 94).

These arguments would tend to underwrite Arnhart's naturalistic approach to morality, but it is not at all clear that this has any particular implications for politics. However, if the discussion is extended to include the philosophy of law, some clear entailments do arise.

The 'progressive' (leftist) approach to jurisprudence is founded on principles such as the extension of equality and universal human rights. But the concept of equality has no foundation in the natural world: it is an abstract mathematical notion, and part of the Enlightenment project to reduce politics to abstract universal principles. Descartes believed that a mathematical description of the natural world (*res extensa*) was possible, but the only way of achieving it was to create a separate realm for human subjectivity (*res cogitans*) on the far side of the 'Cartesian divide'. Descartes banished human desires, thoughts and volitions to a transcendental self, thereby freeing science to reconstruct the physical world along rational lines, free from the constraints that would arise from a naturalistic or embodied approach to cognition.

The concept of universal human rights can in turn be traced to Kant's intellectual wrestling with the 'problem of modernity'. How can we account for our sense of freedom, rationality and personal agency in a deterministic Newtonian universe? Kant's answer was to draw a distinction between the physical world and the spiritual 'kingdom of ends'. As human beings are also inhabitants of the latter, we are capable of acting as autonomous moral agents, beyond the demands and constraints of our ordinary physical and psychological makeup. As the 'kingdom of ends' was not subject to the contingencies of history and culture, this paved the way for a universal system of human rights and values.

Unfortunately Kant vacillates between the view that the transcendental subject is merely a logical necessity and the view that it is a distinct noumenal thing (in the earlier sense of Descartes).[15] Despite this ambiguity, Kant's transcendental philosophy has had profound consequences for jurisprudence. The notion of human beings as

---

[15] This very ambiguity, allied to Kant's tortuous prose style, is one of the reasons why the Kantian project has not been subject to the same attacks as Descartes' legacy. 'Cartesianism' and 'dualism' are now pejorative terms in the philosophy of mind, whereas Kant still lives on through the highly influential writings of John Rawls.

ends in themselves — and the subject of universal human rights — is derived from these uncertain Kantian foundations.

Such rationalist arguments tend to be the province of the left. By contrast, conservatism has its philosophical origins in the naturalism of the Scottish Enlightenment. Michael Oakeshott, the greatest conservative philosopher of the twentieth century, argued that the roots of conservative thought lie not, as widely believed, in the political sociology of Edmund Burke, but rather in the moral psychology of David Hume (Oakeshott, 1956, p. 435). Hume believed that all our beliefs and values are grounded in human passions. There is such a thing as natural moral sentiment, and the task of government is to ensure that society is aligned with this naturalistic source of morality. In the same way that Adam Smith showed how economic order could emerge naturally, through the operation of the 'invisible hand', Hume provided an equally naturalistic grounding for morality.

Conservatives, of course, believe that politics is a highly limited activity and that it's not the business of democratic governments to tell people what to think. Concepts like 'right' and 'wrong' arise out of natural moral sentiment and the business of government is to codify natural moral sentiment into laws. The conservative approach to law in general is not that of a reforming or progressive instrument. Sir James Stephen's *History of the Criminal Law of England* (1883) recounts that the object of criminal law is simply to give formal expression to the natural sentiment of anger produced by an act of wrongdoing. There would be no law without a natural passion for vengeance, just as there would be no marriage without sexual desire. Arnhart also endorses De Waal's ethological theory of the evolution of the concept of justice: 'The human sense of vengeance — the desire to get even — is the earliest and deepest expression of the human sense of justice' (p. 79).

Public opinion in modern social democracies tends to be positioned well to the right of the media and political classes, particularly over issues like law and order, immigration and minority rights. So whenever politicians respond to public concerns over these issues they are accused by their liberal opponents of populism, opportunism and 'saloon bar politics'. This became clear recently in the United Kingdom when the (Labour) Home Secretary — who is often seen as further to the right than his Conservative counterpart — was lambasted in the *Guardian* over his plans to allow victims and their relatives to have some influence over sentencing policy. This is an integral part of Islamic law, but is anathema to the modern western concept of 'justice' as an impersonal force, best symbolised by the blindfolded statue on the roof of the Old Bailey, Britain's most senior criminal court.

The internationalist/universalist agenda is usually associated with the left and is ultimately derived from St. Matthew's transformation of the Mosaic commandment to love your neighbour (and hate your enemy) into loving your enemy as yourself, as 'God sends rain on the righteous and the unrighteous'. Unfortunately this requires 'perfection' (Matt. 5.48), whereas Arnhart prefers the more attainable teaching of Aristotle and Aquinas that love for others, as a natural extension of self-love (born of parental nurturing),[16] is naturally more for those most like us than it is for strangers.

---

[16] Population biologists would explain parental nurturing in *Homo sapiens* as the natural consequence of the pursuit of *K*-strategy — the production of fewer offspring that require the investment of high levels of energy in each (p. 103). Aristotle, Aquinas, Darwin and Arnhart are in agreement that this is the prime source of the natural moral sentiments.

What the right sees as 'patriotic pride' the left dismisses as xenophobia and tribalism. But if Arnhart is right, empathy for those outside the tribe comes from travelling, not Christian asceticism or categorical imperatives.

It would appear to be the case that the implications of Arnhart's case for Darwinism in ethics and jurisprudence are inherently weighted towards the political right. Is this perhaps the reason that Wilson's keynote speech at the 1996 HBES meeting was met by such a hostile response? Many of the theorists in evolutionary psychology, Richard Dawkins being an obvious case, like to view themselves as centre-left politically, and it is hard to see how this would be compatible with ethical naturalism. Arnhart has been described as a 'right-wing neo-Thomist' (private correspondence) and certainly the title of his (2000) paper 'Why conservatives need Charles Darwin' would give credence to this view.

Some thinkers on the left have made an effort to reverse the colonisation of Darwinism by right-wing forces. Peter Singer makes a bold attempt with his new book *A Darwinian Left*, but he can never really break free from Kant. For example, as his reviewer in *Scientific American* says: 'To Singer, and to me, the core of the Left is a set of values, most notably that worth is intrinsic and doesn't depend on success or power' (Van Valen, 2000).

Given Singer's controversial views on the (lack of) intrinsic worth of disabled foetuses this is a strange remark. The other problem is that this view is heavily dependent on the Kantian reworking of Christian principles. The work of Damasio (1994) and others would imply that the 'worth' of an individual (as an autonomous moral agent) is in proportion to the development of pre-frontal cortex and it's hard to imagine what a non-transcendental theory of absolute worth might look like. Singer then concludes with a listing of 'what a Darwinian left would do', much of which could have been culled from any manifesto for 'compassionate conservatism'.

Left-leaning political philosophy has tended to comprise a fusion of the Kantian notion of intrinsic value (and its derivative, universal human rights), together with the Hobbesian and Lockean view that denies a natural human sociality and argues that 'the moral inclinations are utterly artificial products of the social contract' (p. 72). In combination with the twentieth-century move towards collectivism the left tends to rely on the power of the state to construct and enforce morality. It is hard to see how this essentially top-down model can be reconciled with the (Aristotelian and Darwinian) view that human sociality is an indirect product of the natural affection produced by the parent-child bond.[17] In the same way that left-leaning politicians have dismissed the workings of the invisible hand in politics, they have been equally sceptical about the emergence of morality independently of state guidelines. For similar reasons the left tend to support the view that sexual roles are largely socially constructed and the product of power relationships in society. Given the published work by sociobiologists in all these areas it is hard therefore to imagine exactly what a Darwinian left would look like, notwithstanding Professor Singer's efforts.

---

[17] Often the subject of left-wing attack, as in the Kibbutz movement and the failed experiment of the Oneida Community.

80   L. ARNHART

THOUGHTS ON DARWINIAN NATURAL RIGHT:
A RESPONSE TO SUTHERLAND AND HUGHES

Larry Arnhart[18]

Ethics is a product of human consciousness. Animals act in a goal-directed manner to satisfy their desires based on their information about the changing environments in which they live. Animal movement is thus inherently normative or value-laden insofar as animals cannot live without choosing between alternative courses of action as more or less desirable. To the extent that some animals are conscious of themselves and their environment, they can act voluntarily by consciously gathering information related to their desires and then acting according to their conscious assessment of the information in relation to their desires. The greater extension and complexity of human consciousness allows human beings to formulate deliberate conceptions of right and wrong and to act in accordance with those conceptions, which makes human beings the only ethical animals. Only human beings can pursue happiness as a deliberate conception of the harmonious satisfaction of their desires over a whole life, because only they have the capacities for reason and language that allow them to formulate a plan of life, so that they can judge present actions in the light of past experience and future expectations. To fully explain human consciousness, therefore, we must explain ethics as a product of human consciousness.

In *Darwinian Natural Right*, I have tried to explain ethics as rooted in human biological nature. The good is the desirable, I argue, and there are at least twenty natural desires that belong to the nature of the human animal as shaped by natural selection in evolutionary history. I am grateful to Keith Sutherland and Jordan Hughes for the generous way in which they state their general agreement with my position. But as I would expect of such thoughtful readers, they also see some difficulties in my reasoning. I will respond to their three main objections, which concern Hume's account of the is/ought dichotomy, the complex interaction of reason and desire in neurobiology, and the problem of incommensurable desires.

Hume on the Is/Ought Dichotomy

The most common objection to any view of ethics as rooted in human nature is that moving from natural facts to moral values, from an *is* to an *ought*, requires a fallacious inference. This argument is attributed to David Hume, who is said to have shown that there must be a radical separation between questions of what *is* or *is not* the case, which belong to the realm of nature, and questions of what *ought* or *ought not* to be done, which belong to the realm of morality. Part of my response to this objection is the claim that this sharp dichotomy between *is* and *ought* is not Hume's position but Immanuel Kant's. According to Kant, judging what *is* the case belongs to the 'phenomenal' realm of nature, but judging what *ought* to be belongs to the 'noumenal' realm of freedom. He then uses this separation to argue against Hume's naturalistic view of morality by claiming that morality belongs to an autonomous realm of reason that transcends the realm of nature.

[18] Department of Political Science, Northern Illinois University, DeKalb, IL 60115, USA.
Email: arnhart@aol.com

Sutherland and Hughes, however, defend Barry Stroud's claim that Hume really does separate facts and values: Hume argues that we cannot derive an *ought* from an *is* by reason alone, because such a transition requires some experience of the sentiments or passions to bridge this gulf. This Humean view of the fact–value dichotomy is consistent with Hume's ethical naturalism because the transition from facts to values through the moral sentiments is itself natural.

Like Sutherland and Hughes, I now find Stroud's reading of Hume persuasive. So I would modify what I say about this in my book. Although Hume does accept the fact–value dichotomy, his version of the dichotomy differs radically from Kant's; and it is Kant's version that denies ethical naturalism. According to Hume's version, we cannot deduce a moral *ought* from a natural *is* by reason alone, because moral judgments require moral sentiments or passions that are part of human nature. But according to Kant's version of the dichotomy, the moral *ought* belongs to a transcendent realm of human freedom beyond the realm of human nature. I would defend Hume's version and reject Kant's.

In the same section of his *Treatise of Human Nature* where Hume distinguishes *is* and *ought*, he explains his fundamental point about how reason without passion cannot infer moral conclusions by using the example of judging incest. Hume asks 'why incest in the human species is criminal, and why the very same action, and the same relations in animals have not the smallest moral turpitude and deformity?' (Hume, 1888, pp. 467–8). If moral judgment were simply a matter of reason discovering certain factual relationships in the world that are right or wrong, and if incestuous relationships are factually the same in the case of human beings and animals — that is, sexual mating between closely related kin — then animal incest should be just as morally abhorrent as human incest. Although nonhuman animals might lack the rational ability to perceive the immorality of their incestuous relations, those relations would still be immoral as a matter of fact even if they could not perceive that fact. But, of course, we do not perceive animal incest to be immoral, as we do human incest, even when the factual relations are the same, because human incest arouses a moral sentiment of blame that animal incest does not. To move from the factual judgment that some animals are mating with close relatives to the moral judgment that this is wrong, we must feel a moral emotion of disgust towards such mating; and we naturally tend to feel this in the case of human mating but not in the case of animal mating.

There is now a lot of evidence to support a naturalistic explanation of the incest taboo as rooted in the moral emotions of human biological nature. Elaborating an insight suggested by Charles Darwin, Edward Westermarck argued that there was a natural propensity for human beings to feel a sexual aversion towards those with whom they had been associated in early childhood, that this propensity had been favoured by natural selection in evolutionary history as a mechanism for avoiding the deleterious effects of close inbreeding, and that this propensity would be expressed culturally as a moral rule against incest.[19] Although reason is important in formulating kinship rules and in generalizing the emotional aversion to incest as customary or legal norms, it is the emotional aversion itself that originates the sense that incest is wrong. And if Westermarck is right, the propensity to acquire this aversion belongs to

---

[19] See Westermarck (1922), Vol. 2, pp. 162–239; (1932), pp. 246–50. For the extensive evidence and arguments supporting Westermarck's hypothesis, see Wolf (1995). I have elaborated my defence of Westermarck in Arnhart (1998a; 2000a).

the biological nature of the human species as shaped by natural selection. This illustrates how Darwinian biology can support a Humean view of ethics by rooting it in a modern scientific account of human nature. The judgment that incest is wrong arises not as a Categorical Imperative of reason knowable by any rational agent in the universe, but as a natural emotional propensity of the human species. Hume is right. Kant is wrong.

The Neurobiology of Reason and Desire

Although they generally agree with me in adopting Hume's naturalistic view of morality, Sutherland and Hughes criticize Hume for employing 'an artificial dichotomy between passion and reason', which they identify as a mistake common to 'folk psychology'. The falsity of this dichotomy between reason and emotion has been clearly revealed, they believe, in the neurobiological research surveyed by Antonio Damasio and others, which shows 'the interdependence of reason and emotion'. The idea of disembodied rationality as opposed to embodied emotions is false, because all human reasoning is embodied and thus embedded in the somatic and social life of the human animal.

I agree with them that our philosophic view of morality would be improved by a moral psychology that recognizes the mutual dependence of reason and emotion as revealed by modern neurobiology. But to speak of 'interdependence' implies separation as well as union. A science of psychology requires that we distinguish faculties of the mind even when we know that in practice human mental experience is a complex interaction of factors in which everything is connected to everything else. Moreover, as Sutherland and Hughes indicate in their comments on the neurological basis of psychopathy, some people with neuropathological disorders can show 'rational-analytic behaviour' that is dysfunctional because they lack the social emotions that guide normal human behaviour. As Damasio indicates, the problem with such people is that reason has been separated from emotion: 'to know does not necessarily mean to feel, even when you realize that what you know ought to make you feel in a specific way but fails to do so' (Damasio, 1994, p. 211). Ted Bundy 'knew' what he was doing when he brutally murdered his victims, but he could not 'feel' the moral emotions that such brutality elicits from normal human beings. Such extreme cases show that reason and emotion are separable, even as they confirm that healthy human functioning requires the union of reason and emotion.

This sustains Hume's insight that we need to separate the effects of these two parts of the mind for the sake of psychological analysis, but we also need to see that sanity and social life require their inseparable cooperation (Hume, 1888, pp. 415–18, 493). It seems to me that 'folk psychology' — that is, the common-sense understanding of human psychology — recognizes the complex interdependence of reason and emotion, and that progress in neurobiology will only give us technical refinement in what we already know by common sense.

That technical refinement is important, however, in filling in the gaps in our knowledge about how we acquire moral character by habituation and how we learn moral rules by instruction. So, for example, although there is plenty of evidence to support Westermarck's claim that human beings are naturally inclined to learn a sexual aversion for their early childhood associates, which then is expressed culturally as rules against incest, we still need to know how neurobiological mechanisms mediate this process of habituation and instruction.

The Problem of Incommensurable Desires

Sutherland and Hughes rightly identify the deepest problem in my position when they write: 'Arnhart's basic moral principle — that the good is the desirable — is simply inadequate as an arbiter among profoundly incommensurable desires'. It is inadequate 'because we can never establish a single, ultimate standard of desirability against which competing claims can be measured'. People in different situations will disagree about what is 'desirable', and when they disagree deeply about issues that are too fundamental to be ignored, there might seem to be no way to resolve the disagreement except by one party exploiting the other through force and fraud.

Some other readers have seen the same problem in my book. John Hare,[20] a Professor of Philosophy at Calvin College, defends the tradition of Christian Kantianism in ethics, and he criticizes my book as illustrating how a purely naturalistic view of ethics tends to 'reduce the ethical demand' by denying the ethical demand for impartial and universal benevolence as contrary to natural desires. He summarizes my argument as a 'double inference': 'The good, he says, is the desirable, and the desirable is what is generally desired by human beings.' (Hare, 2000)[21] But when these natural desires create conflicts between human beings, he argues, there is no higher principle in a purely naturalistic ethics to resolve such conflicts without coercion or manipulation. For example, if the natural desires for wealth, social status, and political rule lead some human beings to enslave others, the desire of the masters to exploit their slaves comes into conflict with the desire of their slaves to avoid being exploited. If the masters have enough power to suppress the rebellion of their slaves, the masters will prevail by force and fraud; and there is no reason in terms of natural self-interest for the masters to give up the institution of slavery. Similarly, if men can satisfy their natural desires by exploiting women through patriarchal practices, the men have no reason to stop such exploitation as long as they have the power to suppress any resistance from the women. He believes the record of history suggests that both slavery and patriarchy have been widely practised because there was a conflict of desires between exploiters and exploited, and the exploiters prevailed through force and fraud.

When such conflicts in the natural desires arise, is there any ethical principle that would resolve them? Sutherland and Hughes appeal to a principle of Aristotelian naturalism, while Hare appeals to a principle of Kantian idealism. Sutherland and Hughes argue that we must look to 'considerations beyond desirability', which would provide 'objective, quantifiable criteria by which we can adjudicate among competing moralities'. They suggest that such a standard would be found in Aristotle's conception of human happiness (*eudaimonia* in Greek) understood as human well-being or flourishing. This could provide 'measures of health and well-being that apply universally to human beings in virtue of their common biological structures'.

---

[20] The son of the influential moral philosopher R.M. Hare (see above, p. 64).

[21] For an elaboration of his Christian Kantianism in ethics, see also Hare (1996).

Hare argues that we must look to Kant's Categorical Imperative, which says that we should act only on those rules that we can will as universal laws, and we should treat all people as ends in themselves and never merely as means. According to this Kantian (and Christian) version of the Golden Rule, we are morally obligated to respect the equal worth of all human beings by showing a disinterested and universal benevolence towards all people without favouring our own desires over the desires of others.

I agree with Sutherland and Hughes that the general uniformity of human nature supports some universal standards of moral judgment. But I would stress that the application of these universal standards to the particular circumstances of particular individuals in particular societies requires prudence or practical judgment that cannot be reduced to 'objective, quantifiable criteria'. For example, I argue in my book that female circumcision (clitoridectomy and infibulation) frustrates the natural desires of both women and men, and thus we can rightly condemn such practices even where they are deeply rooted in local traditions. But a prudent respect for such local traditions demands that we work for gradual reform rather than immediate abolition. As I suggest in my book, we could employ strategies of reform similar to those used to abolish foot-binding in China. Such an attempt to gradually reform an oppressive tradition implies some universal standard of human welfare, but it also implies respect for human diversity. The moral controversies surrounding practices such as female circumcision illustrate the four sources of moral disagreement: fallible beliefs about circumstances, fallible beliefs about desires, variable circumstances, and variable desires. Moral reform in such cases requires a shrewd judgment of what would be best, given the beliefs, the desires, and the circumstances of the people involved.

As Sutherland and Hughes indicate by their quotation from my chapter on men and women, I think that deep conflicts of interest can create moral tragedies in which there is no principle for resolving the conflict. In the cases of men exploiting women and masters exploiting slaves, I argue that in the long run such exploitation is self-defeating for the exploiters. In most cases of exploitation, the exploiters do not understand that exploitation is not truly desirable for them. In any case, I doubt that there are many cases in which exploiters have been persuaded to voluntarily give up their exploitation even though this was contrary to their interests.

Hare thinks that the way to resolve conflicts of interest is to appeal to a truly disinterested and universal benevolence. I agree that our moral concern can expand to ever-wider circles to include our extended kin, our clan, our group, our nation, all of humanity, and perhaps even all life forms. The experience of 'globalization', in which we experience the advantages of cooperation with strangers and thus develop some sympathy for their situation, illustrates this. But while universal sympathy is possible, it will never be totally impartial, because it will always favour those close to us over those far away. The rhetoric of groups like Amnesty International and Human Rights Watch confirms this. We can feel sympathetic concern for people whose unjust suffering is vividly presented to our imaginations so that we feel some affinity to them, but we do not feel equal concern for every human being regardless of their relation to us. Indeed, people who would care more for strangers in a foreign land than for their own family and friends would be moral monsters. The futility of Hare's appeal to utterly disinterested benevolence as contrary to human nature is indicated by what he says about the 'moral gap' in Kantian ethics: since human beings on their own can

THOUGHTS ON DARWINIAN NATURAL RIGHT 85

never live up to the moral demands of true benevolence, because they are corrupted by 'original sin', they need God's assistance to bridge the gap between their natural inclinations and the moral law. But until such a divine transformation of humanity occurs (on earth or in heaven), it seems that Hare's Kantian morality is not fitted to the human condition.

In contrast to Hare, and in fundamental agreement with Sutherland and Hughes, I defend Darwinian natural right as a form of moral realism. As social animals that cannot live well without cooperating with others, we try to overcome conflicts of interest by finding confluences of interest. We are helped in doing that by our natural propensity to sympathize with the pleasures and pains of others. Sometimes we are inclined to exploit others, but that inclination provokes resistance from our potential victims. Sometimes our moral conflicts create tragic choices in which we must choose between goods or choose the less bad as good. To choose correctly requires a prudent management of our desires in a manner that fits our circumstances. There are no abstract rules that would make this easy. We must use the tools nature has given us — our capacities for thinking, feeling, and learning. We make the best of what we have as we strive to satisfy our deepest desires as mortal creatures naturally adapted for life on earth.

References

Arnhart, L. (1995), 'The new Darwinian naturalism in political theory', *American Political Science Review*, Vol. **89**, No. 2, June 1995.

Arnhart, L. (1998), 'The search for a Darwinian science of ethics' *Science and Spirit*, **9** (1), pp. 4–7.

Arnhart, L. (1998a), 'Westermarck's ethics as Darwinian natural right,' paper presented at the International Westermarck Symposium, Helsinki, Finland, November 19–22, 1998.

Arnhart, L. (2000), 'Why conservatives need Charles Darwin', *First Things* (Forthcoming).

Arnhart, L. (2000a), 'The incest taboo in natural law and Darwinian ethics,' paper presented at the conference on 'Incest, Inbreeding, and the Incest Taboo,' Stanford University, Stanford, California, February 24–26, 2000.

Blair, R.J.R. and Cipolotti, L. (2000), 'Impaired social response reversal. A case of "acquired sociopathy" ', *Brain*, **123**, pp. 1122–41.

Brown, A. (2000), *The Darwin Wars* (London: Simon & Schuster).

Cahn, S.M. and Markie, P. (ed. 1998), *Ethics: History, Theory, and Contemporary Issues* (New York: Oxford University Press).

Capaldi, N, (1966), 'Hume's rejection of "ought" as a moral category', *Journal of Philosophy* **63**, pp. 126–37.

Curry, O. (2000), *What is the Relevance of Evolution to Ethics?* (CPNSS Discussion paper series, London School of Economics), forthcoming.

Damasio, A.R. (1994), *Descartes' Error: Emotion Reason and the Human Brain* (New York: Grosset/Putnam).

Damasio, A.R. (1999), *The Feeling of What Happens: Body and Emotion in the Making of Consciousness* (New York: Harcourt Brace).

Darwall, Gibbard and Railton (1997), 'Toward *Fin de Siècle* ethics: Some trends', *in Moral Discourse and Practice: Some Philosophical Approaches*, ed. Stephen Darwall, Allan Gibbard and Peter Railton (London & New York: Oxford University Press).

Dawkins, R. (1976), *The Selfish Gene* (Oxford: Oxford University Press).

De Waal, F. (1996), *Good Natured: The Origins of Right and Wrong in Humans and Other Animals* (Cambridge, MA: Harvard University Press).

Dickens, P. (2000), *Social Darwinism: Linking Evolutionary Thought to Social Theory* (Milton Keynes: Open University Press).

Frankena, W.K. (1939), 'The Naturalistic Fallacy', *Mind*, **48**, pp. 464–77.

Goldsmith, M.M. & Horne T.A. (ed. 1986), *The Politics of Fallen Man* (Exeter: Imprint Academic).

86   K. SUTHERLAND, J. HUGHES AND L. ARNHART

Hare, J. (1996), *The Moral Gap: Kantian Ethics, Human Limits, and God's Assistance* (Oxford: Oxford University Press).

Hare, J. (2000), 'Evolutionary naturalism and the reduction of the ethical demand', paper presented at a conference on 'The Nature of Nature,' at Baylor University, Waco, Texas, April 13–15, 2000.

Harman, G. (1977), *The Nature of Morality* (New York: Oxford). Repr. In Cahn and Markie (1998).

Hudson, W.D. (1969), *The Is-Ought Question: A Collection of Papers on the Central Problem in Moral Philosophy* (London: Macmillan/St. Martin's).

Hughes, J. (2000), 'A normative science manifesto', paper accepted for presentation at APLS 2000, the Annual Meeting for the Association for Politics and the Life Sciences in Washington D.C., Aug. 31–Sept. 3, 2000.

Hume, D. (1739/1985), *A Treatise of Human Nature* (Harmondsworth: Penguin Classics).

Hume, D. (1888) *A Treatise of Human Nature*, ed. L.A. Selby-Bigge (Oxford: Oxford University Press).

Huxley, T.H. (1894/1989), *Evolution and Ethics* (Princeton: Princeton University Press).

Katz, L. (ed. 2000), *Evolutionary Origins of Morality*, A special issue of *Journal of Consciousness Studies*, **7** (1–2).

Lévi-Strauss, C. (1966), *The Savage Mind* (Chicago: University of Chicago Press).

MacIntyre, A. (1982), *After Virtue* (London: Duckworth).

McShea, R.M. (1986), 'Political philosophy, human nature, the passions', in Goldsmith and Horne (1986).

Moore. G.E. (1903/1959), *Principia Ethica* (London and New York: Cambridge University Press).

Oakeshott, M. (1956), 'On Being Conservative', reprinted in Oakeshott (1991).

Oakeshott, M. (1991), *Rationalism in Politics and Other Essays* (Indianapolis: Liberty Fund).

Panksepp, J. (1998), *Affective Neuroscience: The Foundations of Human and Animal Emotions* (New York: Oxford University Press).

Pears, D. (1984), *Motivated Irrationality* (Oxford: Clarendon Press).

Rolls, E.T. (1999), *The Brain and Emotion* (Oxford: Oxford University Press).

Rose, H. and Rose S. (ed. 2000), *Alas Poor Darwin: Arguments against Evolutionary Psychology* (London: Jonathan Cape).

Rule, A. (1989), *The Stranger Beside Me* (New York: Signet).

Ruse, M. (1986), *Taking Darwin Seriously* (Oxford: Blackwell).

Rutherford, J., 'An Ecological Organic Paradigm: A Framework of Analysis for Moral and Political Philosophy', *Journal of Consciousness Studies*, **6** (10), pp. 81–103.

Segerstrale, U. (2000) *Defenders of the Truth : The Battle for Science in the Sociobiology Debate and Beyond* (New York: Oxford University Press).

Singer, P. (2000), *A Darwinian Left: Politics, Evolution, and Cooperation* (New Haven: Yale University Press).

Stephen, J.F. (1883) *History of the Criminal Law of England* (New York : Franklin).

Stroud, B. (1977), *Hume* (London: Routledge).

Van Valen, L. (2000), 'How the left got Darwin wrong', *Scientific American*, http://www.sciam.com/2000/0600issue/0600reviews1.html

Warnock, G.J. (1967), *Contemporary Moral Philosophy* (London: Macmillan).

Weber, M. (1904/1949). ' "Objectivity" in social science and social policy', in *The Methodology of the Social Sciences*, ed. and trans. E. Shils and H. Finch (New York: Free Press, 1949). Repr. in *Readings in the Philosophy of Social Science*, ed. M. Martin and L.C. McIntyre (Cambridge, MA: MIT Press, 1994).

Westermarck, E. (1922), *The History of Human Marriage*, 3 vols., 5th edition (New York: Allerton Book Company).

Westermarck, E. (1932), *Ethical Relativity* (London: Kegan Paul, Trench, Trubner & Company).

Williams, B. (1985), *Ethics and the Limits of Philosophy* (Cambridge, MA: Harvard University Press).

Wolf, A. (1995), *Sexual Attraction and Childhood Association: A Chinese Brief for Edward Westermarck* (Stanford, CA: Stanford University Press).

Yalden-Thomson, D.C. (1978), 'Hume's view of "is-ought" ', *Philosophy*, **53**.

Name Index

Akerlof, George 117
Alexander, J. 98, 103
Alexander, Richard 6, 17, 178
Anscombe, Elizabeth 205–6
Aquinas, Thomas 328
Aristotle xii, 297, 313–15, 322, 326, 328, 333
Arnhart, Larry xxxvi, 313–16, 319–30, 333
Aumann, R.J. 102
Aureli, F. 10, 12–13, 20
Axelrod, Robert 6, 17, 45–6, 76, 172

Bach, J.S. 181
Barchas, P.R. 12
Batali, John 45
Batson, Daniel C. 82, 131
Bean, P. 17
Bebel, August 285
Beckstrom, John 317
Behe, Michael 181
Bekoff, Marc xxxii
Benedict, Ruth 24
Bentham, Jeremy 180, 219, 287
Bernheim, B.D. 93
Bernstein, Eduard 285
Bernstein, I.S. 12, 14
Bin Laden, Osama 181
Binmore, Kenneth 91, 94, 96, 100, 145
Bjornerstedt, J. 96
Blake, William 314
Blurton-Jones, Nicholas 7, 9
Boehm, C. 5, 11, 14
Boesch, C. 6
Boesch, H. 6
Borgers, T. 97
Boyd, Robert 17–18, 78, 140–41, 213
Bratman, Michael E. 146
Brink, David 213, 253
Broad, C.D. 83
Brosnan, Sarah xxxii
Brown, A. 314
Bundy, Ted 323, 332
Burke, Edmund 328
Buss, David 314

Butler, Joseph 83, 129

Cacioppo, J.T. 19
Cahn, S.M. 318
Camerer, C. 93
Camerer, V. 93
Campbell, J. xxiii, 207
Cannon, Susan 283
Capaldi, N. 319
Carey, Susan 162–3
Carnap, Rudolf 300
Carpenter, C.R. 12
Carroll, Lewis 301
Carson, Thomas L. xxvii
Castles, D.L. 20
Chagnon, Napoleon 293
Cheney, D. 6
Chomsky, Noam 162
Clark, R.D. 82
Clarke, Samuel 319
Clutton-Brock, T.H. 17
Collier, J. xxv
Cordischi, C. 10
Cords, M. 9, 13
Cosmides, Leda 136
Cozzolino, R. 10
Crawford, M.P. 6
Cronk, L. 18
Curry, Oliver 318–20

Damasio, Antonio R. 322–4, 326, 329, 332
van Damme, E. 97–8
Darwall, Stephen 318
Darwin, Charles xi, xii, xiii, xv, 3, 22–3, 57,
    70–71, 74–5, 89, 96, 100–101, 103, 109,
    132, 169–70, 179, 181–7, 283–92, 297–8,
    313–15, 320–22, 329, 331
Das, M. 13, 20
Davies, Paul 182
Dawkins, Richard xvii, xix, xxviii, 4, 23, 73, 107,
    134, 187, 314, 329
Demaria, C. 12
Dennett, Daniel xxviii, 4, 137, 169–83

Descartes, René 327
Desportes, C. 12
Devitt, Michael xxiii
Dodsworth, R. 14
Dostoevsky, Fyodor 206
Duffy, J. 93
Dunbar, Robin I.M. 136
Dworkin, R. 244

Edwards, S. 6
Ehardt, C.L. 12, 14
Eibl-Eibesfeldt, Irenäus 292
Ekman, Paul 109
Elster, Jon 130

Falett, J. 9
Fedigan, L. 19
Fedigan, L.M. 19
Feinberg, Joel 83, 130
Feldman, Marcus W. xxix
Ferri, Enrico 285
Feurerstein, J.M. 6
Fisher, R.A. 77–8
Flack, Jessica C. xxi, xxxii, 3–31
Flew, Anthony 298
Flex, A.G.N. 187
Foder, J. 163
Foot, Philippa 205
Fox, G. 6
Fragaszy, D.M. 6
Francis of Assisi, Saint 304
Frank, Robert H. xviii, xix, xxxiii, 5, 107–18,
    144, 172
Frankena, W. 212
Freud, Sigmund 323
Fruth, B. 6
Fudenberg, D. 97

Gage, Phineas 323
Gale, J. 96, 100
Gauthier, David 91, 144
Gewirth, Alan 302, 304–7
Ghiselin, Michael 152
Gibbard, Alan 94, 132, 172, 226–7
Godfrey-Smith, P. 164
Goldman, A. 153
Goodall, Jane 6, 9, 14, 20, 43
Gould, Stephen Jay xxi, 292, 314, 326
Gouzoules, H. 14
Gouzoules, S. 14

Gruter, M. 11
Güth, W. 99

Haeckel, Ernst 284, 298
Haldane, J.B.S. 124
van Halen, L. 329
Hall, K.R.L. 11
Hamilton, William D. xvi, 6, 71–2, 75–7, 102,
    107, 125, 172
Harcourt, A.H. 14
Hardin, Garrett 4, 11
Hare, John 333–5
Hare, R.M. 314
Harman, Gilbert xxii, xxxv, 208, 212, 318
Harms, W. xxv, 100
Harsanyi, John 91–2
Hatfield, E. 19
Hawkes, Kristen 7–8
Hayaki, H. 19
Hemelrijk, C.K. 9
Henrich, J. 18
Heyes, C.M. 22
Hitler, Adolf 225, 253–4, 304
Ho, T.H. 93
Hobbes, Thomas 4, 41–2, 50, 61, 91, 94–5,
    172–3
Hofbauer, J. 97, 101
Hofstadter, Richard 284–5
Hohmann, G. 6
van Hooff, J. 13–14, 20
Hughes, Jordan xxxvi, 313, 326
Hughes, William 216–17
Hume, David xii, xxxiv, 22–3, 94–5, 139, 172–3,
    185–7, 189, 206–7, 211, 218, 313–15,
    319–23, 328, 330–32
Hutcheson, Francis 319
Huxley, Julian 298
Huxley, Thomas Henry xii, xiii, xiv, xv, xxv,
    xxviii, xxix, xxx, xxxi, 3–6, 186, 297,
    314–15

Irons, W. 293

James, William 302
Jeannotte, Lisa A. 17, 19
Jencks, Christopher 148
Jensen, G. 14
John Paul II, Pope 175
Johnston, M. 207
Jonker, L. 96

Joyce, R. xxii, xxiii, xxiv, xxv, xxxv, 203–22
Judge, P.G. 13, 20

Kagan, Jerome 116
Kano, T. 6
Kant, Immanuel xxvii, 22–3, 180, 195, 198–9,
    204, 237, 299, 303, 309, 315, 320, 323–4,
    327, 329–32, 334
Kappeler, P.M. 12–13
Kavka, Gregory 81, 146
Kevles, Daniel J. xiv
Keynes, J.M. 93
Killen, M. 25
Kitcher, Philip xix, xxii, xxxii, 35–68, 131
Konner, Melvin 109
Kortlandt, Adriaan 7
Kortlandt, J. 6
Kropotkin, Petr 6
Kummer, H. 4, 9, 25
Kuroda, Suehisa 6–7

Lafollette, Hugh 81
Laland, K.N. xxix
Lavoisier, Antoine 214
Lemos, John xxiv, xxxvi
Levine, D. 97
Lewis, David 104
Lewontin, Richard 292, 314, 326
Lindburg, D.G. 20
Lindley, Richard 139
Lorenz, Konrad 71–2
Lovejoy, O. 188
Lumsden, Charles J. 188–9
Luttrell, L.M. 10, 12, 19
Lykken, David 109

Moser, Paul K. xxvii
McDowell, J. 207
McGinn, Colin xxviii, xxxi, 138
McGuire, M.T. 6
Machiavelli, Niccolò 41–2, 50
MacIntyre, Alisdair 305–7, 316
Mackie, John L. xxiv, xxvi, 132, 144, 171, 204–5,
    208, 215, 217, 219–20
Mackintosh, James 288
McDowell, John xxvi
McShea, Daniel W. xxi
McShea, R.M. 327
McShea, Robert J. xxi
Markie, P. 318

Marler, P. 15
Martinsen, David xxiii, 207–8, 210, 212
Marx, Karl xxxvii, 316
Mattern, 292, 294–5
Maynard Smith, John 71–2, 76–8, 96, 291
Mayr, Ernst 292
Meisner, L.F. 19
Mendoza, S.P. 12
Midgley, Mary 4
Milgrom, P. 102
Mill, John Stuart 180, 287
Mitra, D. 6
Moore, G.E. xiv, xxiv, 58, 173, 186–7, 193,
    211–12, 283, 297, 309, 313, 317–19
Moore, James 7
Morgenstern, O. 91–2
Moulin, H. 93

Nagel, R. 93
Nagel, Thomas 83, 138, 244
von Neumann, J. 91–2
Newton, Isaac 169
Nietzsche, Friedrich 216
Nishida, Toshisada 6–7, 11
Nissen, H.W. 6
Nitecki, D.V. 5
Nitecki, M.H. 5
Nozick, Robert 83, 197
Nucci, L.P. 25

Oakeshott, Michael 328
Odling-Smee, F.J. xxix
Oleson, Kathryn C. 131

Palmer, Jack A. xx
Palmer, Linda K. xx
Panksepp, J. 322
van Panthaleon van Eck, C.J. 20
Parfit, Derek 239
Parker, G.A. 17
Pearce, D.G. 93
Pears, D. 323
Perry, S. 6, 8
Petit, O. 14
Pierotti, R. 6
Pinker, Steven 314
Plantinga, A. 164
Plato 194
Plutarch 203
Preuschoft, S. 12

Price, George 71–2, 96
Putnam, H. 212

Quine, Willard V.O. 121
Quintilian 203

Rachels, James xxvii
Railton, Peter xxxv, 138–9, 251
Rappaport, Anatol 76
Rapson, R.L. 19
Raskolnikov, Rodion Romanovic 206, 210
Rawls, John 21, 91, 123, 132, 190–91, 197–9
Reinhardt, V. 14
Ren, M.R. 13–14
Richards, Robert J. xxxvi, 208–10, 283–311
Richerson, Peter J. 17–18, 78, 140–41
Ridley, Mark 134
Ridley, Matt xxi
La Rochefoucauld, François de 123
Rockefeller, John D. 284–5
Rolls, E.T. 322
van Roosmalen, A. 12, 16, 20
Rose, Hilary 326
Rose, L.M. 6, 8
Rose, Steven 314, 326
Rosenberg, Alexander xix, xx, xxxiv, 132,
　　　169–84
Rothstein, S.I. 6
Rottschaefer, William xxiii, xxviii, 207–8, 210,
　　　212
Rowell, T.E. 13
Rule, Ann 323
Ruse, Michael xxi, xxii, xxvi, xxvi, xxvii, xxviii,
　　　xxxiv, xxxv, 172, 177, 185–206, 210,
　　　216–17, 219–20, 289–91
Russell, Bertrand 297–8
Russett, C.E. 186
Rutherford, James 327

Sacco, P.L. 96
Sahlins, Marshall 284
Samuelson, L. 96, 100
Sarin, R. 97
Scanlan, J. 14
Scanlon, T.M. 244
van Schaik, C.P. 12–13, 20
Schelling, Thomas 54, 108
Schlag, K. 96
Schmittberger, R. 99
Schwartz, Barry 132

Schwarze, B. 99
Scott, J.P. 19
Scucchi, S. 10
Segerstrale, Ullica 317
Selten, R. 92, 99
Sesardic, Neven xviii, xix, xxxiii, 121–50
Seton, Ernest 25
Seyfarth, R. 6
Shafer-Landau, Russ 225, 244
Shaw, Laura L. 131
Sicotte, Pascale 16
Sidgwick, Henry 297
Sigg, H. 9
Sigmund, K. 97, 101
Silberbauer, G. 4
Silk, J.B. 6, 10
Simon, H.A. 18, 180
Singer, Peter xxviii, 123, 138–9, 196, 329
Skinner, B.F. 176–8, 316
Skyrms, Brian xix, xxxiii, 91–106
Smith, Adam 22, 328
Smith, John Maynard xvi, xix
Smith, Michael xxiv, 207, 213
Snowdon, C. 6
Sober, Elliott xvi, xix, xxxii, xxxiii, xxxiv, 17,
　　　69–90, 102, 127, 130, 133, 135, 151–6,
　　　158–64, 172
Sommers, Tamler xxxiv, 169–84
Sorokin, Pitirim 316
Spelke, Elizabeth 162–3
Spencer, Herbert xii, xiii, xiv, xiv, xxv, xxv,
　　　xxviii, xxix, xxx, 186, 283–5, 297,
　　　309–10, 316
Spinoza, B. 94
Stahl, D.O. 93
Stalin, Joseph 181
Stampe, Dennis 82
Stephen, James 328
Stewart, R.M. 83
Stingl, M. xxv
Stitch, Stephen xxxiv, 151–65
Street, Sharon xxxvi, 223–80
Stroud, Barry 320–21, 331
Sturgeon, Nicholas xxii, 253
Sugden, R. 94
Sutherland, Keith xxxvi, 313–36

Taylor, C.E. 6
Taylor, P. 96, 186, 199
Teleki, G. 6

Teresa, Mother 194
Thierry, B. 12, 14
Thompson, Paul xxiv
Thrasymachus 216
Thurnheer, S. 13
Tokuda, K. 14
Tooby, John 136
Trivers, Robert L. xvii, xxx, 6, 11, 107, 123, 172, 187

Uchii, S. 23
Uno, H. 19

Vanderschraaf, P. xix, 103
Veenema, H.C. 13, 20
Verbeek, P. 20
Verleur, D. 20
Virchow, Rudolf 285

de Waal, Frans B.M. xxi, xxxii, 3–31, 43, 51–3, 328
Wallace, A.R. 175
Waller, Bruce N. xxi, xxii, xxvi, xxvii, xxviii
Watson, John B. 316
Wedgwood, Hensleigh 287

Weibull, J. 96–7
Weigelt, K. 93
Westermarck, Edward 22, 56, 315, 331–2
Whiten, A. 20
Wickler, W. 25
Wilkinson, Gerald. S. xxi, xxviii, xxxii
Williams, Bernard 139, 318
Williams, George C. xvi, xvii, 4, 23, 25, 71, 73, 76–7, 126, 186, 292, 314
Wilson, David Sloan xvi, xix, xxxii, xxxiii, xxxiv, 17, 69–90, 102, 127, 151–6, 158–64, 172
Wilson, Edward O. xv, 107, 127, 131, 172, 176–8, 187–9, 283, 286, 289–93, 302, 315, 317, 329
Wilson, James Q. 140
Wittgenstein, L. 219–20
Wollaston, William 319
Woolcock, Peter xxiv, xxix, 216–17
Word, L.E. 82
Wrangham, Richard 6, 9, 47
Wynne-Edwards, V.C. 126

York, A.D. 13
Yoshihara, D. 13, 20

Printed and bound by CPI Group (UK) Ltd, Croydon, CR0 4YY

21/10/2024

01777095-0002